DEEPWATER HORIZON 2020

Remembering BP's

2010 Disastrous Blowout—

10 years later

DEEPWATER HORIZON 2020

Remembering BP's

2010 Disastrous Blowout—

10 years later

Includes:

(Book 1) THE SIMPLE TRUTH

(Narrative nonfiction)

&

(Book 2) FROM THE PODIUM

(Nonfiction)

by

J.A. Turley

The Brier Patch, LLC
Littleton, Colorado, USA

Published by:
THE BRIER PATCH, LLC
P.O. Box 184, Littleton, CO 80160-0184, USA

Deepwater Horizon 2020 (April 2020)

ISBN 13:   978-0-9858772-6-2 (Paperback)
ISBN 13:   978-0-9858772-7-9 (eBook)

Introduction

Deepwater Horizon 2020 combines two previously published books about the cause of a true incident—BP's 20 April 2010 Macondo blowout aboard Transocean's *Deepwater Horizon* drilling rig.

The tragedy has been defined as one of the most lethal, costly, manmade, environmental disasters in history.

The author researched depositions, rig reports, company documents, and public information about the blowout, and then wrote and published *The Simple Truth: BP's Macondo Blowout*, in 2012. Though the data-driven book is diagramed, footnoted, and referenced throughout, it is written as narrative nonfiction to tell the story about drilling BP's deep Macondo offshore exploration well at the hands of fictional characters—surrogates for survivors and the eleven witnesses who died that terrible night.

Based on his research for *The Simple Truth*, the author made technical, academic, and keynote presentations around the world about the cause of the blowout. After his 75th presentation in 2016, he wrote and published *From the Podium: The Cause of BP's Macondo Blowout*. The nonfiction work is a full compilation of the author's slides, text, drawings, Q&A, conclusions, and recommendations.

Now, 10 years after the tragedy, with a new generation of students, faculty, and technical and management personnel across the oil-and-gas industry, as well as new ranks of astute readers across the globe, *Deepwater Horizon 2020* combines the previous works, eliminates duplication, and links footnotes, diagrams, and references between the books.

The author's mantra has been and remains: Only if we understand and care about the cause of BP's Macondo blowout will we know why it should not have happened and why it should never happen again.

* * *

BP's 2010 Macondo blowout should not have happened, and no person wants to experience anything like it ever again. To this end, post-blowout **accident investigations**—academic, legal, regulatory, societal, forensic, environmental, financial, technical, presidential—cascaded onto the scene in the weeks, months, and years that followed.

Analysis of the catastrophe evolved on three key fronts:

(1) Among the companies that were collectively charged with safely drilling the deep-water oil-and-gas exploration well, what *pre-blowout* safety, procedural, communications, human-factors, financial, and/or other organizational processes may have contributed to the blowout?

(2) What happened, minute-by-minute, decision-by-decision, action-by-action, on the rig and in the office that allowed the 3-½-mile-deep, high-pressure oil well to erupt without control onto the rig, explode and catch fire, kill eleven, injure and disable dozens more, sink the rig in a mile of water, and dump more than 200 million gallons of crude oil into the Gulf of Mexico?

(3) After the initial blowout and with the skeleton of the burned *Deepwater Horizon* drilling rig settled on the seafloor, the well continued to flow. No quick fix existed. Hence, what physical, mechanical, and procedural measures should have been available *pre-blowout* but had to be developed posthaste to: (a) stem the persistent months-long massive flow of oil from the well into the mile-deep Gulf of Mexico, (b) secure the well, and (c) minimize and remediate extensive environmental damages at sea, along the Gulf Coast, and into the estuaries?

The answers to (1) and (3) above—researched by professional, academic, and business task forces across the globe—define the *Root Causes* of the blowout and led to fines, penalties, sanctions, lawsuits, enhanced regulations, and corporate restructurings, both Macondo-related and throughout the industry.

The answers to (2) above—researched by the author and others—define the *Technical Causes* of the blowout, as revealed by the hours-long chronological stream of decisions and on-the-rig data between the *last* successful-as-planned wellbore activities and the *first* too-late moment that led to the then-unstoppable man-made disaster. Drilling and operations personnel, petroleum engineers, company men, toolpushers, managers, academics, students, investors, industry professionals, and surviving family members want and need to know what happened, and what could have, should have, been done with documented hard data and rig-site leadership decisions that, if handled otherwise in real time, would have precluded the lethal well-control disaster called the Macondo blowout.

This book, *Deepwater Horizon 2020*, does not address Root Causes, finger pointing, attorneys, deep pockets, names, or companies. Rather, it focuses entirely on the *Technical Causes* of BP's 2010 Macondo blowout, throughout both parts: *The Simple Truth* and *From the Podium*.

BOOK 1

THE SIMPLE TRUTH

BP's Macondo Blowout

———————————————

Narrative Nonfiction
by

J.A. Turley

The Brier Patch, LLC
Littleton, Colorado, USA

Published by:
THE BRIER PATCH, LLC
P.O. Box 184, Littleton, CO 80160-0184, USA

ISBN 13: 978-0-9858772-1-7
ISBN 10: 0-9858772-1-9

This is a work of narrative nonfiction based on a true incident—BP's 2010 blowout aboard Transocean's *Deepwater Horizon* and the resulting oil spill in the Gulf of Mexico. Names and characters are products of the author's imagination and are used fictitiously, and any resemblance to actual persons, living or dead, is entirely coincidental. Public data about companies, equipment, the well, and the rigs, though modified by the author for ease of reading, form the setting for the story. The Epilogue is nonfiction. Opinions and errata are the author's.

FOR JAN

Forever My Love,
my best friend,
my CFOOE.

ACKNOWLEDGEMENTS

During the project's two-year evolution, its working title morphed from NOWHERE TO RUN, to SPILL IN THE GULF, and finally to *THE SIMPLE TRUTH*. A number of people with mixed skills—prolific readers, published writers, engineering professors, oil-and-gas executives, world-class offshore drilling consultants—took time with the full manuscript in order to help me write a better book. They are joined by a long list of others who contributed editing skills where they could do the most good.

Apologies ahead of time if I've missed naming significant contributors—even the mean ones who constantly tested the thickness of my skin. My list includes: Roger Abel, Daven Anderson, Ian Ballard Bonnie Biafore, Nikki Baird, Andrea Catalano, Chris Devlin, Ben Ebenhack, Jim Ewing, Lizzie Funk, Steve Hagen, William R Hagen, Elizabeth Hall, Jennifer Harrelson, Ed Hickok, Angie Hodapp, Michelle Hoff, Martha Husain, Beckie Kagan, Mary Ann Kersten, Linda LeBlanc, Bill Madison, Anne Mini, Mindy McIntyre, James Norris, Mike Oldenburg, Kathy Reynolds, David Thyfault, Nat Tilander, Kevin Paul Tracy, Mike Walker, Bill Wall, Chris Wineki, Kevin Wolf—plus other anonymous-by-choice friends and family around the world.

A special thanks to the geoscience and petroleum engineering faculty and students at Marietta College who endured my guest-lectures 2010-2012 even as ever-evolving investigative data revealed and confirmed new blowout-related information not previously made public.

Cover credits and my personal thanks go to Scott Baird—Baird Enterprises, LLC.

Kudos also to Brian Schwartz—there's a reason he's called the Kindle Expert.

Photo credits: (1) The photo of Transocean's *Transocean Marianas* (Diagram 1) and the pre-Macondo photo of Transocean's *Deepwater Horizon* (Diagram 7) were supplied by Transocean, permission on file.

(2) The public-domain photo of the burning *Deepwater Horizon* (Diagram 21) was taken by Richard Braham, U.S. Coast Guard, on 21 April 2010.

TABLE OF CONTENTS

The Spindletop discovery well, near Beaumont, Texas, blew out on 10 January 1901. Also called the Lucas Gusher, the well spilled as much as 100,000 barrels of oil per day for nine days.

Spindletop

INTRODUCTION

Blowout—I, the author, am appalled by the word and its potential consequences.

From the Internet, one can download photographs of the infamous Spindletop blowout, near Beaumont, Texas, in 1901. Nobody in the oil-and-gas industry is proud of such reminders of its clumsy, greedy history. Times have changed. Educated and experienced engineers and geoscientists work in harmony with the physical laws of nature and the need for environmental stewardship. Sophisticated technologies and safe operating procedures abound. Front-line leaders take responsibility for the tasks at hand. All of which means the industry can drill, as it has for decades, even the most intractable wells. And with none of Spindletop's theatrics—oil and gas roaring up from the depths, out of control, pouring onto the land or into the sea.

But if that's the case, what about a horrific blowout on 20 April 2010 in the Gulf of Mexico, defined by fire, deaths, and oil on a rampage for months on end? How is it possible a world-class energy company, BP p.l.c. (formerly British Petroleum Company) and an equally world-class offshore drilling company, Transocean Ltd, lost control of the well they were drilling *109 years* after that miserable, ancient-technology photo-op at Spindletop?

Day-one publicity about BP's Macondo blowout (BP's well, Transocean's rig) focused on eleven tragic deaths, devastated families, and a spectacular fire at sea. Then came a second round of publicity about the loss of Transocean's mammoth drilling rig, the *Deepwater Horizon*, and about the involved companies' failure to prevent such a disaster. Finally, a media blitz about the when-will-it-ever-end months of oil and gas flowing from the seafloor and the billions of dollars in social damages and untold environmental harm to the Gulf and its coastline and estuaries. Beyond BP and Transocean, the entire oil-and-gas industry, especially offshore in the Gulf of Mexico, found itself under intense media and political magnifying glasses.

Federal investigators, though handicapped without the eleven frontline workers who had died, wasted no time. They subpoenaed a bounty of 24/7 drilling data that had been sent wirelessly for months to BP offices in Houston, Texas. They deposed and questioned-under-oath key employees from BP and from Transocean and from all subcontractors involved with the well. They dug into designs, actions, data, decisions. They hired independent contractors to perform forensic analyses of equipment salvaged from the remains of the well. They questioned if the water was too deep, Mother Nature too onerous, the technology too demanding. Though investigators issued final reports in September 2011, BP's blowout will be defined by billion-dollar, finger-pointing lawsuits for years to come.

The effects of the tragedy were no doubt horrendous, but an energized, disappointed, angry public wants to know what *caused* the blowout. By early 2011, journalists and authors had crafted almost a dozen nonfiction books about the disaster. A few authors addressed rumors and suspicions about *cause*, but all jumped on *effect*—their research fueled by the media frenzy that had tracked, documented, and photographed for months every tragic nuance of the event, from containment booms to beaches to boardrooms.

Far from satisfied, I leaned on my own education and work history to dig deep into the wealth of investigative data with one goal—to answer the question about *cause*. Having taught petroleum engineering and been directly responsible for decades of offshore drilling operations in the Gulf of Mexico and around the world, I expected and confirmed the answers were far from mysterious—they were black and white. I wanted to share those simple-truth answers fully knowing that only if we understand and care about the cause of BP's Macondo blowout, will we know why it should not have happened and why it should never happen again. But I also knew a dozen technical one-liners about the cause would be meaningless to those readers with little knowledge about life offshore, the good people who work there, and the kinds of decisions made on a daily basis.

Hence, this book.

But what's a *Macondo*? It's BP's "code name" for the deep geologic structure they were drilling into, hoping they would discover oil and gas. On the streets and in coffee shops, the well, too, was called Macondo. BP's Macondo well. Hence, BP's Macondo blowout, in perpetuity.

THE SIMPLE TRUTH invites readers aboard the two rigs that drilled BP's deep-water Macondo well in the Gulf of Mexico. Transocean's *Transocean Marianas* started the well but was damaged by a late-2009 hurricane and never returned to the site. A new contract called for the Macondo well to be reentered and finished in early 2010, using Transocean's *Deepwater Horizon.*

For convenience and without prejudice, the named characters in the story are fictional. Such characters occupy real positions. They represent a collection of *any, all, none* of the key people who worked the Macondo project, including, with respect, the eleven good men who died on the job. The offshore setting, gross drilling data, engineering procedures, operational decisions, key actions and inactions, and the tragedy that follows—though I simplified and nested them into the story—are based on comprehensive investigative data.

The resulting story is fictional. The answers to the question about cause are not.

Footnotes and diagrams throughout the novel define key terms and source material.

In the extensive nonfiction Epilogue, the reader will find references to a number of post-blowout investigations, published books and data on the topic, and the human and technical causes of the disaster.

To all readers of *The Simple Truth*: Welcome aboard and enjoy the ride.

J.A. Turley—2012

Perhaps the Macondo story went something like this .

Book 1

PROLOGUE—The End

20 April 2010—9:49 P.M.

Jessica Pherma's world exploded. Not the planet, but the *Deepwater Horizon*, her half-billion-dollar, drilling-rig home in the Gulf of Mexico. The event started fast, but only after hours of warnings, unheeded until too late. From under the rig an ugly growl thundered, echoed in her chest, a mere prelude to the black-geyser eruption of mud and water that blew through the rig floor and dwarfed the 24-story-tall derrick.

"Shut it in!" Jessica screamed, tally book in hand, waving her arms toward the men on the rig floor, toward the deluge that threatened to drown them all. "Shut it in, now!" she bellowed again.

A new noise—gas roaring from a vent line attached high in the derrick. Natural gas and atomized oil—a lethal concoction—its thick, acrid aroma unwelcome on any rig. The black cloud grew, swallowed the derrick, blanketed the rig floor. Jessica, an experienced geologist with a life-long fire phobia, feared the worst—the end of life, a spark away.

"Open the diverter!" Earthquake-size rumblings racked the rig and shook her body, and she wondered if the words had even left her mouth.

Men ran—some from the fury, others into the maelstrom.

Jessica, too, ran. Grabbed a life vest. Slid one arm in—

A blinding fireball of reds and oranges and yellows filled the night sky, the sound and concussive force beyond movie magic. The blast—a mix of fire, steel, and hardhats—slammed her body and drove her across the deck and against a four-inch-high drainage rail. She grabbed what she could and held on tight. Eyes locked open, looking over the edge and into the abyss, all she could see was water. Black water—the sea 60 feet below. And her tally book, dropped and falling, its white wings flapping on the way down, beckoning a vision of her dad's face, framed on a white pillow in his open black casket.

Another chest-crushing blast scooted her down the rail, covered her in debris. Something heavy, metal, on her back, pushed her down. Visions of death filled her skull, oozed from pores in baking skin. With brute strength, she commanded hands, elbows, knees to take over, lever her up, free the rubble, and shake it off. Dizzy and numb, she turned to face the billowing blaze that engulfed the giant derrick as if kindling in a campfire.

Words flashed. Daddy. Death

No. Not yet. Not this horrid night. Not by fire . . . though the radiant heat cooking her face and penetrating her body to the core of her bones threatened to incinerate her on the spot.

More people, silhouettes, running, stumbling, in silent slow motion, away from the inferno.

Jessica stayed low. Crawled away from the edge. Away from the fire, its brutal roar behind her growing louder, as if hungry, carnivorous, closing the distance, daring her to slow down. She found a stairwell, up to the helideck. On her knees, gasping for air, she scaled first one step, then another, her own flesh and bone and blood on steel.

A wailing man, a face she knew, stepped from the crowd, helped her up. The man, his face defined by anger and blame, grabbed her shoulders and shook hard. "You're BP," he howled. "Do *you* know what happened?"

Jessica yanked herself free, his question less important than the next minute of her life. She hugged her chest. Piercing heat attacked her back, while adrenaline chills tormented her body, her torso a conflict of fire and ice. She turned as if flipped on a barbecue grill and welded her gaze onto the growing inferno consuming the rig, fifty miles from shore, perched on top of water deep enough to drown a nation, its late-night background as dark as death.

Unable to stand the heat, she turned back to the helideck crowd, her legs weak, the horizon tilting. She studied faces. Faces of men. Fewer women. Illuminated by fire. Huddled in terror. Colleagues. Acquaintances. People she hadn't met.

She wondered about missing friends—Barry, Tanker, Daylight, others. No doubt fighting the battle. Maybe their last.

Her dad. She missed him. Would miss him forever.

And Mom and Sissy. She would always love them, though they were so wrong about her dad's death.

Like last rites. Something important to say. To confess. To get off her chest.

And one more thing. The truth. The simple truth.

While someone bellowed commands about mustering, and lifeboat stations, and abandoning ship, she found the man who'd asked her the question. A question that deserved an answer.

"Yes," she told the man, her voice weak, her body crashing. "I know . . . *exactly* what happened."

While able crew members helped Jessica and others to lifeboats under fire-lit skies, she found her stamina increasing, driven by a passionate goal: stay alive, tell the story.

OCTOBER 2009

CHAPTER 1—The Beginning

The helicopter's transparent-plastic floor dared him to look straight down. Not wanting to puke in the cockpit, Barry Eggerton gazed out the windshield and occasionally to his right to make sure the pilot was still awake. By mid-morning, he spotted his target, a gnat-size speck, halfway to the horizon.

Barry rode shotgun in a bright-yellow PHI chopper as it caught up with the *Transocean Marianas*,[1] framed by blue sky above choppy seas forty miles south of the Mississippi Delta. The 200-foot-square drilling rig, encompassing almost an acre, rode silently behind two oceangoing tugs. Even though the tow rate was only two knots, white water boiled behind the workboats and around the massive rig's buoyant flotation columns.

Barry had the pilot circle the rig counterclockwise, which allowed him to take cellphone pictures through the port-side window. His youngest daughter, Bailey, sixteen, collected the pictures for her special photo album—Daddy's drilling rigs. He wondered how long her interest would last. How long before his ex and her asshole-hubby of twelve years got tired of him dropping by. How long before Bailey would be so busy with life he'd need an appointment to give her a hug. Or hand her a picture of a rig.

As soon as the helicopter landed and the pilot gave him the okay, Barry disembarked, thankful for the solid footing of the helideck. Drilling deep, demanding wells from floating drilling rigs was his forte. Chopper rides—no thanks. He grabbed his duffel bag and backpack, then ducked under the chopper's still-rotating blades and made his way down the exit stairs to the main deck.

The rig was quiet—no drilling activity, no cargo lifts, no personnel transfers. No horns, sirens, rumbling engines. A soft sea breeze carried the odor of new paint wafting from a crew of roustabouts with scrapers and brushes working on handrails, the results glossy gray.

Barry cared nothing about paint. As BP's company man,[2] he cared only about managing the well. *His* well—safe, on time, on budget. Anxious to start drilling, he made his way to the quarters building. Just inside, he turned right and entered an office with a big picture window that faced the main deck and the substructure of the rig and derrick. He caught a whiff of fake pine needles. The paint had smelled better.

**Diagram 1 (Photo)
Transocean's
*Transocean Marianas***

Note helideck
(forward starboard
corner)

Note
anchor lines
on corners

Source & Permission
on file

Diagram 1

A big guy at one of two desks in front of the window hung up a phone and uncoiled from his chair. He wore magenta Transocean coveralls. Barry knew the code. Magenta, on duty. Otherwise, Transocean blue. The guy's smile crinkled his enormous face, entirely proportional to his massive body, a solid 3XL-tall.

Barry looked up and introduced himself and met Transocean's toolpusher.[3]

"Myron Forster," the big man said. "My daddy's also Myron, so please call me Tanker."

Tanker pointed to the second desk that faced the window. "That one's yours," he rumbled, "unless you want the one that faces the wall. Keys are in the drawer in case you need to lock her up. Your room's just down the hall, first on the right, across from mine."

Barry settled in, both men with their feet on their respective desks. They compared one-liners to see where their paths might have almost crossed—North Sea, West Africa, Indonesia. Wherever they'd been, Tanker managed Transocean rigs and Barry managed BP wells. Tanker's three decades of offshore drilling had overlapped a number of times with Barry's own twenty years. Fully engrossed in dredged-up memories, they recalled people, hurricanes, rig disasters.

"No disasters on *my* rig," Tanker declared.

"Nor with *my* well," Barry said.

Tanker picked up a sheet of paper, dropped his feet to the floor, and faced Barry. "Mississippi Canyon Block 252, well number one—that's quite a mouthful. What's your well plan look like?"

"We call the well *Macondo*.[4] It'll be a quickie. Fifty-one days to almost 20,000 feet in a mile of water." He didn't mention the well cost—almost $100 million for a dry hole, even more if the well discovered oil and gas.

Tanker squared his shoulders. "The *Marianas* can do 25,000 feet in 7,000 feet of water."

Barry liked that—a toolpusher proud of his rig. The rig that would drill BP's well.

A service-hand wearing scruffy red coveralls stepped into the open doorway and focused on Tanker. "Surveyor says we're one hour from location. Said you wanted to know."

Barry snapped a look at his watch, the passage of time his enemy. Especially rig time, at hundreds of thousands of dollars a day, which added to well costs. And influenced performance evaluations. And year-end bonuses—for himself and the entire food chain of bosses above him.

One hour to location. He willed the tugs to pull harder, faster.

Barry considered it good news, though not unexpected, that Tanker came across as a toolpusher he could trust to run the rig, to keep his crews working around the clock, drilling ever deeper. Barry had no doubt they would work well together. And all the big man had to do was stay out of the way when it came to Barry managing the well.

* * *

During the next twenty-two hours, Tanker and his crew filled the rig's flotation chambers with seawater and ballasted the rig down to its drilling draft. Anchored and stable, the *Transocean Marianas* was ready to drill Macondo.

So was Barry. He pointed out the window. "Driller got any coffee up there?"

Tanker grinned and led the way. They trekked across the main deck and around the pipe racks, stacked with 40-foot-long joints of 36-inch-diameter structural casing that would start the well. The beginning of Macondo.

Barry patted and ran his hand along one of the joints of smooth steel pipe that would soon be buried in the seafloor, never to be seen again. Twenty-two steps led them up to the rig floor and its doghouse, traditional home of the drilling crew's coffee station.

Barry took a small cup, a cardboard demitasse, and poured himself an inch of the viscous brew. He sipped and shuddered. As he'd expected, the rig-floor coffee, in contrast to the superb galley coffee, was midnight black, sickeningly sweet, and served from a used coffee can rather than an electric pot—a potential source of spark ignition disallowed on open rig floors.

Tanker introduced Barry to the driller,[5] David "Daylight" Stalwart. Hidden under a Transocean hardhat, dark safety glasses, gloves, and magenta coveralls, the about-average-size driller with a West Texas accent was classically incognito. From the color of his cheeks, Barry guessed he was either a brownish white guy, or a toffee-colored black guy. Or maybe Hispanic.

Daylight, as did Tanker, called him *Mr. Barry*.

Barry had worked smaller rigs, where each driller's station had included a chair, a few gauges, a brake, and a doghouse. Not so on the *Marianas*. A Plexiglas wall surrounded Daylight and a bank of sophisticated instruments. The clear wall gave the driller an open view of the rig floor and up into the derrick. It also shielded him from roughnecks armed with high-pressure, wash-down hoses. Barry was in favor of anything that made life easier for the driller—the frontline supervisor on the rig floor, the guy fully in charge of making the drill bit drill deeper, the guy Barry and Tanker depended on to first see drilling problems and react accordingly.

After the introduction, Tanker left the rig floor to check on an arriving chopper.

Daylight didn't slow down. He and his crew used the rig's hoisting equipment to lift another joint of structural casing through the vee door and raise it vertically into the derrick. Then with Daylight's expert guidance, the crew finessed together the pre-installed male-female connections, locking the new joint to the one hanging below the rig floor.

Barry checked his watch, willing it to slow down. "Remember when we used to weld each joint?" he said. "Took damn-well forever."

Daylight nodded, spit tobacco juice, sipped his coffee, and raised his cup.

It was time. Couldn't be avoided. Barry tipped his own cup to his lips. The sugar-thick brew moved like syrup. He willed his eyes to stay open, his throat to not gag. He sucked in the lukewarm fluid. Swallowed. Held his blinks. Swallowed to clear his throat.

Daylight again hoisted his cup, as if it were a shot of fine scotch.

Eyes watering, Barry nodded. "Fine brew. Now let's run some pipe."

"Yes, sir, Mr. Barry. We good to go." Daylight raised the casing a few inches, his crew pulled the slips[6] , and he lowered the connected joints through the rig floor, headed to the sea.

Barry gave him a thumbs up and stepped back into the doghouse, a simple storage shed for hand tools, O-rings, and buckets of pipe-thread lubricants, nested in the aromas of new grease and bad coffee. A plastic Ziploc bag held an unopened container of Skoal—smokeless tobacco, long cut, classic. He wondered if adding a bit of Skoal to his coffee would improve it. Probably.

He wrote notes in his tally book[7] and began a sketch of the well yet to be drilled. A small square for the floating rig. A horizontal line for sea level. Another line for the seafloor a couple inches down the page. Blanks for numbers to go on computerized drawings in the office.

Tanker leaned into the doorway. "Mr. Barry, got somebody who'd like to meet you."

A woman stepped forward. Wearing black boots, dark slacks, and a blue windbreaker, she spoke from under a snow-white BP hardhat. Lightly shaded safety glasses covered her eyes. "Jessica Pherma, BP onsite geologist."[8] She held out her hand and shook Barry's, her grip firm.

"Miss Jessica, I'll leave you two to chat," Tanker said. He touched the brim of his hardhat and headed toward Daylight.

"Name's familiar," Barry said. "Have we met?"

"I've logged BP wells for the past ten years, but we haven't worked together."

In Jessica's distinct Southern accent, her word *wells* came out as *whales*, but her strong words and staccato pace came across bigger than her Russian-gymnast body.

"Might've met in a meeting," Barry said, "or maybe a Christmas party. Regardless, if you're the assigned geologist, you're days early. We just started running structural casing."

"I'm early on purpose. I don't usually get to see this part of the well as it's being drilled, so I've got lots of questions. One of your bosses told me you're the guy with all the answers."

"Whoa, who says—"

"Mr. Barry," Daylight boomed from his padded chair, "if y'all can give us some room, we fixin' to pick up more pipe."

"That's our exit cue," he told Jessica as he nudged her out of the way.

"Page me before you start jetting," he told Daylight.

"Me, too," Jessica told the driller.

Side-by-side, headed to the stairs, Barry wondered how long it would take his new shadow, a geologist, to get bored following him around, asking him questions.

Double-timing his steps, she asked, "What's jetting?"

About a day, he reckoned.

---

[1] *TRANSOCEAN MARIANAS*— Source and photo credits on file. The *Transocean Marianas* is a moored deep-water semisubmersible drilling rig. Sedco 700 design. Built in 1979, rebuilt in 1998. BP commissioned Transocean (rig owner) to use the *Marianas* to drill the Macondo well.

[2] COMPANY MAN—usually a petroleum engineer; works for operator (BP). Responsible for managing the well—engineering, logistics, data, decisions, services, costs, executing procedures. On Macondo, a number of company men shared the job, often supported by drilling engineers, all working 14-day overlapping schedules. Fictional character Barry Eggerton represents, by himself, BP's entire offshore Macondo drilling staff, from the beginning of the well until the end.

3 TOOLPUSHER—(TP) works for Transocean, which owns the rig. TP manages the equipment and crews who drill the well per company-man instructions. Transocean's crews worked a 21-day rotating schedule. Fictional character Tanker Forster represents, by himself, Transocean's TPs, Offshore Installation Managers, and Masters on Macondo.

4 MACONDO—BP's code name for its Mississippi Canyon Block 252 deep geological prospect, and the nickname for the well (officially MC 252-1). Original "Macondo" reference from a 1967 notable work of fiction—*One Hundred Years of Solitude*—by Gabriel García Márquez.

5 DRILLER—works for the toolpusher. As Transocean's frontline supervisor on the rig floor, he makes the drill bit drill new hole. Fictional character "Daylight" Stalwart represents all day drillers, night drillers, and assistant drillers on the 21-day schedule.

6 SLIPS—mechanical wedges that trap and hold the uppermost joint of casing or drillpipe into the rotary table, with the rest of the pipe (the string of pipe) hanging below. Allows another joint to be screwed on (or unscrewed and taken off). Slips have to be "pulled" before the pipe can be moved up or down.

7 TALLY BOOK—An 8-inch tall, 4-inch wide, lined, hard-cover booklet. Makes record keeping easy. The book is designed to fit in a pocket, especially of coveralls.

8 GEOLOGIST (petroleum)—works to understand the type and structure of underground rocks and formations. The petroleum geologist assesses geology of drilled formations by examining drilled cuttings, and by running hi-tech electronic tools (called well logging) in the open-hole wellbore. Fictional character Jessica Pherma is BP's onboard geologist, 24/7.

CHAPTER 2—Breaking Ground

Jessica kept pace with Barry's long legs. Had he asked, she would have said the cool sea breeze was perfect jogging weather, and she was good for an hour at full pace. And no rig was that big.

But he didn't ask. He was a big man, but not as big as Tanker, and wore clean camo coveralls like preparing for a duck shoot. But office scuttlebutt said he was a deep-well expert, an all-business kind of guy. Exactly why she wanted to get inside his head. Preparing for the day when she'd get his job.

She followed him into the quarters building, then an office. Other than sunlight pouring in through a picture window, a dismal, macho office. Three putty-colored steel utility desks and matching file cabinets. An array of wall-mounted LCD screens showing drilling parameters. Lifeboat stations highlighted on a rig schematic. A plastic plant in a bucket for ambiance.

Barry pointed to the only desk facing a wall. "That's yours. The one with the junk is Tanker's." He sat at the third desk. "Are you settled in yet?"

"My gear and briefcase are with the steward.[2] I'm supposed to find him in the galley."

"Then the galley it is," Barry said, "and I'll let you buy me a cup of coffee."

Jessica joined Barry, side-by-side down a short hallway, toward the galley, while whiffing the pleasant aroma of dark-roast coffee combined with the sweet bouquet of hot butter and toasted garlic. Soon, she told herself. Soon she'd start the process to get everything she could out of Barry Eggerton before they finished drilling the Macondo well.

* * *

After her safety orientation, Jessica hung up her travel clothing and emptied her suitcase into a three-drawer chest bolted to the floor in her six-by-ten quarters. A twin bed, tucked snuggly into a corner, sported a hard pillow, brown blanket, and white sheets—military tuck. Steel door. No window. The steel walls supported a bed-head lamp, an air vent near the acoustic-tile ceiling, and a round mirror near the door. Bullet-proof, green-brown carpet. An in-suite bathroom with a jetted tub would have been nice, but at least the unisex facility she'd seen across the hall had a high-volume shower head.

She inhaled, closed her eyes, leaned her head back, and sniffed again. Almost fresh, except for a hint of food, paint, diesel. From a travel-size bottle, she sprayed a single shot of apple fragrance toward the vent. Inhaled again. Pictured trees. Fresh flowers. Apple blossoms.

The top of the dresser seemed the perfect place for a small family picture—mother, sister, father, herself. She focused on her dad's image, the last photo taken before—

No, she scolded. Work to do. She slipped off her jacket. Unbuttoned her blouse. Unzipped her boots. Stripped down to panties and mid-neck sports bra.

Though BP generously offered to supply coveralls to offshore employees, she'd found nicer styles on-line. Cute and tailored to fit, her one-piece, long-sleeve outfits were tough enough to wear to the rig floor, and they could survive being thrown into an industrial-spec washing machine and a fire-hot dryer. She flipped through a stack of five and picked lavender for her first day on the rig. The tough fabric felt cool against her skin as she finished dressing.

A sharp tapping on her steel door announced her first *Marianas* visitor. Guessed Barry, Tanker, maybe the steward. She opened her room to a man dressed in white. A galley hand. Thirty-something. Built like Barry. A six-footer. Big chest. Black eyes. Scraggly black hair. A tray of hot chocolate-chip cookies would have been nice, but his hands were behind him.

"Rig floor called," the guy said, his eyes moving fast—left breast, right breast, as if she were nude. "Somebody lookin' for you," he added, followed by a crooked-tooth, tobacco grin.

"Thanks," she mustered, as a wave of revulsion crawled down her back.

He walked away, message delivered, then glanced back, met her eyes. Grinned again.

She flipped the rude creep a mental bird. Only rarely had she ever overturned a bad first impression, and it wasn't likely to happen with him.

She picked up her hardhat, squared it in the mirror, and closed the door behind her.

* * *

Jessica found Barry and Tanker collaborating with the driller. She ignored them, went straight to the doghouse, and poured herself a slug of aromatic dark roast. One sip. Sweeter than cactus candy, just the way she liked it. She joined the trio.

Tanker introduced her to the driller—Daylight. Nice guy. Good manners.

Barry grinned. "How's the coffee?"

"Great, though it's a little cool and needs a touch more sugar."

Daylight called to one of his floor hands, hiked his thumb over his shoulder, pointed to the doghouse. "Need a refill on the coffee. Make sure she be hot and don't forget the sugar."

Jessica hoisted her cup toward Daylight and nodded. He returned her salute.

"You called," she said to Barry. "What's up?"

He explained they were getting ready to bury about 250 feet of structural casing into the seafloor. "The short length casing is hanging below almost a mile of drillpipe."[10] He pointed to a small television monitor on Daylight's instrument console. "This live video, fed by an ROV, shows the bottom of the casing right at the mud line."

Jessica knew the term—*ROV*. Remote-operated vehicle. A small unmanned submarine loaded with bright lights, cameras, and tools. Boy toys for the drillers, not needed by geologists.

"How deep's the water?" she asked.

"Near 5,000 feet to the mud line—we'll get a better measurement later."

Jessica reached to her back pocket, found it empty—no tally book. Not good. She was a data person. Notes galore. She made a fist, mad at herself.

Tanker, huddled with Daylight, said, "Mr. Barry, we're ready to start jetting."

Barry gave a thumbs up.

Daylight pushed buttons, turned knobs, kicked in the mud pumps. The video monitor showed a cloud of muddy water billowing from the seafloor at the bottom of the casing.

"That's just a trifle short of fascinating," Jessica said to Barry. "What's happening?"

"As Daylight lowers the casing, he's pumping seawater down the drillpipe where it exits through jet nozzles at the bottom of the casing. The jetting action blasts a hole in the mud below the casing. The weight of the casing pushes it into the hole, nice and snug, so we disturb the bottom as little as possible."

"Because . . ."   .

"Because the structural casing is the start of Macondo—the foundation for the entire well. It'll be important when we run future casing strings,[11] especially the one with the subsea wellhead."

"Then I can hardly wait," Jessica said. She'd never seen a *subsea wellhead*, but since Barry said it was important she wanted to see it, touch it, commit it to memory.

* * *

Other than during a short break to retrieve her notebook, Jessica shadowed Barry. Not much to see—Daylight pumping the brake handle and lowering the drillpipe and its attached casing to keep up with the deepening hole.

Barry seemed satisfied to watch the slow-moving drillpipe.

Jessica wasn't—she scanned the rig floor. She'd worked in mudlogging units adjacent to rig floors her entire career but had never stopped to just look. To enjoy the view. To hear and feel the dull rumble of motors and the rattle and clang of stands of drillpipe—two or three joints of drillpipe screwed together—all mixed with sounds and vibrations that came from nowhere, everywhere. The driller's station, complete with gauges and charts, surrounded by its protective wall. The massive drawworks, never quiet, reeling steel cable in and out, raising and lowering the lifting blocks, which raised and lowered the drillpipe, footage and speed as directed by the driller. As directed by *Daylight*, she corrected herself.

Looking up—way up—she took in the derrick. Reminded her of a truncated version of the Eiffel Tower. Major truss beams. The four corners, 40 feet on a side. Rig specs, posted on the wall in the office, noted it was 174 feet tall. That it could lift almost 700 tons.

Hell for stout her dad would've said.

And off to one side—in stark contrast to the steel drilling rig—a small wooden floor, scarred and splintered. The wood cushioned the threaded pin-ends of row after row of stands of drillpipe that stood together like soldiers waiting for Daylight to call them into service. She'd been on rigs where drillpipe connections had washed out—the high-pressure mud slicing through the steel like butter. She'd learned the failures may have been the result of the tiniest ding to the threads from improper handling. The wood solved the problem by keeping the threaded pins off the steel deck.

Beyond the wood deck, she took in the view out the tall vee-door opening in the bottom of the derrick, through which all materials were moved from the pipe deck to the derrick. Actually, an upside-down *vee*, more like an *A*, it was nothing more than a steel-beam opening in the derrick. Out the vee door, fifteen feet down to the main deck, beyond the pipe racks filled with casing and drillpipe, the office window reflected more gray steel.

"That be it, Mr. Barry," Tanker said, pulling her out of her reverie. "On the mark."

Barry took notes in his own tally book. Jessica peeked. He'd drawn a stick picture. He flashed the page at her. "The top of the casing is at the mud line, the bottom's buried 254 feet deeper, near 5,300 feet.[12] We'll get the water depth as soon as Daylight confirms his tally. It'll be put on the IADC form, plus my daily drilling report."

The noise level on the rig picked up—the rumble of the drawworks. Daylight, focused, worked the drillpipe. Picked up. Slacked off. Picked up again.

"What's he doing?" Jessica asked Barry.

"We've got a drill bit and bottom-hole assembly—BHA—attached inside the bottom of the structural casing. When we pumped seawater, the jet nozzles in the drill bit blasted the hole for the casing. As soon as Daylight gets unhooked from the drilling assembly, the casing will stay where it is, and we'll drill about 1,000 feet before we set more casing."

Jessica held back her joy—a thousand feet represented another five percent of her 20,000-foot goal, with less than 15,000 feet yet to go. Of course the first 5,000 feet, she recollected, being comprised of seawater, had gone rather fast.

* * *

Jessica made her way from the rig floor down to the main deck, crowded with pipe racks and equipment. Halfway to the office to make a phone call she spotted the rude galley hand. He leaned back, big arms draped casually over the handrail, his eyes tracking her approach. An exaggerated smirk on his face dared a slap, but she had a more-practiced weapon, a present from her dad. She made a right-handed fist, held it by her gut where he could see it, and headed his way—

A horn blasted. She jumped, recognized the standard warning for an overhead crane operation, then flash-scanned her surroundings, ready to run or stay put.

No need to worry. In full view over the side of the rig, a twenty-foot-long, bright-white, Sperry Sun mudlogging trailer climbed another foot, lifted by one of the rig's monster cranes.

She waved her arms to get the attention of the crane operator perched in his one-seat cabin on top of the crane's 30 foot tall pedestal. "Be careful with that," she yelled. "It's my home for the next couple of months."

Jessica wasn't a mudlogger.[13] Never had been. But the mudlogging unit would be her working office, her lab, for most of her tenure offshore. While the mudloggers did their job, monitoring drilling data on BP's behalf—for Barry—their unit was where she'd do her most important work, her forensics. Not for clues to a crime, but about decomposed life in the form of hydrocarbons—oil and gas. Diagram 2

The crane operator laughed and cupped his hands to his mouth. "I'll treat her gentle as a butterfly, Miss Jessica." Beautiful laugh. Deep bass voice. Enormous shoulders.

She threw him a thumbs up.

Typical rig. By the end of her first day, everybody knew her name. *Miss Jessica. Miss Pherma. Ma'am.* A hundred volunteers who would defend her in an instant against the foolhardy singleton who might try to cross the line. A hundred good men.

Plus the jerk from the galley, no longer in sight. She'd already sized him up. Big body, little brain. Ready for him, yet glad to have him off her radar, she had a more important agenda.

* * *

Minutes later, alone in the office, Jessica called a number in Houston for Ranae Morgan, her closest friend. Ranae, a Ph.D. geophysicist,[15] called herself Blasé Ranae because all she did was work. She had a vested interest in Macondo—she had supervised the seismic interpretation and proposed the exploration well to BP's executive committee. Though Jessica didn't report to Ranae, the senior manager had helped her secure the Macondo rig assignment.

As soon as Ranae answered, Jessica spit it out. "I'm on the rig, and we just spudded your well.[16] That means I'm within two and a half miles, less than 15,000 feet, of paying you back."

Ranae's gentle laugh filled the phone. "You don't have to do this."

"Not negotiable. I owe you, and a discovery with your name on it is the only way you'll ever get paid. We should be done by mid-December, certainly by Christmas."

A beat of silence. Too long. Jessica feared where Ranae might go.

"It's been more than a year, Jess. We're long ago even."

Belligerent memories found Jessica. Her dad's dead eyes. The shock of false claims, wrongful charges. Her mom's refusal to talk about the accident. None of which Jessica wanted to discuss with Ranae by phone.

And certainly not with a living soul on the almost-woman-free rig. Her fingers found the blue indented scar on her left cheek.

"Two years ago this Thanksgiving," she said. "I wouldn't have, couldn't have, made it without you."

"Then I know you'll do your job well," Ranae said. "You always do."

With small talk and warm goodbyes behind them, Jessica broke the connection. Yes, she would do everything she could to make a discovery for Ranae. Perhaps an early Christmas present, waiting only for Barry to drill it and Jessica to find it.

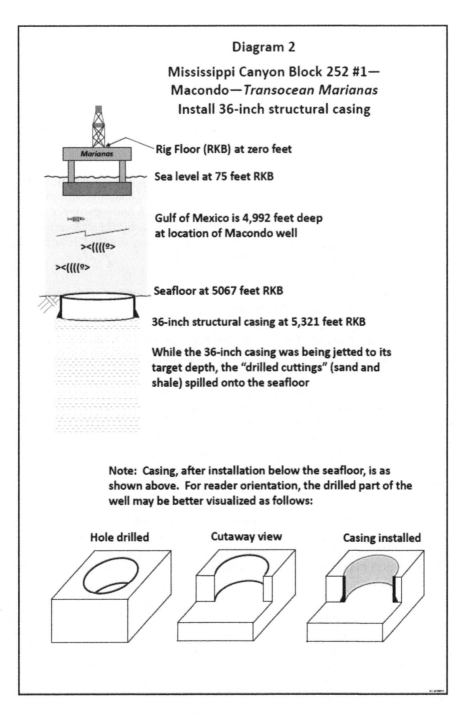

Diagram 2

Mississippi Canyon Block 252 #1—
Macondo—*Transocean Marianas*
Install 36-inch structural casing

Rig Floor (RKB) at zero feet

Sea level at 75 feet RKB

Gulf of Mexico is 4,992 feet deep
at location of Macondo well

Seafloor at 5067 feet RKB

36-inch structural casing at 5,321 feet RKB

While the 36-inch casing was being jetted to its
target depth, the "drilled cuttings" (sand and
shale) spilled onto the seafloor

Note: Casing, after installation below the seafloor, is as
shown above. For reader orientation, the drilled part of the
well may be better visualized as follows:

Hole drilled   Cutaway view   Casing installed

Diagram 2

⁹ STEWARD—in charge of the galley, meals, and quarters. It pays to be nice to the steward.

¹⁰ DRILLPIPE—(or drill pipe)—a joint of drillpipe is a single length of steel tubing (normally 5", 5-1/2", 6-5/8" diameter), usually 30 feet long. Drillpipe has larger diameter couplings (tool joints) for added strength and wear resistance. A *stand* of drillpipe is usually three joints screwed together (90 feet long). A *drill string* is the entire length of many joints of drillpipe, plus the BHA (see below), plus, ultimately, the drill bit on the bottom end.

¹¹ CASING STRING—The entire length of casing, comprised of many joints of casing (each about 40 feet long) screwed together. As the well gets deeper, a number of casing strings will be run and cemented in place. Each successive casing string must necessarily be smaller diameter than the already-installed casing string it fits inside.

¹² See **Diagram 2**, page 35 herein—(MC 252#1—Macondo—*Marianas*— Install 36-inch Structural Casing). This series of drawing will follow the progress of the well, starting with large-diameter structural casing, followed by deeper drilling. Each new drill bit has to fit inside the previous casing. Then the next casing also has to fit inside the previous casing. This means that as the hole gets deeper, subsequent drill bits as well as installations of casing will be smaller diameter.

¹³ MUDLOGGERS—Companies like Sperry Sun, hired by the operator (BP), supply mudlogging staff and equipment to monitor the well and all drilling activities 24/7, with communications directly to the company man and geologist. Though critical to the drilling of the Macondo well, Sperry Sun personnel are not named in the story.

¹⁴ DISCOVERY—The majority of sedimentary formations around the world contain water, and only few contain commercial quantities of oil and gas. Geophysicists and geologists decide which exploration-well location will maximize the chance of a discovery, which can be made only by drilling the well. A discovery means the well contains commercial quantities of recoverable oil and gas. If not, it's a dry hole.

15 GEOPHYSICIST—specialist geoscientists and managers who assess seismic data, develop geologic prospects, create subsurface maps, and build cases for exploring underground geologic structures. Fictional Ranae Morgan is one of a number of BP's onshore geotechnical managers. Though more senior than Jessica, she is not Jessica's boss.

16 SPUD—By industry tradition, a well is spudded the first time a drill bit drills new hole. Engineers say, "We spudded the well." Geologists say, "We spud the well." Neither side will ever yield.

CHAPTER 3—The Mentor

Barry scrambled down the rig-floor steps, Jessica on his mind. A breath of freshness. A sight to remember. A geologist on the rig floor, watching, talking. Not normal. Geologists more often hid in dark corners. Studied maps. Looked through microscopes at bits and pieces of rock. Counted fossils. Minded their own business—just the way he liked it.

But Jessica was different. She'd asked him questions no geologist had ever mentioned. He wondered if she might be a reporter. Incognito—like Daylight. Preparing a feature article about life offshore. Maybe a story about a company man. Like himself. Probably not.

He found her in the office. "Dinnertime. Great spread. At least a dozen entrees."

Without looking up from her laptop, she said, "Go ahead. I'll be there in ten."

So he found Tanker. Barry, a steak-and-potatoes guy, filled his plate with steak and potatoes. "You're missing out," he told Tanker. "Can't beat a rib-eye steak on a rig at sea."

"No red meat for me," Tanker said as he finessed his knife and fork to gain ground on a platter heaped with deep-fried jumbo shrimp. "Bad for my cholesterol."

Minutes later, Jessica placed her tray on the table and parked herself next to Tanker, across from Barry. "Mind if I join you?"

"No problem," Barry said.

Tanker wiped his mouth with a napkin and said, "Please do, Miss Jessica."

Barry eyed her plate. Steamed vegetables. Mashed potatoes. "They run out of steak?"

Jessica speared a piece of broccoli. "Nope. When I walked by, there were a half dozen, juicy and hot, for anybody who wanted one."

Barry sliced off another chunk of beef. He chewed slowly and gazed for the first time at the geologist's soft blue eyes under spiky blond hair. Cute. But with an interesting scar.

"That from a right hook?" he asked, pointing his fork her direction.

She sat up. "Pardon me?"

"The scar. Your left cheek. Right hook?"

She reached up and gently touched the scarred area. "Actually, no. Had it been, I'd be serving time for manslaughter."

"I wouldn't blame you." Barry ignored Tanker's chuckle and used a finger to ruffle his own right eyebrow. "I've got the same kind of scar, but it's hidden a lot better than yours."

Jessica crunched raw cauliflower, lips tight, her eyes on his.

"By the way," he said, "I didn't see your name on the personnel manifest."

She swallowed. "Probably because of the silent P."

He paused. "A silent P? In Jessica?"

A smile cracked her face. "P-H-E-R-M-A. Like a hard Italian mattress—very firma."

Before Barry could answer, a wall phone rang. He couldn't ignore it any more than he could a siren and flashing lights in his rearview mirror.

The steward answered and gave Tanker a nod. He took the call, then came back and excused himself from the table. "I'll be on the rig floor," he told Barry before making his exit.

"Do we need to go with him?" Jessica asked.

Barry dipped a bundle of extra-crunchy French fries into a puddle of ketchup. "We? No. Tanker would have told me if the problem involved the well."

My well, he thought, as he bit off the red-headed ends of the fries.

* * *

Barry agreed to show Jessica the well program. No geologist had ever bothered to ask, and he didn't mind. He unlocked his desk and handed her the quarter-inch-thick *Initial Exploration Plan*. "It's all public information—won't help me drill the well." Then he handed her a notebook. "This is the well program—less pages, more details. Sections include the objective of the well. Schedules and costs. Design criteria for casing, cement, drilling mud. Plus information about formation pressures, fracture gradients, and even a depth-versus-days chart."

She flipped pages, punctuating her scanning with questions.

He answered with explanations about the high rig cost, the sizes of planned strings of casing, the date she'd first get to see drilled cuttings.

Before her next question, he swiveled his office chair, faced her, and gazed into eyes that shared no secrets. "You writing a book?"

She placed the notebook on her desk and leaned forward, feet and knees together, hands folded in her lap, an answer on her lips. An answer that didn't come.

"So you *are* writing a book."

"No. Sorry." A determined look crossed her face. "I need to tell you something, but it's confidential. Which means I need to be able to trust you."

Barry mimicked her posture. "Macondo's a tight hole.[17] Open acreage in every direction. Everything we see out here is confidential, on a need-to-know basis only. You and I represent BP. We're it. If we can't trust each other, nobody can."

She glanced toward the open doorway. "I'm leaving BP next summer. Going back to school." Her face was red; her voice, quiet.

"Studying what? More geology? An MBA?"

"Petroleum engineering. A Master's." She snapped another look at the empty doorway. "I didn't want to tell you because I don't want my boss to know before it's time."

"Jessica, ease up—I'm not going to tell anybody. Why petroleum engineering?"[18]

She exhaled and settled deeper into her desk chair. "Offshore engineers always have a lot going on, and their responsibilities increase with experience. I need the challenge."

"And you're picking my brain for the get-ready part?"

She grinned. "Everything you know would be a good start."

Right. Fat chance. "What school?"

"Colorado School of Mines, or maybe Texas A&M. Where'd you go?"

"Penn State," he said. "My first job was as a reservoir engineer, in the office, but I hated the drudgery. You interested in drilling too?"

"Absolutely. If I'm going to be on a rig the rest of my life, I want your job."

"It's good work," Barry said. For the right person, he thought. "Have you considered Louisiana? They've got the best drilling program you can find."

"LSU? Absolutely. Good drilling, good sports. Definitely on my radar."

"I hope you know, petroleum engineering will be tough, especially for a geologist."

"*Especially* for a geologist? What's that supposed to mean?"

"Most geologists don't like numbers," Barry said. "They couldn't care less about drilling operations, how something's designed, or being responsible for the well."

Jessica blinked, fast, as if she had something in her eyes. "That seems a mighty bigheaded thing to say, since you probably don't have a clue about what geologists do."

He held up his hands. "I'm just saying geology's not a very big job."

Book 1

Her entire face puckered around her nose and mouth. "It takes both of us to drill the well. You do part A, I do part B."

"Yeah," Barry said, "you sort through bags of drilled cuttings for hours on end and try to figure out the geology, what kind of rocks are in the hole. Not a lot of work scope."

Jessica gave him a twist of a smile. "Ever fish for bass under sunset-red skies? Throw a top-water lure along the weed line or under a dock, let the ripples settle, then twitch the lure?"

"All the time, but what's that got to do with geology?"

"You know the feeling you get when you expect the water to explode? That's me each time I look through the scope at a new set of samples. The tiny pop of fluorescent colors from an oil show is every bit as exciting to me as a double-digit bass taking my top-water lure."

Barry couldn't imagine looking at samples from even one ten-foot interval of hole, let alone from ten *thousand* feet.

"You're shaking your head," Jessica said, "like you've got something besides coffee out here that keeps you perpetually excited."

"Hey, there's no script for making hole—drilling deeper. If we drill a foot of rock and all goes well, then the well's safe and I'm happy. But what happens when we drill that next foot? Maybe nothing. Or the bottom drops out of the well, and we lose circulation. Or we get stuck.[19] Or we twist off the drillpipe. Or the well kicks.[20] Or all the above at the same time. Or none of the above. And what about the next foot? Like twitching your lure—better be ready."

"And that's why we're both out here," Jessica said.

"With all due respect, I can drill the well without you."

"I bet you can, "but where will you drill it? What's your target? How deep? And how will you know what you've found, if anything?"

"I'm not saying we don't need geologists, it's just that—"

Jessica turned her chair sideways and swung her feet onto her desk, ankles crossed. "Just so you know, I'm a geological engineer, BS and Master's. Which means I've suffered through thermodynamics, strength of materials, differential equations—just like you. I think I can handle Drilling 101.[21] That said, I'll ask just one time. Will you help me?"

Barry picked up his coffee cup—cold and empty. An ugly flash of memory reminded him of the time his ex had asked if he would quit working offshore to save their cold-and-empty marriage. He'd answered honestly—not a pretty sight. And now, for the first time in years, somebody actually wanted his help. Felt kind of good. "I'm glad you're going back to school," he said, "and of course I'll help you."

"Thank you."

Book 1

"When I'm not busy."

She cocked her head. "For your information, when my boss assigned me to Macondo, I asked if I could go out early. Not a problem, he said, as long I stayed out of the way of the drilling department. That would be you, because he doesn't know jack about who you are."

"Where's this going?" Barry asked. "I said I'd help."

"Since we both have full-time jobs and neither of us works for the other, I won't bother you unless there's something exciting going on."

"*Exciting*," Barry said, "is not a good word out here. We want calm. Boring. Predictable. No mistakes. Safe. On time. Under budget."

"Nice speech, but I mean activities I don't normally get to see. I've been around drilling mud and casing, and I've seen things that can screw up progress and junk a well. But I've never participated in drilling decisions, or running casing, or fighting a kick—the kind of stuff you're responsible for every day. That's what I want to see and learn about. With your help."

"Works for me," Barry said. "Except kicks, which we can do without."

"Granted, but if the well does kick, I want to be there, start to finish."

Barry shrugged, but his heart wasn't in it. Kicks were serious business. Formation fluids—oil, gas, water—in the wellbore. Clock ticking. Zero tolerance for failure, which could cost him the well. Or worse. And the last thing he'd need during a kick was a bunch of show-and-tell questions from an excited geologist.

". . . mutual benefit," she said, her soft accent pulling him back into the conversation. "Like if I can help you with either mudlogging or wireline logging, just let me know."

"Count on it," Barry said, unable to think of anything about her business he didn't already know. "And you have my word on the grad-school thing."

He stood and stretched, checked his watch. Counted hours in his head. "If you're interested in the casing job, you better get some sleep. I'll have the steward call you when there's something *exciting* to see."

"As soon as I get the call," she said, "I'll head to the rig floor."

Watching her leave, Barry considered the upside of their agreement. Whatever happened in the days to come—good days, rough days, terrible-horrible-bad days—he'd at least have a BP colleague to talk to. Somebody who understood offshore and had expressed interest in the importance of his job.

Or somebody who could quickly drive him nuts.

---

17 TIGHT HOLE—Slang term for a proprietary exploration well. Data is confidential. Common when adjacent acreage is not yet leased.

18 ENGINEER—uses math and science to make projects faster and safer, and more productive, cost efficient, and environmentally friendly. Petroleum engineers apply geo-science and engineering (mechanical, structural, civil, geological, electrical) to the petroleum industry, whether designing, drilling, and managing wells, or producing oil and gas (O&G), or managing O&G reservoirs.

19 STUCK (drillpipe and drill bit)—if you can't pick up the drill string, go deeper, or rotate, you're stuck. Caused by cave ins, gumbo, tight hole (not the proprietary kind defined above—here it means the hole is physically restricted). On a rig, the term "fishing" refers to trying to get the pipe unstuck. If you can't get the pipe (the fish) free, you get to plug the old hole with cement and drill new hole (sidetrack) around the abandoned wellbore and its "fish."

20 KICK—If formation fluids (oil, gas, water) enter a wellbore otherwise full of drilling mud, the new fluids will expel (kick) mud out the top of the well (it overflows). Kicks require immediate attention, often referred to as well control. More on this topic later.

21 DRILLING 101—Slang term for the first-semester course in drilling engineering as part of petroleum engineering curriculum. Topics include drilling equipment, drill bits, rock mechanics, drilling fluids (mud), wellbore hydraulics, casing design, cement, well control.

CHAPTER 4—High and Deep

Barry spotted Jessica standing at the handrail, looking west. Clouds hid the late evening horizon, but she looked fresh and cheerful, dressed the color of sunshine. Like a pit-crew poster girl, waving a flag in a NASCAR winner's circle. Barry wore the same army-green coveralls he'd put on the day before. They looked like he'd napped in them, because he had.

But coveralls weren't his problem. He hadn't thought about his ex-wife for months, when he'd visited his daughters at Easter. Yet the previous evening he'd thought of her a number of times, compared Jessica to her. Not healthy. He didn't even know Jessica. But, hell, he apparently hadn't known his wife either.

He stood next to Jessica and leaned his elbows on the rail.

"Evening, Barry. Now what do I need to know about the hole you just drilled?" [22]

He enjoyed her sing-song soft drawl, wondered how it would be in a normal conversation, like in a restaurant, sharing a bottle of wine. In another life.

"Join me on the rig floor," he said, "and I'll show you." He led the way.

"The drilling assembly," he said, pointing to the components, "consists of a drill bit and reamer, which together drill a 32-1/2-inch hole. We drilled to 6,217 feet, which is 1,150 feet below seafloor. Now that we're done with it, we'll store the entire assembly until needed on another well."

Jessica stepped back a pace. "I've seen lots of drill bits but never that big. When I normally first get out to a rig, you guys are using drill bits about half that size."

"They'll come later, you can be sure of that."

"So what's next?"

"The casing crew.[23] They're rigging up to run 28-inch casing."

"Long or wide?"

"That's the outside diameter of the pipe," Barry said, then caught her grin.

"I couldn't help myself," she said. "I'm not usually concerned with the casing." She studied a page in her tally book. "I've got a lot of numbers—

but our depths don't match, since all my structural-mapping depths are based on sea level." [24]

"We use RKB, the rotary kelly bushing, right in the middle of the rig floor."

"I understand RKB," Jessica said, "but why can't you reference from sea level?"

"Two reasons. If the driller hangs 1,000 feet of pipe below the rig floor, we say he's run pipe to 1,000 feet RKB. Doesn't matter that the rig floor is 75 feet above the water. Second, once we run the drilling riser, all our pressure calculations are based on that elevation, with the rig floor being zero depth."

"Zero depth. Interesting term. Did you ever get the true water depth?"

"The seafloor's at 5,067 feet RKB, so you tell me."

A beat. "It's the difference in that and RKB. Which means it's 4,992 feet deep." [25]

"Our numbers match," Barry said.

Jessica wrote a note. "My, but it's good the water's not 288 feet deeper, because that'd be a mile. And there's no way I could ever jump off a rig this high or swim in water that deep." Her words were soft, but a shiver crossed her shoulders.

"Then I hope you never have to," Barry said, wondering just how much she feared heights, and how much she feared water, on a rig far from shore, in open ocean a mile deep.

———————————

[22] DRILLING—The process of making hole (getting deeper). Drill bit is attached to the heavy bottom-hole assembly (BHA), on top on which is the drillpipe. To get the drill bit to drill, the driller rotates the drillpipe, which rotates the drill bit. The driller also slacks off on the brake, which applies a portion of the weight of the heavy BHA onto the drill bit. The driller pumps mud down the drillpipe, where it exits through holes in the drill bit, cools the rotating drill bit, and mixes with the cuttings. The mud and cuttings return to the surface up the annulus. At the surface, the cuttings are screened from the mud (for assessment by the geologist), and the clean mud goes into the mud pits for reuse. As a well is deepened, drilling is often accomplished with a down-hole mud turbine added to the BHA. The turbine rotates the drill bit without the drillpipe having to be rotated.

23 CASING CREW—Contracted by the operator (BP), the crew makes up (connects the joints) and runs the casing into the wellbore. Though casing crews will come and go to run casing throughout the Macondo well, the workers are not named in the story.

24 DEPTHS—Geological records are referenced to sea level, even in Colorado. Drilling-related depths are based on RKB—the Rotary Kelly Bushing. RKB allows easy calculation of how much pipe it takes to reach bottom (i.e., the bottom of the well).

25 From Reference (3)—BP's Internal Investigation—pp 19—includes all referenced depths throughout the well (and as shown in subsequent references).

CHAPTER 5—Cement Job

Come morning, Jessica found Barry in the office, apparently talking to his boss, a tradition she'd observed forever—the drilling department's not-to-be-interrupted morning report. Didn't matter, she knew how to get to the galley.

She had two English muffins, a poached egg, and an earful of idle chatter from the adjacent table. The casing crew had finished their job before sunrise and were ready to go to the house.

After breakfast, alone in the office, she read the BP morning report. No surprises. Next stop, the rig floor, where she knew she'd find Barry and company.

Yep. Barry, Tanker, and Daylight. Barry invited her to join him for a cup of doghouse coffee.

They sipped together.

He gagged.

She smacked her lips.

"So you've got a little more than 1,000 feet of 28-inch casing," she said to Barry, "standing inside structural casing that's three feet across. 26 Won't it try to tip over?"

"No. First, because it's got nowhere to go. Second, we built in all kinds of safety factors for burst, collapse, and tensile strength. The pipe we just ran weighs more than two hundred pounds a foot, and—"

"Barry, I'm teasing. And since I'll learn a lot about casing design in school, how about just a brief summary?"

"The pipe's really strong," he said.

"Better. So now we cement the casing?"27

"Yep." Barry checked his watch, then raised his hand and got Tanker's attention. "Call when you're ready," he told the toolpusher. "Jessica and I will be at the cement unit."

Barry took off. Jessica, excited, stayed in his wake. She'd never even been on a rig during a cement job, as those occasions had always represented a day or three off, time to go to the office, do laundry, rejuvenate.

She followed Barry into the bowels of the *Marianas*. Places she'd never seen on any offshore rig because she'd never needed or wanted to go there. Everything made of steel, painted gray, labeled with words and arrows in white and red. Piping and pumps and tanks and walkways and handrails along drop-offs. She was intimidated by nothing she saw, except the drop-offs. Drop-offs, like cliffs and mountain roads, high places in general, she didn't like.

Though there were bright lights everywhere, with not a dark corner to be seen, a six-foot-two overhead beam caught six-foot-tall Barry's hardhat with a resounding thump, followed by his guttural "Sonofabitch."

Jessica's hardhat missed the beam by several inches.

Through Barry's introduction, Jessica met and shook hands with a Halliburton cement operator and his supervisor, both in red coveralls and red hardhats.[28] She looked down at her own pair of butter-yellow coveralls and wondered if she owed it to herself to order a red pair. But not Halliburton red, she told herself. Too bright. Perhaps something a little softer.

The cement unit—she'd been afraid to ask Barry to describe it in less than a thousand words—appeared anything but user-friendly. She'd seen massive, truck-mounted cement systems on the road and parked at land rigs on a number of occasions, but never related to her scope of work. The unit in front of her was tiny by comparison. The skid-mounted equipment had been painted bright red, perhaps recently, as no blemish marred its surface.

The cement operator stood on a raised platform in front of a control panel overlooking two rectangular tanks full of water. Behind him, an area crammed with what appeared to be large pumps and giant pistons—silent reminders of courses she'd taken in fluid dynamics.

Barry took out his tally book, so Jessica did the same. The Halliburton supervisor referred to a clipboard. Barry read numbers from his notes, comparing his to the supervisor's. Every time Barry read a number, the cementer said, "Check." Casing sizes with both inside and outside diameters and weight per foot. "Check." The depth to which the structural casing had been set. "Check." The depth to which the 32-1/2-inch hole had been drilled. "Check."

The cementer took a phone call, nodded to Barry, hung up.

Barry resumed the numbers. The footage of 28-inch casing that had been run, plus the dimensions of the mile-long landing string on which the casing hung. Jessica would have called the landing string *drillpipe*, since that's what it was.

"Check." "Check." "Check."

Jessica thought about challenging a number, maybe just yelling "Twenty-six" and seeing which guy crapped his pants trying to figure out how he'd gone wrong. She had dribs and drabs of most of the numbers but was missing too many to know whether she should add, multiply, or divide. She played a mental game trying to keep up, until the guys started talking about cubic feet per foot and sacks and barrels.

Barry and the cement supervisor grunted and nodded, which Jessica reckoned to be some kind of mental contract. Checks and balances. Two brains. Zero tolerance. Sure enough, Barry made a phone call and the cementer pushed buttons to start pumps and began doing something to the water in the bright-red tanks. Within minutes the water turned gray and got thicker, then much thicker, then much more gray.

She got it. Cement. Sacks of cement. All her years on rigs, and she'd never stood by a cement unit actually mixing cement. She gave herself a pat on the back for having learned a lot without Barry's help. No mentoring. Learning by osmosis. Simple confirmation by example that when he was busy she needed to respect his job. Stay out of his way.

Hah. Those were the parts she'd want to see, ask questions about.

Barry gave her a come-with-me nod. He led the way, walking fast, backtracking the route they'd come. He ducked under the beam and kept walking. Jessica tiptoed as high as she could to see if she could tag the beam with her hardhat. She missed, by a lot.

Barry was all business on the rig floor. He spoke to Tanker, then to Daylight. The three most senior on the rig, in tune with each other, all focused on the task at hand.

Jessica, invisible.

No rig-floor activity—just the opposite of what she saw on the seafloor, courtesy of Barry's submarine and its headlights. The video camera showed the top of the 36 inch casing, inside of which the 28 inch casing hung. Swimming creatures speckled the entire area. Small stuff. Nothing that looked like a proper fish. All in perfect silence, a mile down. She'd still not seen the unmanned submarine, but she was getting good at squinting her eyes just right to see the pictures the small vessel sent from the seafloor up to the rig.

The casing hung from the landing string, which stuck up through the rig floor. A red device made of valves—probably Halliburton's—had been connected to the top of the landing string, above the rig floor. Connected to the red device were a number of short, flex-joint steel pipes—*chiksans*—that crossed the rig floor and disappeared below.

The noise level from rig machinery ratcheted up, as the chiksan pipes rattled, jumped, and thumped, albeit fractions of an inch, as if they were alive. Filled with wet cement, Jessica guessed. Driven by those big Halliburton pistons on the other end of the line.

As soon as she caught Daylight looking her way, Jessica pointed to the vibrating line and shot him a thumbs up.

Always the gentleman, he spit in his cup, grinned, and returned the gesture.

The video screen changed. Water, slightly muddy, billowed at the seafloor from the top of the casing—all in shades of gray, lighted by the submarine.

She glanced back at the trio of men. All busy. Daylight, always working, was up from his seat, wiping down his instrument panel with a rag. Barry and Tanker, perhaps sensing her presence, reading her thoughts, looked right at her. Tanker sent her a thumbs up, and Barry followed. A double thumb job.

Not to be outdone, she raised her thumbs—nails manicured—toward the two honchos. Nodded for good measure. But they didn't need her for their jobs. Maybe didn't even want her around—except for her geological work—in spite of her good intentions. Perhaps by the end of the well they'd feel different about her.

She set her mind to making that happen.

––––––––––––

26 See **Diagram 3**, page 51 herein—(Depth of 28-inch Casing). Note that the well is now deeper and the casing is smaller diameter.

27 CEMENT—All casing strings placed in drilled holes need cement to keep them in place and to isolate rocks, fluids, and pressures. Cement slurry is pumped down the inside of the casing and up the outside (the annulus). By adding chemical retarders, the cement slurry is kept pumpable (called green cement, with the consistency of a chocolate malt) for a pre-determined number of hours, to allow time for pumping and for resolving problems.

28 CEMENT OPERATOR (and supervisor)—Once casing is run, specialist contractors like Halliburton (hired by BP) mix and pump cement down the casing and up the annulus that surrounds the casing. They also execute other necessary pumping jobs (squeeze cement into casing shoe, set packers, set cement plugs, pressure test for leaks). Though involved throughout the drilling of the Macondo well, these positions are not named in the story.

Book 1

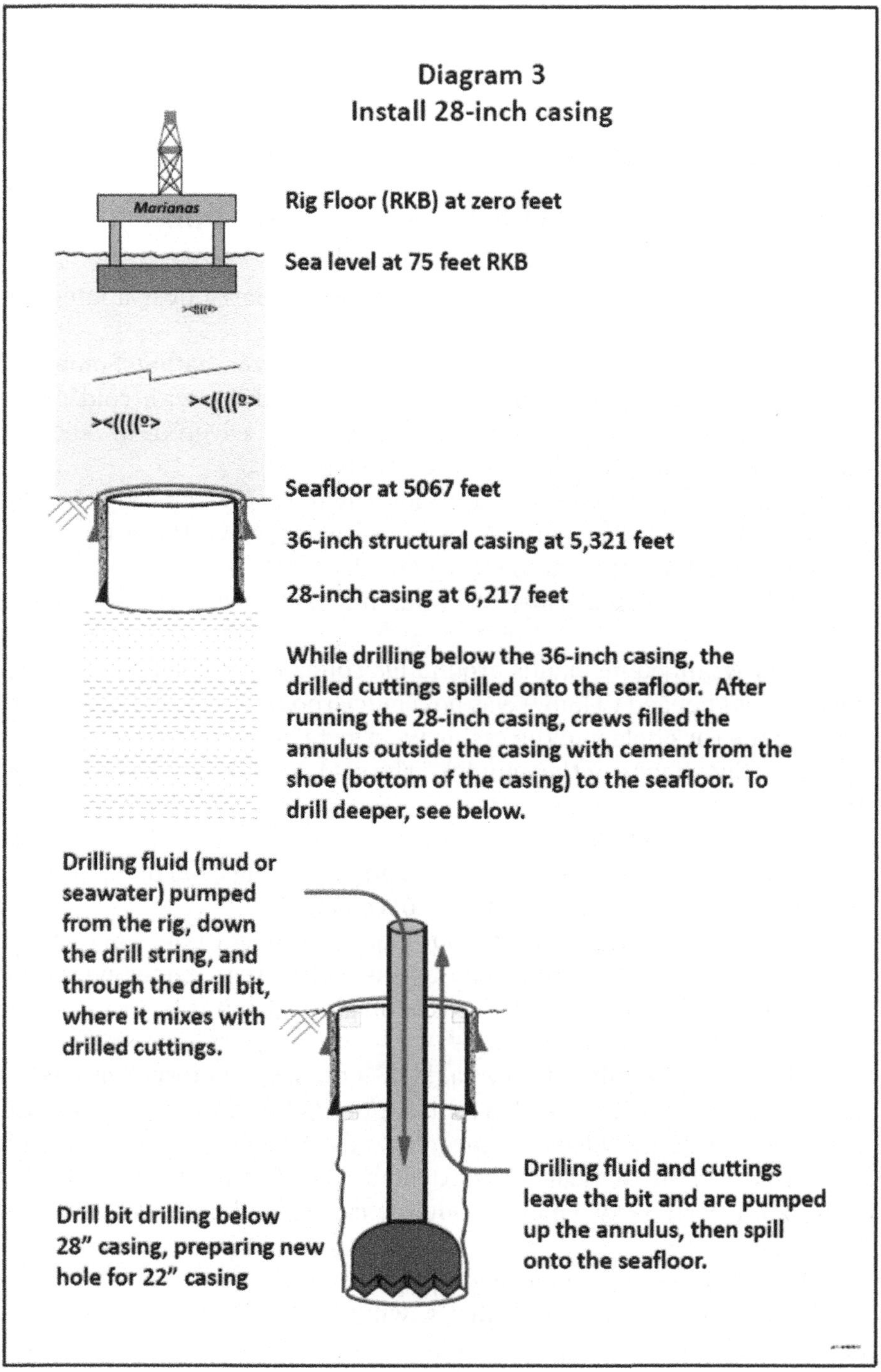

Diagram 3
Install 28-inch casing

Rig Floor (RKB) at zero feet

Sea level at 75 feet RKB

Seafloor at 5067 feet

36-inch structural casing at 5,321 feet

28-inch casing at 6,217 feet

While drilling below the 36-inch casing, the drilled cuttings spilled onto the seafloor. After running the 28-inch casing, crews filled the annulus outside the casing with cement from the shoe (bottom of the casing) to the seafloor. To drill deeper, see below.

Drilling fluid (mud or seawater) pumped from the rig, down the drill string, and through the drill bit, where it mixes with drilled cuttings.

Drilling fluid and cuttings leave the bit and are pumped up the annulus, then spill onto the seafloor.

Drill bit drilling below 28" casing, preparing new hole for 22" casing

Diagram 3

CHAPTER 6—The Foundation

"So I'm no longer a cement virgin," Jessica told Barry over a late-night snack in the galley.[29]

They weren't alone. Three guys in line. A dozen eating. Some with pancakes and eggs and giant cinnamon rolls. Others with cold cereal. Another with a massive pinkish-purple steak covered in ketchup. Reminded Jessica of how her dad used to eat.

"Cement virgin?" Barry said. "Not anymore. By the way, I didn't include you in the calculations—way too noisy down there for lots of numbers, and I didn't want to scare you off."

"Numbers don't scare me," she said, "but tell me what you did. The big picture."

Barry wrapped his hands around his coffee cup. "The quick version, then. What I needed to know was how much room there was between the walls of the rock hole and the casing we'd just run."

"Like the volume of the annulus," she said, having been on a rig a few times herself.

He nodded, said nothing.

"It seems like the tricky part," she said, "is getting from diameters and lengths of pipe, to sacks and volumes of cement."

"You got it, but it's easier than it looks. We've got tables and charts, books of conversion factors, and no-brainer software that does everything except prevent garbage answers based on garbage input."

"Example?"

Elbows on the table, Barry said, "A haunting memory. I missed an input number—strokes per barrel for a pump—and my little computer gave me a correctly-calculated incorrect answer, which I used to tell the toolpusher when to take the next step. We ended up with about a thousand feet of hard cement inside the casing, and not near enough in the annulus."

"I bet that was a fun morning report."

"Not that I recall. And that's why I do all the checking and crosschecking with the cementers."

"So for this job you made your calculations and then confirmed them with Halliburton. Then what? And I don't need the numbers."

"No numbers? And you want to be an engineer?"

"I am an engineer."

"Sorry, *petroleum* engineer?"

"Thank you. And it's still yes. No numbers. Just the highlights."

Barry shrugged. "Won't seem natural without numbers, but I'll try. The cement crew mixed the calculated sacks of cement to the agreed density, then pumped the slurry down the landing string and through the inside of the casing. We followed the cement with enough barrels of seawater to make sure the cement exited the casing and came up around the outside of the pipe, in the annulus. We left about fifty feet of cement inside the casing."

Jessica held back her smile. "That was good, Barry. A little wordy, but a lot better than reading a textbook. But why'd you leave cement inside the casing?"

"Because when the drill bit finds hard cement inside, we'll know it's also hard on the outside—which is what we want."

"So in the annulus," Jessica said, checking her notes, "the bottom of the cement is at the bottom of the pipe—the shoe. Where's the top?"

Barry pointed an index finger toward the ceiling. "As high as we could pump it."

Jessica played the game. "And that might be where?"

"You saw it. The ROV camera, when the flow from the casing changed from muddy water to what looked like lava. That means the top of the cement is at 5,067 feet, which is at the top of the 28-inch casing, which is at the seafloor. We can't pump anything higher than that—at least not yet."

Jessica did the math. "So we've got cement from the seafloor down to the bottom of the casing. Seems like overkill, more than a thousand feet of cement. That's a heck-of-a-thick patio. Why so much?"

"We're getting ready to drill deeper," Barry said, "a lot deeper, and we accomplished two things. First, we covered up about a thousand feet of weak rocks that'll never bother us again. Second, and perhaps more important, we finished a major portion of a strong foundation that we'll need for the rest of the well."

"Then I have two questions for you," Jessica said. "First, are all the casing and cement jobs going to be this easy? Because if they are, I certainly don't need to witness another one."

"The calculations won't change," Barry said, "but the cement formulations will. Lots of variables. Well geometry. Depth, pressure, temperature. And especially if we've got hydrocarbons, lost-circulation zones, or both."

Hydrocarbons. Jessica's favorite topic. "Okay, make a note. Since I'm out here to help discover a huge reserve of oil and gas, I may want to watch a future cement job. Now second question. You said we'd completed a portion of the foundation. What foundation?"

"After we drill the next section of wellbore," Barry said, "to about 8,000 feet, which is 3,000 feet below the seafloor, we'll run 22-inch casing to bottom. It'll get the same kind of annular cement job, from the bottom of the casing up to the seafloor. That casing, with a special subsea wellhead attached at the seafloor, will finish the foundation, and it'll be one strong mother."

Jessica scanned her notes. "So we'll have 22-inch steel casing that's surrounded by cement, inside 28-inch steel casing that's surrounded by cement, all inside 36-inch steel casing that's buried 250 feet into the seafloor. Is all that maybe a clue as to the importance of the *special subsea wellhead*?"

Barry leaned forward, started to speak.

Jessica gave him a time-out sign and picked up their empty coffee cups. "I'm buying this time,[30] because I want to hear the rest of the story before I give you the news about my schedule."

She was at the coffee pot, scooping sugar into her mug, when the scraggly-haired galley hand stepped up beside her, his shoulder against hers. Nothing serious. Nothing aggressive. But in her space. With no escape other than departure. Like trying to avoid a block of ice in a hot tub. She stood her ground.

He used a clean white towel to wipe up nonexistent stains around the pitcher of cream. "You quite the coffee drinker, Miss *Jessica*," he said, pushing harder against her. "Does a mug in each hand and all that sugar mean you a sharin' person?"

"Sharing?" she said as she stepped away and turned to face him. "No. One's for drinking, and the other's for backup. You know, in case I spill one, or toss it, or need to swing it like a hammer."

Big grin. "Yeah, right," he said, his eyes flashing from her face to her chest, back and forth, up and down. He chuckled, then left her standing and disappeared into the kitchen.

Bastard. Not what she wanted. She added cream to Barry's mug and returned to the table.

"Who's your snuggly friend?" Barry asked.

"Not a friend—just somebody trying to be funny," she lied. "I told him he was out of line."

Not a muscle in Barry's face moved, though his eyelids seemed to harden. "If he gives you the tiniest pinch of shit, I want to know."

Jessica nodded.

"Say it," Barry said.

"It's nothing, but I'll tell you if he doesn't back off."

Barry looked toward the kitchen, then back at Jessica. "We don't need to talk about the wellhead for a few days, but you mentioned your schedule. Anything I can help with?"

"Depends on when I can start collecting cuttings.[31] Any idea?"

"Like the leg bone connected to the knee bone," Barry said. "From the bottom up—we'll run the next string of casing with the wellhead attached. That will allow us to run the blowout preventers. With the BOPs in place, we can run the marine riser.[32] And we need the riser before we can drill with mud returns back to the rig. And you need mud returns before you see your first cuttings."

"And all this takes place after we drill more hole and run another string of casing. How long will that take?"

"Eight to ten days," Barry said, "unless we have hole problems."

Jessica thought about the days until she'd be able to collect her first geologic sample. She had two choices: stay on the rig during the interim and learn what she could, or go back to the office and her own cozy apartment for a week. Two choices: Q & A with Barry, or R & R at home.

She turned her empty coffee mug upside down. "If you can get me on the manifest tomorrow, I'll go to town for a few days. I've got plenty to do in the office and you don't need me bugging you. If your time estimate changes, please get word to me, because I want to see and touch the wellhead before it disappears forever."

Barry picked up both coffee mugs. "Then enjoy your days off, because it'll be busy when you get back." He returned the empty cups to the service counter, and they left the galley together, headed toward the office. "There's a chopper going to the beach at nine in the morning," he added, "right before the weekly safety meeting, which I know you'll hate missing."

"You think I should stay for the meeting?"

"I was teasing."

"So was I."

"There'll be no problem getting you a seat."

They said their goodnights, but Jessica heard something tired in his voice, saw a slump in his shoulders. Maybe she'd feel that way, too, when she matched his 20-plus years as a company man. Twenty years of fourteen-day hitches, drilling agonies, safety meetings, two-hour catnaps, and a constant deluge of calls to and from the office.

She returned to the galley, alone, and picked up a clean coffee mug. A clean, heavy, earthenware coffee mug. For a middle-of-the-night sip of water, she would have said, had she been asked.

In her room, she sat on the edge of her bed, pulled off her boots, and began her mental checklist of the outfits she needed to buy online. Rush delivery. Three days maximum.

She made sure the door to the hallway was shut tight, then collapsed into bed, with two fingers of her right hand locked into the handle of the heavy coffee mug under her pillow.

---

[29] GALLEY— Open around the clock, serving four hot meals a day, no charge, ever. The impeccably polished stainless-steel buffet line serves the hundred-plus personnel on board. A number of personnel work from six until six. Others from twelve to twelve. Supervisors and service hands eat when they can. With 24/7 drilling operation, someone is always working, sleeping, getting up, going down, or just plain hungry. Dinners and breakfasts overlap, with unlimited cookies, salads, eggs, pie, steaks, desserts, soup, fries . . . available around the clock.

[30] BUYING—Slang term, since there's nowhere on a rig to spend even a nickel. *I'm buying* means let's go eat, or get a cup of coffee.

[31] CUTTINGS—rock chips generated by the drill bit . . . actual samples of the rocks being drilled.

[32] MARINE RISER (or Drilling Riser)—Large-diameter (21-inch OD) pipe that connects the floating rig to the blowout preventers (BOPs), which are attached to the wellhead at the seafloor. The well is drilled down through the riser and BOPs. Drilled cuttings are pumped up to the rig through the drillpipe-riser annulus. Until this equipment is installed, the geologist sees no cuttings.

CHAPTER 7—The Wellhead

Six days later, the staccato beat of chopper blades got Barry's attention. He'd seen the manifest of names. One stood out. Silent P.

Finish the job, he told himself.

He'd begun the day down below the rig floor with Tanker and Daylight's crew testing the blowout preventers. After each function and pressure test, he and Tanker had check-marked and initialed their approvals. All systems go. Barry signed the test sheet and closed his notebook.

"Gotta go," he told Tanker.

He climbed steel-grating steps up to the pipe deck and watched Jessica Pherma disembark. Convinced himself he needed a cup of coffee. In the galley. Now. Before she got there. Which she would. Soon.

Having taken a shortcut to the galley, he was seated at a table, mug in hand, when she walked in, all smiles, choppy blond hair bouncing.

"Well, look who's here," she said, encased in pristine baby-blue coveralls that matched her eyes. "Seems I left you in the same seat a week ago."

"Hey," Barry said, casual, macho. He stood and shook her hand. "Welcome back. Grab a cup—tell me about your days off."

Sipping and listening, he learned she'd slept-in most mornings, worked late in the office most evenings, shopped at the Galleria, eaten at some of Houston's finest, and had given eBay a number of quick-delivery challenges.

A half hour later she refilled their coffees—his with cream, hers black with three sugars.

"I worked around the clock," Barry said. "Spent millions of BP dollars but not a dime of my own, ate four meals a day within twenty feet of this table, and called my folks. Twice."

Jessica, two palms up, radiant smile. "And the well?"

"We're getting ready to run 22-inch casing and the wellhead. Drink up and get your gear—it's show-and-tell time."

* * *

They huddled at the rig end of the pipe rack.[33] Shaded safety glasses and hardhats shielded their eyes. Barry pointed and prodded, describing details as Jessica asked questions.

"When this last joint of 22-inch casing is run," he said, slapping the horizontal body of the pipe, "the two goodies on the end will be looking up. The lower unit's the wellhead, and the upper is a connector that will receive the blowout preventers."

"Interesting." She opened her tally book, clicked her pen.

"The connector's important because its matching upper half will be attached to the bottom of the blowout preventers. The two halves mate together to attach the BOP to the wellhead."

Jessica stuck her head inside the connector and yelled, "Hello."

An echo returned through the casing: "Hello, Miss Jessica." She looked up and waved to a grinning roustabout standing at the other end of the pipe.

Barry wondered if everybody on the rig knew her name. Probably.

"Tell me about the wellhead," Jessica said, still peering inside, "since it's so almighty important to the foundation you mentioned. I'm seeing sloped surfaces, ring grooves, a number of profiles." She stepped back and adjusted her hardhat. "I'm sure all those details are important, but surely there's a nice, simple explanation. You know, just the big picture, in words I can use with my fourteen-year-old nephew."

"You want my help," Barry said, having thought about his volunteer mentoring job for most of the week, "but you're taking the fun out of this."

"Barry, I've worried about sessions like this for almost week, and I can feel it coming. You can't wait to tell me all you know. I'm duly impressed with your breadth of knowledge, I really am, but the fun is all for you, not for me. I need the big picture first. Details later." She gently rubbed the scarred area on her cheek, her face in deep-thought mode. "Have you ever explained your job to your parents?"

"Of course," he said, "in detail, because they want to know."

"Or because they want to show their interest, keep you happy? Do you tell them about profiles and seals and barrels? Speak in acronyms? Tell them you ran a NPQ on the RDU, then pumped TXP until you ran out of gas?"

Barry recalled a number of good stories he'd entertained his folks with that sounded similar. But this was Jessica.

He patted the wellhead and connector. "Then let's do it like this. After we run and cement the 22-inch casing in the well, with all this *stuff* attached to the top, we'll be ready to run the BOP. The bottom of the BOP will attach to this connector." He patted the brute once again.

"That was good," Jessica said. "Keep going."

"Later, after we drill more hole, there'll be a time when we want to run casing all the way from the seafloor to the bottom of the well. To help us, we'll attach a donut-shaped device to the top of the casing. When that *donut* gets to this contraption—" He patted the wellhead. "—it'll be too big to go through, and therefore the casing won't go any deeper."

"I can picture a donut. Does it have a name? A function?"

"Yes," Barry said, "a nephew-size name." He found a grin, proud of himself. "The donut is called the *casing hanger*, because the casing—after we cement it in place—will hang from the hanger." He paused, waiting for even the whiff of an accolade.

"So the donut and casing just hang there forever," stoic Jessica said.

"Yes, but with a lockdown seal ring."

Jessica gave him the look.

"Okay, for your nephew. If, and that's a big if, this well ever needs production casing, a lockdown *device* will be installed on top of the *donut* to lock it in place."[34]

"That was good. Did you leave out anything?"

"A ton," Barry said, "which we won't worry about unless the well's a discovery."

"Oh, *discovery*, such a delicious word." She added a number of notes to her tally book, then peered again into the throat of the heavy equipment.

"Should I guess," she said, "that the remainder of the grooves in the wellhead are for additional casing strings, each with its own donut?"

"You got it. The well program includes no less than seven strings of casing, though they'll be a mix of long strings and liners.[35] The long strings of casing are the ones that need to be hung from the wellhead with donuts."

"Long strings and liners," Jessica said. "How do you decide which to run?"

"Let's rain-check the answer until it's time to make the decision. It'll mean more to you—and to your nephew—than anything I can tell you now."

Jessica opened her tally book and wrote a note. She read it to Barry: "*Get liner versus long-string explanation. From Barry. As promised.*"

"Agreed," he said, not remembering the word *promise*, though he liked the spirit of her note.

<u>33</u> PIPE RACK—frames and vertical stanchions on the pipe deck that accommodate storage of multiple joints of drillpipe and casing, stacked in layers, so each joint can be accurately measured before being fed through the vee door (triangular opening at base of derrick) to the rig floor.

<u>34</u> CASING HANGER (donut) & LOCKDOWN SEAL RING—shown in **Diagram 17** (page 222 herein). Will mean more later in the story.

<u>35</u> LINER (or casing liner)—a short section of casing (a short casing string) that runs from the bottom of the previous larger-diameter casing (with a couple hundred feet of overlap) to the bottom of the hole.

CHAPTER 8—Waiting

That evening Barry left Tanker on the rig floor with the casing crew and went down to the Sperry Sun mudlogging unit.[36] Jessica invited him in and asked him if he was drilling yet.

"Not yet," he told her. "Still have to run and cement the casing, install the BOP and riser, drill out and test the shoe, and change out the mud system."

She threw him the time-out sign and gave him a short tour. Showed him laboratory equipment, an array of chemicals, a gas chromatograph, and a stereoscopic binocular microscope. "For viewing marker fossils and cuttings in 3D," she told him.

He nodded.

Hands on hips, she said, "I need this stuff to watch for shows,[37] which of course won't happen until you start drilling."

He ignored the bait. "Nice rundown, but I've been in mudlogging units thousands of hours."

"That's why I gave you the elementary tour," she said. "I didn't want to bore you with esoteric details."

He liked the flash of a tease that crossed her face and twisted her grin.

And besides, he and Jessica jointly depended on the mudloggers, though with different goals. He depended on them for backup drilling information, as received from rig-floor and cement-unit monitors and gauges. Key information included how fast they were drilling, the amount of torque required to turn the drill bit, and the amount of drag—friction—when the driller picked up the drill bit or shoved it to bottom. The mudloggers were also good for formation pressures. Gas readings. Geology—sand or shale. And high on the list, the mudloggers kept a volumetric count of all drilling fluids in the active system, both in the wellbore and in mud pits on the rig. Their job, as paid contractors, was to plot data, watch trends, and raise the alarm—to the company man, Barry—if they observed anything out of sorts with any drilling-related parameter.

Jessica, as BP's on-site geologist, wasn't paid to drill the well. Instead, her job was to focus on lithology—depths and rock types and microfossils coming out of the wellbore on a foot-by-foot basis. Importantly, her data and constant communications with her boss in Houston allowed her to make depth adjustments to deep, structural, subsurface maps that had been derived from seismic data months before the well was approved. And the bottom line was simple: the Approval for Expenditure—AFE[38]— had been authorized for an exploration well to be drilled to assess the geology, fine-tune the subsurface maps, and confirm or deny the presence of commercial hydrocarbons, all on Jessica's shoulders. For these activities, she was the on-rig boss. Of course she couldn't do anything without the well, for which Barry was responsible. The well Barry would consider successful if he were able to drill it to the AFE depth on time and under budget.

If Jessica and Barry were lucky, she would identify at least one good *show*, the tiniest trace of oil and gas, [39] which they would then pursue with vigor, hoping to confirm a major discovery.

Barry had no role in the search for shows, except to provide drilled cuttings to Jessica. Though he'd always been thrilled to drill a discovery well—like maybe his shares of BP stock would jump in value— hydrocarbons in the wellbore significantly added to the complexity and cost of the well. His well. His job.

* * *

Hours later, Barry again looked in on Jessica. One of the mudloggers directed him to the BP office. "Miss Jessica said she wouldn't be back until she had some drilled cuttings to look at. Didn't seem too happy when she said it."

"Tomorrow," Barry said.

Which left him time to have dinner with Jessica. Maybe.

Which didn't work. After looking for her, and not inclined to knock on her quarters door, he noted her tally book, abandoned on her office desk. He spent ten minutes flipping through pages. A repeated acronym—*SLD*. Then a dozen quotes—things she and he had discussed.

Cement.

The wellhead.

Plus pages from before Macondo.

And famous quotations.

Right-of-way's something you have . . .

Trust your life to nobody because . . .

Doors plus drawers . . .

Yak, yak, yak.

Footsteps in the hall.

He replaced the booklet and about had a panic attack trying to get settled in his own chair.

Wasn't her.

---

[36] MUDLOGGING UNIT—off limits to all personnel on the Marianas, with the exception of the mudloggers, BP staff, and the toolpusher. Mudlogging, wireline-logging, and drilling data were highly protected, shared on the rig, and transmitted to the beach, 24/7, on a need-to-know basis only (tight hole). The mudlogging unit—a sophisticated, air-tight, air-conditioned trailer—was wired with almost two dozen computer screens, monitors, gauges, and charts, a number of which replicated the half-dozen monitors in the driller's cabin. Though the two systems provided full redundancy, the mudloggers were contracted by BP to monitor and record drilling parameters.

[37] SHOW—evidence of oil and gas obtained from drilled cuttings, or as seen in the mud returns. A show, most often first seen by an onsite geologist, can prove to be either "nothing" or the key indicator of a potentially successful exploration well.

[38] AFE—Approval for Expenditure. The Macondo AFE called for a 51-day exploration well to a depth of 19,650 feet, for a dry-hole abandonment cost of $96 million. See Reference (1).

[39] OIL & GAS IN ROCKS—Oil & gas (O&G) are found primarily in sedimentary rocks like the solid rocks that make up the walls of the Grand Canyon, not in pools or caves as often seen in young-adult books and the media. A key criteria for sedimentary rocks is porosity—a measure of the space between the grains that make up the rock. Another is permeability—a measure of a rock's ability to transmit fluid. Shale has some porosity but near-zero permeability; hence, no flow. Sandstones can have high porosities (so can hold lots of water, oil, or gas) and high permeabilities (which allow O&G to be extracted). Most O&G in the Gulf of Mexico is found in sandstone reservoirs.

CHAPTER 9—Expectations

Barry saw Jessica only occasionally during the 22-inch casing and cement job. He'd expected her to watch the running of the blowout preventers, though once the BOP splashed—entered the water—there hadn't been much to see. Well, except for the latch-up on bottom, seen through the video cameras. Now *that* had been exciting. Too bad she missed it—he didn't expect to see the BOPs again until the end of the well, likely in December.

He was working on the last few bites of a fine chunk of prime rib when Jessica joined him for an early-afternoon dessert, the first time he'd seen her all day. She targeted carrots and cookies—he joined her for a cookie.

Afterwards, she opened her tally book, asked him about the setting-depth of the casing.

"I recall you don't want numbers," he said, aware that if she'd been with him, she could have witnessed every aspect of the job and noted the depths herself.

"Some I do, for my records."

"If you want depths, they're listed on the IADC report." [40]

She closed her tally book, gave him the stare.

No emotion on her face. Thinking. Reminded him of the pre-divorce looks he used to get any time he broke the news he had to go to the office on one of his days off. But that was then—so he aimed for a compromise between bitter and sweet. "The wellhead, as I recall, is ten feet above the seafloor. And below the wellhead, we ran almost 3,000 feet of casing. That would put the shoe near 8,000 feet."

A long beat. "You need to talk to the steward," she said, continuing her hard stare, "because I've got the feeling your bloody steak was a quart low."

"Meaning?" Barry asked.

"You're talking ABC, thinking XYZ. My skin's thicker than that—spit it out."

"You missed watching us run the blowout preventers."

"I didn't know it required an audience."

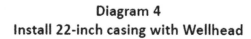

Diagram 4
Install 22-inch casing with Wellhead

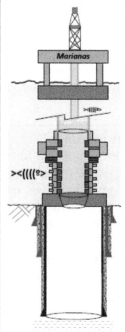

Rig Floor (RKB) at zero feet

Sea level at 75 feet RKB

Marine Riser (connects rig to blowout preventer and wellhead)

Blowout Preventers (BOPs) on top of Wellhead at 5,057 feet

Seafloor at 5,067 feet

36-inch structural casing at 5,321 feet

28-inch casing at 6,217 feet

22-inch casing at 7,937 feet, with wellhead permanently attached at 5,057 feet

While drilling below the 28" casing, the cuttings went to the seafloor. After running the 22" casing and the wellhead, the 22" casing's annulus was filled with cement from the casing shoe to the seafloor.

This critical step allowed the BOPs (18-3/4" inside diameter) to be installed on top of the wellhead, and the riser (21" outside diameter) to be installed on top of the BOPs. Note: scale of drawings is exaggerated

New drilled cuttings from below the 22" casing will be circulated up the drillpipe annulus (through rock, casing, BOP, and riser) to the rig

Diagram 4

"None necessary. But you won't learn squat about BOPs in school, and there's nothing like standing next to one, five stories tall, hundreds of tons, to give you a feel for one of the most-important safety devices you'll ever depend on when you're running a rig."

"Which means there are other equally important safety devices that require field trips? Like what?"

"There are many," he said, "some you can touch, others you can't—including the simplicity of a column of correct-weight drilling fluid. But whether it's mud or the BOP, neither means anything without a good brain making educated decisions. Which means you need to know how everything works."

Jessica, arms folded, stretched at least an inch taller. "You can't push me into a show-and-tell. I'll see the BOP the next time we run it."

"Jessica, I don't work for you, nor you for me. But you asked for my help. If there's to be a show-and-tell, it's because you asked for it. So until I hear otherwise, from you, I'll be on your case, telling you what you need to know, showing you what you need to see. Like the BOP. Why? Because when you go to school, or if you ever want to describe a BOP to your nephew, it'd be nice to say you actually walked by one."

He expected an argument. Instead, she leaned forward in her chair. "Sorry," she said, her voice soft. "I don't expect a full-size BOP in a grad-school lab, but there'll be lots of drawings, maybe videos."

He appreciated her words, her attitude—the difference between lemons and lemonade. "Good idea. I have a BOP schematic in the office. You interested?"

She hesitated, then, "Yes. Thank you. I'll fill our coffees; you get us a stack of chocolate-chip cookies."

Right, he thought, a BOP tea-party.

---

[40] See **Diagram 4**—(Depth of 22-inch Casing). See page 65 herein. This drawing shows the BOP installed on top of the wellhead, and the drilling riser on top of the BOP. The IADC form (International Association of Drilling Contractors) is the official morning report for most drilling operations around the world. Used to record all daily well activities on a 30-minute basis, 24/7. Includes all equipment into and out of the well, drilling mud data, casing setting depths, names of personnel, etc. The Daily Drilling Report (DDR) is the operator's (i.e., BP's) official record of the well. The DDR includes IADC data plus costs, trouble time, logistics, and proprietary geological information (lithology, shows, etc.)

CHAPTER 10—Blowout Preventers

Barry put the drawing flat on his desk. "One Cameron, 18¾-inch, 15,000-psi BOP." [41]

Jessica rolled her chair next to his and took a look. "Holy crap, it's huge."

"Fifty-four feet tall and 400 tons."

She stared at the drawing, broke a cookie in half, then half again. Popped a piece into her mouth and took a sip of coffee. Chewed. Swallowed. "Okay. Ready."

"The upper half of the BOP is called the LMRP—the lower marine riser package."

"Makes no sense. How can the *upper* be the *lower*?"

"It's called the *lower* because it's on the lower end of the marine riser."

"So is there an upper package, a UMRP, at the top of the riser?"

"Good guess. There's a lot of gear on top of the riser, tucked under the rig floor, including the mud-diverter assembly,[42] plus the slip joint and riser tensioner." He looked at his watch.

"Sorry," she said. "So where do we start?"

"Let's do the bottom first. Remember the 22-inch casing, wellhead, and BOP connector we saw on the pipe rack?" He pointed to the bottom of the drawing. "They're just above the seafloor. The BOP stack is installed on top of the wellhead—connector to connector. There's another connector on top of the stack."

"Which connects to the lower riser package," Jessica said.

"And which is already attached to the riser," he added. "And on top of that, there's the diverter system."

Jessica, hands up—a traffic cop. "Sorry, I get it. Back to the BOP."

Barry took a manly bite of cookie, about half, and swigged cool coffee. He wiped crumbs off his mouth and pointed to the drawing. "When the LMRP is attached to the lower half of the BOP, the combination is referred to as the BOP stack, or just the BOP."

"I guess that's better than the L-M-R-P-B-O-P combo." She grinned. "And now that everything's connected together and installed on the seafloor, you can tell me which parts of the blowout preventer actually do the preventing."

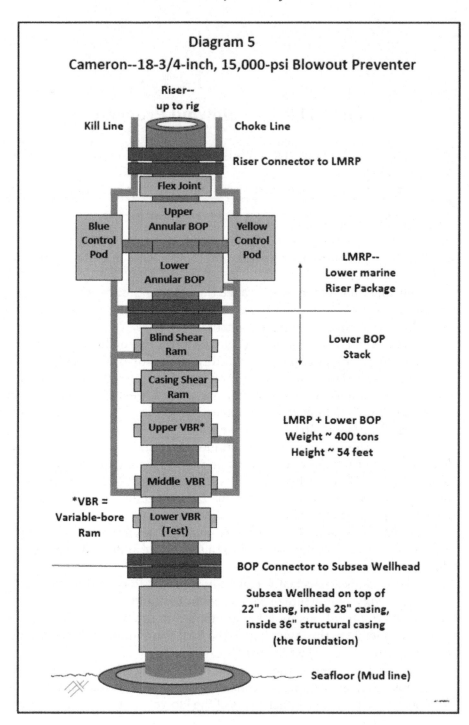

Diagram 5

He leaned back in his chair.

"Don't get too comfortable," she said. "I don't have all night."

"I'm just regretting you're not standing beside the brute, feeling its size, complexity, the entire package, because it's so much more than shown on the drawing."

"Next time," she said. "But for now, the drawing."

"Then just the basics. First, BOPs are designed to stop fluid flow, whether we have drillpipe or casing in the hole, or maybe no pipe at all. Which is why we have four kinds of BOP units to stop the flow."

Jessica pointed to the top of the drawing. "Like the annular BOPs."[43]

"Yep. They're like heavy-duty rubber donuts. They can close around anything—any drillpipe, any casing, or even nothing."

"Why would you want to close around *nothing*?" she asked.

"In case the well kicks after we've pulled all the drill string and the drill bit out of the hole. All we'd have left is the drilling riser, the wide-open BOP, casing, and an open wellbore."

"Duh. Sorry." Then, more pointing. "Why two annular BOPs?"

"Because it's the workhorse, and we love redundancy."

She ran her finger down the page. "And shear rams?"

"The blind-shear rams.[44] They're like cage fighters—they take no prisoners. The two sides of the blind-shear rams are designed to close against each other. If pipe is in the way, so be it. The pipe gets cut, and the shear rams close tight and seal the wellbore."

"When would you ever cut the drillpipe? That seems kind of extreme."

"If the top end of the drillpipe, above the rig floor, is open ended when the well kicks, closing the annular BOP would force all the flow to exit through the drillpipe. Not good—the well's still kicking. We've got a small device called an inside BOP that we can try to stab onto the top of the flowing drillpipe, but if that's not successful, we'd have to cut the drillpipe to seal the well."

"And the casing-shear rams do the same with casing?"[45]

"Good guess. They're the big boys. And I'm glad to say I've never needed them in an emergency."

Jessica tapped a finger on the drawing. "Wow. I can't wait to find out what a VBR is, but I need a potty break."

"And I need a quick trip to the rig floor," Barry said. "I'll have a coffee for you when you get back."

* * *

A half-hour later, Barry returned to the office, empty handed.

Jessica pushed a cup of hot, creamed coffee his way. "Daylight called me, said you were on your way."

Barry sipped his coffee. "Damn, that's good. And in the meantime, Tanker, the mud guys, and I fixed a little mud problem. Ready to go back to work?

"Yep." She pointed to the BOP drawing. "What's a VBR?"

He pointed to the footnote. "Variable-bore ram."[46]

"Oh, I knew that. Right there on the drawing."

"Right," he said. "The variable-bore rams are important because most well-control activities involve pipe in the wellbore. They're like the shear rams, except they've got semi-circular faces that self-adjust around whatever size pipe they find. They're mighty convenient when you want to shut-in the well without cutting the drillpipe."

Jessica studied the drawing. "I think I'm getting this whole BOP thing, but it's the first time I've ever seen color designations on a rig—your yellow and blue control pods that are attached to the BOP."

"Again, we like redundancy. The BOP control panels on the rig are also blue and yellow, as are the drums of multiplex lines that run down to the control pods. The colors allow us to know which is which when it comes time for operations or maintenance."

Jessica pointed to the top of the drawing. "Two lines—choke and kill. Sounds like more cage fighting."

"Without the choke and kill lines, the BOPs are nothing but valves and shearing devices. The choke and kill lines are built into the riser, along with all the control lines. When we connect the riser joints together, one joint to the next, the choke and kill lines are automatically connected, pressure tight."

"I hear you, but it's hard to visualize."

"Not a problem. We've got several riser joints still on the rig, if you want to see one. And I can show you the moon pool.[47] It's down under the rig floor. It's the closest you can get to the water and still be on the rig. "

"Uh . . . no," Jessica said, her face suddenly pale. "I'll wait until I can see the entire menagerie—pods, shears, VBRs, the whole BOP. But you can still tell me about the cage fighters—choke and kill."

"Sure, back to the drawing. The kill line is designed to allow us to pump drilling mud or other fluids *into* the wellbore below the BOP stack, especially below a closed ram or annular BOP."

"Where's the mud go?" Jessica asked.

"If we're just testing the casing, it goes nowhere—just a few barrels pumped in to raise the pressure, then we get the barrels back. Otherwise, it'll go up the riser if the BOPs are open, or down the wellbore, into open hole, if the BOPs are closed."

He studied the drawing. "Huh. There's actually a third line, not shown on the drawing, called the riser boost line. We use it to help clear fluids from the massive riser, since the riser holds a tanker-load of mud."

"Barry, you're drifting off the page. Tell me about the choke line."

"Right. The choke line is just the opposite of the kill line. It allows us to *release* pressurized fluids from under a closed BOP. It's the primary discharge line when we're in a well-control situation."

"Speaking of well control, all this stuff looks good and keeps us out of trouble. But what happens when something fails, then something else fails, or the kick's too big. What's the worst that could happen, your opinion?"

Barry's mind flashed on old headlines—blowouts, storms, fires. The Ixtoc calamity offshore Mexico, the Ocean Odyssey in the North Sea, the Ocean Ranger in the North Atlantic. "It's hard to imagine any problem so bad it couldn't be worse," he said to Jessica, though he knew the answer to her question. "A rank blowout would be bad for sure. And with a fire, losing the well, maybe the rig, and especially people . . . "

Jessica tensed in her seat, her face again pale. "In case of fire, wouldn't you just get off the rig? Like to a helicopter? Or a boat?"

Her words helped him to fade the image. "Actually, I'd hope that wouldn't be necessary. We have a system on the rig, as part of the BOP stack, called the EDS—emergency disconnect system. In an extreme emergency, we hit a few buttons on the BOP control panel and two things happen: the blind shear rams go into action, closing off and sealing the well, and the LMRP releases from the lower BOP stack. Those actions allow us to get the rig away from the well without damaging the BOP or dropping the riser."

"How often does that happen?"

"As caused by kicks, never in my career, never with one of my wells. Though it's certainly happened to others."

"Why not you?"

"I'm nothing special, but I'm convinced that understanding basic well-control principles and procedures will carry the day when it comes to managing well problems, including kicks."

"I need to know about kicks," Jessica said, excitement in her voice, "which you're going to show me, if one happens, for sure. And that's why we're looking at the BOP, so I can be prepared."

Barry cringed at the thought of the well taking a kick. And he damn sure didn't need—

"You're drifting on me," Jessica said. "Just nod if you're actually going to let me be on the rig floor if the well kicks. And I want to be at the BOP control panel, watching you work, doing whatever you need me to do."

Barry swallowed hard. Nodded. Thought of a cold day. And hell.

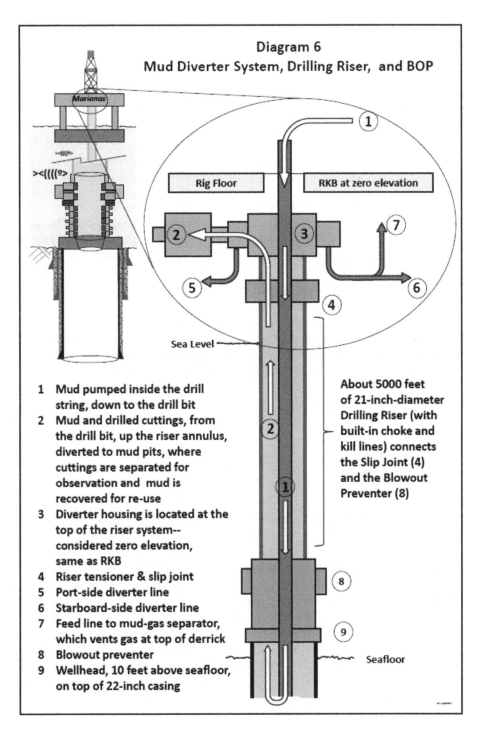

Diagram 6
Mud Diverter System, Drilling Riser, and BOP

Rig Floor

RKB at zero elevation

Sea Level

1   Mud pumped inside the drill
    string, down to the drill bit
2   Mud and drilled cuttings, from
    the drill bit, up the riser annulus,
    diverted to mud pits, where
    cuttings are separated for
    observation and mud is
    recovered for re-use
3   Diverter housing is located at the
    top of the riser system--
    considered zero elevation,
    same as RKB
4   Riser tensioner & slip joint
5   Port-side diverter line
6   Starboard-side diverter line
7   Feed line to mud-gas separator,
    which vents gas at top of derrick
8   Blowout preventer
9   Wellhead, 10 feet above seafloor,
    on top of 22-inch casing

About 5000 feet
of 21-inch-diameter
Drilling Riser (with
built-in choke and
kill lines) connects
the Slip Joint (4)
and the Blowout
Preventer (8)

Seafloor

Diagram 6

---

[41] See **Diagram 5**—BLOWOUT PREVENTER (BOP)—Cameron, 18¾-inch, 15,000-psi BOP. The massive BOP is attached to the Wellhead, at the sea floor, and is connected to the rig via the Marine Riser (also called Drilling Riser). The BOP system is described in the text. See page 68 herein.

[42] See **Diagram 6**—DRILLING RISER, BOP, & MUD DIVERTER SYSTEM. This system is tucked just below the rig floor, where the Mud Diverter acts as the top of the riser (zero elevation). A low-pressure packing in the Diverter forces mud returning up the riser to be "diverted" to the mud pits. Kick fluids, especially gas, can be further diverted overboard through large-diameter flow lines . . . or, if the gas volume is smaller and more manageable, to the Mud Gas Separator. The Riser Slip Joint and Tensioner system allows a constant pull on the riser to keep it upright even as the rig heaves in stormy seas. See page 73 herein.

[43] ANNULAR BOP—Rated for 10,000 psi, driven by hydraulic pressure. By adjusting the "closing pressure," the driller can force the donut-shaped BOP to maintain its pressure seal around the drillpipe, while flexing just enough that even a fat tool joint can be pulled up (or pushed down) through it. This is called stripping the drillpipe. Drillpipe cannot be stripped through any other single BOP, nor can it be stripped through the annular BOP if the closing pressure is too high.

[44] BLIND SHEAR RAMS—two steel blades facing each other. When actuated, they close, cut, and seal the wellbore. Rated for 15,000 psi. They'll cut any drillpipe, but will not cut through a tool joint. To ensure the blind shear ram will work when needed, the driller keeps track of the depth of critical tool joints (in the area of the BOPs). If necessary, for example during a kick, the driller will raise or lower the drill string to get tool joints away from the blind shear rams.

[45] CASING SHEAR RAMS—Designed to cut casing. Though the body of the CSR is rated for 15,000 psi, these rams close and cut, but do not seal the wellbore.

[46] VARIABLE BORE RAM (VBR)—Rated to 15,000 psi. Similar to shear rams, except they have semi-circular faces that self-adjust around whatever size pipe they find. The pipe is not cut. Closing pressure does not need to be adjusted, as it does for annular BOPs. Drillpipe, especially tool joints, cannot be stripped through a singular VBR. But, by using two VBR's, drill crews can strip drill pipe by alternating which VBR is open/closed, and pulling (or lowering) the drillpipe until the next tool joint reaches the closed VBR.

[47] MOON POOL—Large opening through the lower deck—directly under the rig floor and surrounded by handrails—through which the BOP and other tools are lowered into the sea.

CHAPTER 11—Weak Link

Barry dined with two mud engineers, then invited them down the hall and around the corner to his office. Jessica sat at her desk, staring at her laptop screen, slapping her thigh with her tally book. She looked refreshed. Hair damp. Wrinkle-free red coveralls. Kind of a faded red, but not pink.

"Jessica," he said, "you need to meet these guys—they're M-I Swaco, the mud daubers."[48] He did the honors with names, followed by handshakes. The two guests sat on the edge of Tanker's desk—neither wanted his chair.

Jessica and the mud men did a mental scavenger hunt for brethren they might have worked with in the past, both for M-I and BP. The guys named a dozen BP geologists, most of whom Jessica knew well. Jessica, too, recalled mud-engineers' names, rigs, locations and dates, funny stories, problem wells.

Impressed that Jessica could remember details he'd never bothered to lock in deep memory, Barry tested himself—couldn't recall if his last steak had been a sirloin or a New York strip, though it'd been good.

One of the guys named an experienced mud engineer—Jessica knew him too—who'd become a shrimper to avoid two ex-girlfriends who'd staked out the shore base. "The guy should've been run off long before that," Jessica said. "He couldn't make mud in a swamp." That ended the search for names.

No sooner than Barry and Jessica were alone, she turned back to her laptop and punched keys. "Did you know the *Marianas* used to be called the *Tharos*?"

"Trick question?" Barry asked.

"No." She leaned toward the screen. "Before the *Marianas* was a drilling rig, it was a combination floating-hotel, firefighting, hospital vessel. In 1988, it was anchored alongside a North Sea production platform called the Piper Alpha. Some kind of major disaster, a fire, on the platform. A hundred and sixty-seven killed."

"Oxy Piper." A hazy memory unfurled. "A drilling professor, Penn State, said something about the *irony of Piper*."

"And the *irony* was . . . what?" Jessica asked.

"I don't recall—just the *irony of Piper*—something about the fire, and drilling. And I didn't know about the hotel rig, the—"

"*Tharos*," Jessica said. "It was later converted to a drilling rig, worked under a couple of other names, and then became the *Marianas*."

"What's this about?"

Jessica turned in her chair. "I was researching the *Marianas* and found the reference to the fire. Not my favorite subject." Her face formed a question mark. "Would a drilling rig ever need a fireboat alongside?"

"I wouldn't think so," Barry said. "Mobile rigs like this one are drilling only, no production. All we're required to have is a vessel in the area, in case somebody falls overboard. A big North Sea production platform, or a floating production facility like Thunder Horse,[49] both designed to produce hundreds of thousands of barrels of oil a day, that's a different story."

Jessica's contemplative look dissolved into a hard stare, focused on Barry's eyes. "Stay out of my tally book." She whacked the booklet in her open palm—once, twice, three times, then slipped it into her back pocket.

Barry managed a short nod, convinced he had a right to fear a woman with total recall *and* ESP.

"And when do I get to see drilled cuttings?" she asked.

Glad for the non-lethal segue, he swiveled his chair toward hers. "Crews have been taking on mud all day, working with the mud daubers. As soon as we test the casing and get a leak-off test, we'll convert the mud system and drill ahead, which means you'll get your cuttings."

"What kind of mud?"

"Our drilling fluid of choice—synthetic oil-based mud."[50]

"I understand synthetic oil mud—good for the environment. Tell me about the leak-off test."

"It's a simple but necessary test," Barry said. "Using seawater as our drilling fluid, we're going to drill out the cement from inside the casing shoe, plus five or ten feet of new hole. For the test, we'll close the annular BOP and use Halliburton to slowly pressure up the system. For the LOT,[51] the leak-off test, we're looking for the test pressure at which something down hole opens up and takes seawater."

"And you're doing this . . . because?"

"We're looking for the weak link, either rock or cement."

"Whoa," Jessica said, "like you *want* a weak link?"

"Not the way you think. We run casing to cover low-strength rocks, and the cement gives us the isolation we need when we drill deeper. If during the LOT we're able to pressure up beyond the weakness of the formations behind the casing, then the cement is good. Then later, if we lose mud while we're drilling deeper, we'd expect the loss to be in the new hole we just drilled, rather than behind previously-cemented casing."

"Convenient, but how do you know where the mud's going?"

He unlocked and opened his desk drawer, pulled out a notebook, removed a computer-generated graph. He laid the page on Jessica's desk. "This is a pore-pressure, fracture-gradient diagram for the Central Gulf. The diagram shows us how much pressure a shale formation at a given depth should be able to withstand without fracturing." He shrugged. "Within reason of course."

"So why test the cement, since you know it's a lot better than the weak rocks behind the casing?"

"Good cement is what we want," Barry said, "but the deeper our casing job, the less confident we can be the cement job is perfect."

"I would have thought cement is cement," Jessica said. "You mix it, then you pump it, and then you let it harden."

"Maybe for a patio, but we're a long way from where we need our cement. We can plug in well data—temperature, pressure, hole size—and mix the best possible cement. Then all we can do is pump it down the casing and up the annulus and hope it does its job."

Jessica turned the chart where she could read it. "So what does this tell you about the 22-inch casing shoe?"

He used his pen to track lines on the chart, showing Jessica how to read it. "If we have good cement, the shale should be able to withstand about ten pounds per gallon."

"And if the cement's bad?"

"We'll get a much lower leak-off test, because we'd be pumping into weaker rocks farther up the hole, outside the casing. A lower-than-expected test result means we need to squeeze the shoe."[52]

Jessica sat back in her chair, her hands flat on her desk. "So both answers have meaning, just different outcomes."

"Yep, but I want the ten-pound-per-gallon result," Barry said, "because I don't want to spend time fixing cement."

"Understood, but that begs a question. I've studied fluid dynamics, and I understand fluid pressures and pounds per square inch, but where in the world did drilling engineers come up with *pounds per gallon* to talk about pressures?"

"Important topic," Barry said. "So let's start with a simple equation."

Jessica raised her hand as if to stop a kid on a tricycle.

"Don't get all excited," Barry said. "It's the only one you need for the well, but you'll use it a hundred times before we're done. It's also the most important equation you'll use in Drilling 101." He wrote on a sticky pad and gave it to her.

"Just a minute." She opened her breast pocket and retrieved a small object. "My iPod," she said, twirling her thumb over its surface. "Goodbye, Jason Mraz—hello, Vivaldi." She adjusted a single earpiece. "Okay, I'm ready."

"What's that about?" Barry asked.

"If you get to talk shop, I get to pick the music."

"You can't wear that on the rig floor."

"Not a problem." She waved the note. "The equation?"

"It's important, but if you're not interested, I've got better things to do."

Jessica adjusted the ear bud. "It's just music. Quiet classical music that helps me concentrate. Helps me learn." She locked eyes with his, spoke softly. "But if you've got better things to do, please don't let me stop you."

He considered the gauntlet. A point to argue. Music distracted him. As did television. And what was good for him should be good for everybody. Well, his wife had argued otherwise. His wife who was now his ex.

----

48 MUD ENGINEERS—contracted by BP from M-I Swaco to manage the makeup and maintenance of the drilling fluid (mud) during the drilling of the well. Often referred to as mud daubers in honor of the wasp that builds its nest from mud. Though critical to the drilling of the Macondo well, these positions are not named in the story.

49 THUNDER HORSE—BP's Thunder Horse facility is a world-class semisubmersible (floating) PDQ (production-drilling-quarters) facility, located in 6050 feet of water. The facility, designed to produce 250,000 barrels of oil per day, is located about 30 miles south of Macondo in Mississippi Canyon Block 778.

50 DRILLING FLUID (Mud)— The simplest drilling fluid is water—but water doesn't carry cuttings very well. Increased viscosity (like with potato soup) helps carry cuttings from the well. Mud density (pounds per gallon) provides fluid pressure—called hydrostatic head. Water-base muds use water, clay for viscosity, and barite for density, as minimum ingredients. Synthetic-oil-based muds (SOBM) are environmentally superior muds synthesized from vegetable oils, sugar alcohols, or sugar glucose. In complex geology and deep wells, SOBM is often necessary to make the hole "slicker" to help keep the drill bit from getting stuck.

Book 1

[51] LOT—Leak off test. Pronounced L-O-T. Like BOP is B-O-P. Like all other oil-field acronyms, spelled out.

[52] SQUEEZE CEMENT—A squeeze job forces cement slurry into a leak area, like the shoe, and calls for slowing the pumps until the injection pressure rises (cement hardening), then forcing more slurry into the leak and repeating the process until no more cement can be squeezed, thus sealing the leak.

CHAPTER 12—Simple Math

Barry said, "I call it the mud-weight equation.[53] It allows us to calculate the pressure anywhere in a column of fluid, like drilling mud. We start with the mud weight, in pounds per gallon, and multiply it by the depth of interest, in feet. Multiply that answer by a constant—0.052. The answer is the pressure in pounds per square inch—psi."

"Then you're right—it is simple math." Her words were soft. "So why the big deal?"

"Depends. Geologists who work with rocks rarely need to calculate pressures in the wellbore. Company men need to know pressures from spud through abandonment."

"Me too," Tanker said, his big body on its way to his desk chair. After greetings, he said, "Don't mind me—I got a couple of reports to fill out." He sat and went to work as if alone.

"Okay," Jessica said to Barry, "it's a simple equation that's important for company men and toolpushers. Which means you're about to tell me why."

"Three points," he said. "Management of the riser, surface pressures, and overbalance. All related to the equation. And I'll even explain about pounds per gallon."

Out came her tally book. "Talk slow so I can take notes."

"First, the pressure effect of weighted mud in the riser gets critical when we have to pull the riser, or when it's damaged in a storm."

Head down writing, Jessica said, "An example would help, preferably without a bunch of numbers."

"Sure. How would you calculate the pressure at the seafloor below the *Marianas*?"

"So much for no numbers," she said, calculator in hand.

"No," Barry said. "Tell me in words."

"You mean a calculation without calculating?"

"You're the one who told me you don't want numbers."

Tanker glanced over his shoulder at Barry, then at Jessica. He gathered a stack of papers. "I'll be in the galley," he said. "You want me to send coffee your way?"

Barry shook his head.

"Why thank you, Tanker," Jessica said. "I never turn down a sweet cup of coffee."

"On its way, Miss Jessica," the big man said as he made his exit.

"He's a very nice man," Jessica said.

Barry shrugged. "Back to how you'd calculate the pressure at the seafloor."

"Simple. I'd multiple the depth of the water—5,067 feet—times the weight of seawater—you said 8.6 ppg. That answer times the constant— I'll have to look at my notes—tells me the pressure."

"The missing constant is 0.052," Barry said, "but your answer's too high."

"Too high? Crap, I didn't even give you an answer."

"How deep's the water?" he asked.

"Under the rig—5,067 feet."

"Wrong. From the rig floor to the seafloor is 5,067 feet."

A beat. "Crap. Water's 4,992 feet deep."

"So why was you first answer too high?" he asked.

"Because there's 75 feet more water inside the riser than outside the riser." Her face spelled deep thought. "But the two pressures differ by only 30 or 35 psi, which seems kind of trivial."

"Agreed, but what's the pressure difference inside and outside, at the seafloor, if the riser's full of, for example, 15-pound mud?"[54]

"That's easy, but I'm using my calculator—don't try to stop me." She punched numbers. Made a note. Punched more. Another note. "The difference is 1,720 psi."[55]

Still seem trivial?"

A galley hand carried in a mug of coffee. "I believe this is for Miss Jessica," he said to Barry. "Three sugars—extra, *extra* sweet." He leaned over her shoulder and placed the mug on her desk, by her hand, and promptly left the office.

Jessica reached for the mug, hesitated. Barry imagined a faint tremble in her hand. Then, as if the coffee were days old, she pushed the mug toward the wall.

She turned and looked Barry squarely in the face. "Is 1,720 psi trivial? Not at all. But I'd guess the riser's built for mud heavier than 15 pounds."

"You can be sure of that," he said, his mind on the galley hand— Jessica's snuggly friend who wasn't a friend, "but the riser's not the problem."

"And I suppose you'll tell me why."

"The 1,720 psi is important," Barry said, while mentally moving the galley hand up his list of priorities that needed attention, "because if we lost the riser and its 15 pound mud in a winter storm, and for some reason the EDS had failed—"

"Remind me."

"EDS—the emergency disconnect system. If the EDS failed to close-off the wellbore, the pressure exerted by the mud at the bottom of the well would instantly decrease by 1,720 psi, which might result in a bigger kick than you'd ever want to see."

"Wow, give me a minute." She pulled out her ear bud and hung it over her shoulder, then wrote more notes. "I'm going to write a short story— *Top Three Reasons to Love the Mud-weight Equation*. I just don't know which is more important—number one, two, or three."

"There's no hierarchy of importance, though the second point is an easy one, and it's this. When the BOPs are open, the wellbore pressure at the surface is zero and increases with depth. But when the BOPs are closed, if any pressure is added to the mud column, for example 500 psi at the surface through the drillpipe, then all pressures throughout the wellbore, regardless of depth, will increase by the same 500 psi."

"You're right," she said. "That's an easy one. Next point."

"Wrong. You're supposed to ask me how something so simple can be important."

She did.

"Back to our equation," he said. "If we close the wellbore and increase the surface pressure from zero to 500 psi, the entire wellbore sees the 500-psi pressure increase."

"Perfectly logical."

"Now convert that pressure to equivalent mud weight at 10,000 feet."

"How heavy is the mud in the well?"

"Doesn't matter."

She sat back in her chair. "It *has* to matter."

"Trust me. Use the mud-weight equation. Five hundred psi at 10,000 feet. What's the equivalent mud weight?"

"Pressure equals mud weight, times depth, times the constant." She scribbled a note. "So, we rearrange the equation, and now the mud weight equals pressure, divided by depth, divided by the constant." She punched numbers. "Five hundred, divided by 10,000 equals, with that answer divided by 0.052." She studied the screen. "Mud weight equals 0.96. Makes no sense."

"Ah, but it does—0.96 pounds per gallon. The 500 psi creates an *equivalent mud weight* of almost one pound per gallon at 10,000 feet."

She tilted her head—reminded him of a pinball machine—then nodded.

"So the 500-psi of increased pressure causes the mud weight at 10,000 feet to *act* a pound heavier than it actually is."

Barry faux clapped.

"But why should we care?" she asked.

"Why? Because a 500-psi kick at 10,000 feet means we need another pound of mud weight—above whatever we already have—just to balance the pressure of the kicking formation."

A grin found Jessica's mouth. "That is so cool."

"Ah, but what's the 500-psi answer at 8,000 feet?"

Jessica did the numbers. "I get 1.2 pounds."

"And your conclusion based on your extensive research?"

Jessica studied her numbers. "Makes sense. The equivalent mud weight of any fixed surface pressure is more significant at shallower depths than deeper in the wellbore."

"And now do you care?" Barry asked.

"It depends on how much open hole we have and the depth of the shoe." More thinking. "And the leak off test."

"So shut-in surface pressures really do matter," Barry said. "And if you want to quantify the effect—good or bad—there's a simple way to do it."

"The mud-weight equation." A beat. "And I hope that takes us to number three."

"It does. And I'll even answer your question about why we use pounds per gallon."

Jessica jumped from her chair, did three steps of a jig, and sat down. "Can't wait. Sock it to me."

Barry held his grin.

"What?" she asked, studying his face.

"I'm just trying to picture what you do when you're really excited."

"Make a discovery and I'll show you. Now, number three."

"In a word, pounds per gallon is *convenient*. If I mention 9-pound mud to a driller, he knows it's barely heavier than water. Twelve pound mud is a lot heavier. And 16 pound mud—a double-heavy gallon jug—that's really heavy stuff. It would exert tremendous pressure down hole, and we'd never use it unless the formation pressures were very high."

"What's the heaviest mud you've used?"

"We used 21-pound mud on a gas well in Syria. We made it heavy with powdered galena because barite wasn't heavy enough."

"Good well?"

"Yes, but the Syrians didn't want gas—they needed oil."

"Bummer. But I'm still looking for the conclusion to my short story."

"Then this part of the program is dedicated to concluding the mud-weight equation," he said, enjoying her humor, "and understanding pounds per gallon."

She held up her pen. "Don't wait on me."

"Now, always, forever," he said, "we need to be *overbalanced*—more pressure in the wellbore than in the exposed rock formations. But since we don't know the pressures in the formations, which change by the foot, we need a way to keep up with those increasing pressures. And that's what using *pounds per gallon* does for us."

"Example."

"This is a test: If we're drilling a 500-foot interval of rocks pressured to the equivalent of 13.3 pounds per gallon, even if the actual formation pressures, as expected, increase a little with each foot we drill, how much mud weight do we need to ensure we're overbalanced all the way."

"Simple—anything heavier than 13.3 pounds per gallon."

"Right. And the downside of breaking the rule?" He stopped. Waited.

Jessica nodded ever so slightly. "Formation fluids enter the wellbore—the well kicks."

"Right. If you don't want the well to flow, to kick, then being *overbalanced* is mandatory every foot of the way. While drilling. While pulling pipe. During pressure tests. During well control. While abandoning the well. While drinking coffee."

Jessica looked up from her notes. "Coffee?"

"Sure," Barry said. "Come on, I'll buy." He turned and headed to the galley—pictured her shaking her head, rolling her eyes.

* * *

Barry was at breakfast the next morning when Jessica walked in. "We tested the shoe this morning," he said, "in case you care."

"Care? Of course. I was in the mudlogging unit when you ran the test." She pulled out her notes, straightened her back, stood tall. "I watched the test pressures—625 psi on top of seawater, with the casing shoe at 7,937 feet. Your fancy little equation tells me that's a 10.1-pound-per-gallon leak-off test, which means we're good to go." She flipped shut her tally book and shoved it into the left back pocket of her lime-green coveralls, then headed to the coffee pot.

Barry munched a crisp slice of bacon while she garnered the usual *Good morning, Miss Jessica*, nods, and smiles from most everybody who passed her way. She selected a clean coffee mug and—

Then, something else. Somebody else. The snuggly galley hand who the night before had brought her the coffee she wouldn't drink. Leaning over the serving counter, talking to her—at her—and grinning.

Barry hadn't liked what he'd seen the first time the guy sidled up to her. Nor his too-familiar coffee-delivery service the previous evening. And he damn sure didn't like the morning's two-way heated action at the counter.

Jessica got in the guy's face. Said something perhaps acidic, her entire body a hundred-twenty-pound snarl.

Barry couldn't make out her words.

The guy did. He held up his hands. Stepped back. Shit-eating cackle. Leaned toward her and said something only she could hear.

She whirled toward Barry. Red faced. Returned to the table. Slammed down her coffee mug. Sat hard.

* * *

Jessica had wanted to slug the guy, and now she could barely breathe. Had the big bastard touched her, she wouldn't have held back, knowing few men with broken fingers would reach a second time.

"What was that about?" Barry asked, his voice quiet, his eyes signaling concern.

"I can handle it."

A beat. "Jessica, I know you're a problem solver. But out here there are rules, some written, some not."

"Please, dammit, it was nothing."

"Aggression, threats, harassment—real or implied," Barry said. "Not out here. Zero tolerance."

She swallowed hard. Held it in. Thought of her dad. Refused to be weak.

Barry scanned her face, as if looking for nuances that would betray her thoughts. Then, "Stay here," he said. Decision made. An order. Not negotiable.

He got up from the table, went to the wall phone, made a phone call. Short. Few words.

All Jessica could do was watch.

He made another call. Also short. He stayed by the phone. Thirty seconds. It rang. He picked it up before the ring finished. Again, few words, then he hung up.

Another half minute. Tanker glided his big body into the galley, over to Barry. The steward emerged from the galley office. Barry conferred with both, away from the buffet line. The trio went into the kitchen.

Not what Jessica wanted.

Two long minutes. She tried not to look. Icy tendrils down her back.

Three men left the galley, the steward watching their departure. The galley hand, shoulder-to-shoulder between Barry and Tanker, wasn't too happy. None glanced at Jessica. They disappeared around a corner.

Minutes later, Barry returned. Alone. Nonchalant. He sat in the seat he'd left only minutes before.

"I hate it when that happens," he said.

"What's that?"

"Some guy quits. Decides to ride the workboat to the beach, in about an hour. Leaves the steward short-handed."

"I could've handled it."

"What, being a galley hand? Don't sell yourself short—you're too good a geologist for that."

"Thank you," she mouthed, because the words stayed inside.

"It's over," he said. "Forget about it." He tipped her mug toward her. "But you forgot to get your coffee."

She grabbed her empty mug and left Barry at the table. Good mentor, she thought. Honest guy. Hard worker. Trustworthy. And protective, though she didn't need it.

Need. The word triggered a flash of thought. The only thing she *needed* was for her mom to answer the damn phone.

* * *

"That's better," Barry said as soon as Jessica returned to the table, steaming coffee cup in her hand, the skin tone of her face close to normal.

He returned to their earlier topic. "And what's the 10.1 pound leak-off test tell you about drilling ahead?"

Seconds passed. She exhaled softly, as if meditating. Then, "If we try to use 10.1-pound mud or heavier," she said, seemingly gathering resolve, "we'll probably break down the hole and start losing mud."

"And that rule will apply to each casing string we set," he said. "At each new shoe, there'll be a tested upper limit to the mud weight we can use to drill ahead. Any mud or equivalent mud weight, in pounds per gallon, heavier than the leak-off test will likely break down the wellbore and result in lost circulation."

Jessica double-handed her coffee mug. "How much longer before we drill ahead?"

Barry checked his watch. "Another half hour to get the oil-based mud in shape. I hope you're ready."

"I've been ready for what seems like weeks."

Barry stood. "Then I say let's go finish our well, you and me, and see what we've got."

53 MUD WEIGHT EQUATION: P = MW x D x 0.052 Here, P is pressure, in psi. MW is mud weight, in ppg. D is depth in feet below the top of the mud (important to use true vertical depth). The constant—0.052—keeps all the units of feet, ppg, and psi straight. The equation allows calculation of the pressure-induced equivalent mud weight at a given depth: MW = P / (D x 0.052)

54 MUD—Note: 15-pound mud is slang for mud that weighs 15 pounds per gallon (ppg).

55 MUD WEIGHT CALCULATION—Goal: Calculate the pressure difference inside and outside the riser, when the riser is full of 15-ppg mud. First: outside the riser, the Gulf, at Macondo, is 4992 feet deep. Using the equation (P = MW x D x 0.052), the pressure at the seafloor is the "mud weight" (8.6-ppg seawater) times the depth (4,992 feet) times 0.052, for an answer of 2,230 psi. Second: the top of the riser is 75 feet above sea level (5,067 feet above the seafloor). Inside the riser is 15-ppg mud. Therefore the pressure inside the riser at the seafloor is 15 ppg, times 5,067 feet, times 0.052, for a total pressure of 3,950 psi. The difference inside and out (3,950–2,230) is 1,720 psi.

CHAPTER 13—A Lady Named Ida

In Barry's book, drilling through the riser proved to be anticlimactic. He never saw Jessica; she spent her days in the mudlogging unit, apparently thrilled by sacks of cuttings gathered from every ten-foot interval of new hole drilled.

Barry and Tanker cussed and discussed the calendar. The date had a red star, penned by Tanker days before.

"On the schedule," Barry said. "We have to do it."

Mandatory BOP tests—red starred on the calendar. Regular as church on the beach. Matching the repetition of obligatory safety meetings and fire drills, every BOP component went through a weekly function test. Every two weeks, the same components also went through pressure tests.

Daylight had barely drilled five hundred feet of new hole when Barry and Tanker gave him the news. Time to function test the BOPs, Tanker told him.

Drilling stopped.

Jessica came out of her cave.

Barry filled her in, then added, "Not to worry. We'll be back to drilling in a few hours."

Barry and Jessica dined in the galley. He enjoyed her company almost as much as the jalapeno meatloaf slathered in catsup.

She checked her watch, often, like she had something better to do.

"Relax," Barry said, also counting minutes, adding up down-time dollars. "We'll be drilling before you know it."

Tanker slid into the seat next to Barry, his entry atypical—no food, no coffee.

"What's wrong?" Barry asked.

"We've got a VBR that won't close. No pressure response. No signal. Nothing. Tried a dozen times."

Barry knew the answer. Couldn't find the words.

"We need to pull the stack," Tanker said. "Daylight's making a short trip, getting ready to pull the drill string."

"Oh, that's good," Jessica said. "We just got started and you guys want to pull the BOP. How soon?"

"Now," Barry said, resolved, and without an alternative. "With open hole," he said to Tanker, "we'll have to set cement plugs. I'll get with Halliburton."

"Wait," Jessica said. "One variable-bore ram doesn't work? Aren't there two more? Plus two annular BOPs and two shear rams? How many do you need?"

"All of them," Barry said. "Not negotiable. We need everything to work before we drill another foot."

Jessica didn't seem too happy when she left the table, alone.

* * *

A quiet day later, Friday, just before midnight, Barry and Tanker collaborated about the ongoing stump test on the repaired BOP.

Jessica charged to Barry's side and asked, "You have any enemies named Ida?"

"No. Her name's . . . not Ida," Barry said, trying to erase an ugly vision of his beautiful ex. "Why?"

"The National Hurricane Center just issued a weather advisory," Jessica said, her words snare-drum fast. "Hurricane Ida, which started way south, near Panama, is headed north." Out of breath, she added, "Toward the Gulf of Mexico. Toward us. Moving fast."

"Then we got work to do," Tanker said. "Hurricane Rita got the *Marianas* in September 2005. Tore up the anchor system and drove her up the coast almost 150 miles before she grounded in shallow water." He excused himself, said he had a call to make.

Not what Barry needed. "ETA?" he asked Jessica, noting Tanker was already on the phone.

"Landfall, central coastline, mid-day Monday."

"Which gives us tomorrow and Sunday to get out of here."

"But maybe it'll go somewhere else," Jessica said.

Barry headed to the office, Jessica at heel. "Can't even think about that," he said. "If there's a hurricane in the Gulf, regardless of the forecast, it can get wherever it's going in less than two days. And we generally need every bit that much time to secure the well and get everybody to the beach."[56]

"So we get out of here, go home, then drill after we get back. That'd be good. No sense being out here in a hurricane. Me, I'm ready to go whenever you need me to. Unless you need me to help. You know, count people—"

"Whoa, Jessica, take a breath—this is *not* an emergency." A soon as she seemed to focus on him, he added, "Listen, we're way ahead of schedule. Because of the BOP problem, the well's already temporarily abandoned with two cement plugs, which means all we need to do is ready the rig, solve a few logistics problem, and abandon ship."

"Abandon? Abandon ship? Like the Titanic? Oh, my God, Barry, that's bad." Her voice cracked. "When do you want me to go? I can be on the first chopper. And it has to be a chopper, because there's no way I'm climbing on a frigging boat in a frigging hurricane."

"Wrong word," he said, trying to calm her. "We'll disembark, not abandon. When you hear a P.A. announcement *Abandon Ship, All Hands Abandon Ship*, that's for an emergency. Which this isn't."

They entered the building together, Barry tight lipped, Jessica hyperventilating. Had Barry been trying to comfort his daughters, his mother, his dad or brother, or his ex in the same situation, he would have included hugs—giving and getting. Hug Jessica? Well, no.

Inside the office, he sat at his desk; Jessica, hers. "I'll get you on an early flight tomorrow," he told her, keeping his voice as calm as possible, "but for now I need to call Tanker, then my boss. I recommend you get some sleep, and I'll see you in the morning."

"No way could I sleep," she said. "I'll just hang with you, help you, help Tanker, fix the rig, get ready, in case, you know, the chopper comes early."

She reminded Barry of a red-headed six-year-old neighbor boy who'd had a jabberwocky mouth. "Then how about a cup of coffee while I make my call?" he said, to give her something to do. "Extra cream, please."

Jessica looked at him like he'd farted Taps, then picked up the phone. Dialed. Waited. "Hi, this is Jessica. Can you have somebody bring the company man, Mr. Eggerton, a cup of coffee in his office. Double cream, no sugar." A pause. "Well, thank you—that would be nice. Black, three sugars. Thanks." She tapped the hook. Handed Barry the phone. "Coffee's on the way—now you can make your calls."

* * *

"The well's secure," Barry told his boss.[57] "We've got cement plugs at the shoe and just below the mud line. We should be out of here tomorrow morning."

His boss asked him about the rig.

Standard hurricane procedures called for de-ballasting the rig, to make it stand taller in the water, to give it more air gap, to better ensure storm waves went under the rig, rather than through and over it.

"With the entire BOP and riser stored topside," Barry told his boss, "it looks like the rig will ride out the storm in the drilling-mode draft, a little deeper than we want, with the drill floor at 75 feet."

* * *

Two thirty Saturday morning. Barry and Jessica still in the office, Tanker at his desk too.

Barry, amazed Jessica hadn't crashed from adrenaline withdrawal, agreed with Tanker they needed to go over the POB list, identify nonessential personnel.

"POB?" Jessica asked.

"Personnel on board," Barry said.

"We'll start evacuating nonessentials in the morning," Tanker said. "There'll be a lot of competition for choppers and boats, and the longer we wait the worse it'll get."

Barry said to Tanker, "I'd like to see service hands go first, since the well's secure and they're on my payroll until they leave the rig."

Jessica twitched in her seat—body language Barry understood. "I've also got Jessica on an early flight," he told Tanker, "and there are still a few seats. Free ride to the shore base for anybody you want to send in early— just let me know."

Jessica mouthed, *Thank you.*

Barry mouthed, *Welcome.*

"I hate leaving five hundred feet of open hole behind," Barry told Tanker, "but if we're not gone too long, we should be able to get back to drilling soon after we mobilize."

"How long will that be?" Jessica asked.

"I'd guess probably Tuesday," Barry said.

"Could take longer," Tanker said. "Never know what kind of problems we'll see."

"Like what?" Jessica asked.

Barry willed the big man to zip it.

Tanker crossed his ankles and leaned back. "Rig damage, mainly. But there could be damage to workboats, the shore base, roads, pipe yards."

Barry realized the noise he heard came from his grinding teeth. Neither the BOP repair, the hurricane, nor having to abandon the rig had been on his radar. His thoughts were on the approved well costs—the AFE. He didn't want the downtime. The extra rig days. Expensive rig days. With cost overruns. And a delayed schedule.

Jessica stood from her chair. "I'll leave it with you guys—you've finally worn me down." To Barry, she said, "I'll be up early for breakfast—ready for my ride." To Tanker, she said, "If he doesn't get me a ride first thing in the morning, I may need your help hiding his body."

After she left, Barry glanced at Tanker.

The big man said nothing, but one eyebrow inched to attention.

Barry checked the manifest.

* * *

"See you in a couple of days," Barry told Jessica before she boarded her 9:10 A.M. flight. She tossed him a thumbs up.

He and Tanker spent the morning and early afternoon on the disembarkation process. Two crew-boat runs, three chopper flights, and a one-way trip with the workboat.

That evening, the steward put on a gourmet meat-fest for the small remaining crew. Barry, a carnivore in pit-grill heaven, chomped his way through the smorgasbord—steak, prime rib, brisket, coconut shrimp—and thought about Jessica. Probably eating a bouquet of broccoli at one of Houston's fancy salad restaurants in the Galleria area.

Come morning, Tanker checked the tension indicators on each of the anchor lines. Barry hoped the lines wouldn't be put to the ultimate test—the ferocity of Mother Nature at sea, in the form of Ida. As a last step, Tanker and his chief mechanic turned off all rig power except one emergency generator.

The skies above the Marianas were cloud free when Barry, Tanker, and three others disembarked at noon on Sunday, 8 November 2009.

---

56 The BEACH—offshore slang for anywhere on dry land. Chopper to the beach (airport). Boat to the beach (shore base). BP bosses on the beach (Houston).

57 BOSS—All onshore BP management and executive management positions—superintendents, departmental supervisors, drilling operations managers, logistics managers, drilling engineering managers, drilling executives, vice presidents—are collectively identified in the story as Barry Eggerton's boss or manager.

CHAPTER 14—The Glitch

Barry watched the news the next day and tracked the storm through the National Hurricane Center. The NHC had downgraded the hurricane to Tropical Storm Ida, based on maximum sustained winds of 70 miles per hour. Ida followed a path toward the Gulf's marshy shoreline, the same area notched by Hurricane Katrina in 2005.

According to Barry's pen-and-paper plot, the storm blew right over the *Marianas* about dinnertime that Monday. The NHC forecast announced the slow-moving storm was expected to make landfall that night or early Tuesday morning.

From Barry's late-evening perspective, enjoying a cold beer from his south-facing deck in Houston, Texas—450 miles west-northwest of the *Marianas*—he was unable to see a single symptom of a hurricane in the Gulf.

He called Transocean's office the next morning. An operations manager advised that a physical survey of the *Marianas* and other rigs was in progress as they spoke. The ops manager guessed Barry should plan to board the rig the following day, Wednesday. Barry called his boss, told him what he'd heard.

The Wednesday plan didn't happen. The Transocean inspectors found severe damage to the rig. Barry was told repairs might take a week. He called Jessica's office and gave her the news.

"Let me know," she said. "I'm ready to go back as soon as you raise the flag."

"Plans for lunch?" Barry asked. Friends. Direct approach. Testing the water.

A beat. "Sorry, I can't," Jessica said. "My plate's pretty full, but I'll let you buy me a Macondo salad when we get back to the rig."

They said their goodbyes. Friends—sort of.

Three long weeks later, tugs towed the damaged *Marianas* to a shipyard for repair.

* * *

Weeks passed. No rig.

Barry drove from Houston to Lafayette, Louisiana, Christmas Eve morning and took his daughters to lunch. His eldest, Buie, had accepted admission to LSU for the fall semester.

"Petroleum engineering?" Barry asked, hopeful.

"No, Daddy—not everyone wants to work offshore. General business, for now."

Face in pain, he grabbed for his heart, but before he could answer, Brandy, his high-school senior, said, "I like LSU, too, but Chico's trying to talk me into going to ULL."

"ULL? What and where's that?"

"Oh, Daddy—the University of Louisiana at Lafayette. The Ragin' Cajuns."

Barry laughed, felt the generation gap. He would always think of the school as USL, the University of Southwestern Louisiana, its name before the 1999 transformation.

"More important," he added, "who's this Chico guy?"

"Chico Espinosa. He grooms horses at Evangeline Downs—wants to be a jockey."

Conflicting thoughts fought for attention, but considering his middle daughter's five-foot-seven elegant stature, he said, "Aren't those guys usually short?"

"Of course," Brandy said, "and that's what's so perfect—Chico's tiny."

Barry swallowed hard. "What's your mom think?"

Bailey, the youngest, jumped in: "Mom thinks he's cute, but Mervin calls him *Chihuahua*, since he's so skinny."

"That's not nice," Barry said, chalking up a rare accolade for the kids' stepfather, asshole Mervin.

Barry noticed Buie tapping the tip of her finger on her wristwatch. She reminded him of Jessica, though Buie was more subtle. Her tiny nod confirmed she needed to go.

"Hey, I've got something for you guys," he said, "but the packages are for Christmas." He pulled a loaded shopping bag from under the table. "But first—" He opened his notebook and gave Bailey a high-gloss, 8 x 10, aerial view of the *Marianas*. She was thrilled and jumped from her chair to him a hug.

He told the girls about the hurricane.

"Wow."

"Rad."

"Awesome."

They exchanged gifts before they headed to the parking lot. Two cars. His and theirs.

Barry never cried when he saw his daughters during visits, or when he hugged them goodbye, or when he drove away. Nor did they.

* * *

On New Year's Day Barry made the trek to San Antonio to watch football with his folks. His dad called out every play, presumably because he thought Barry's mom was blind. Didn't matter what the announcer said, his dad repeated the words and added his own, as if his mom were deaf too. She was neither—nor was Barry. Didn't matter. His dad announced ball carriers. Sacks. "Hold the bastards." "Dumb sonsofbitches." Touchdowns. Field goals. "Ten to seven—they lead by three." "Ten to fourteen, down by four." Poor Mom, Barry thought—blind, deaf, and apparently illiterate too.

Barry's mom would just smile, nod a lot, clap occasionally. He'd get her another iced tea—three sugars, two slices of lemon—and his dad another beer, more chips, more dip.

Barry took a break about four, brought back take-out barbecue, which the threesome ate in the living room during half time. The last time they'd shared a whole day together, they figured, was in 2008, at Thanksgiving. Football games, Barry recalled—announced in stereo.

Gnawing a baby-back rib, he remembered his discussion with Jessica about him telling his folks technical stuff. About how they'd enjoyed his descriptions. His acronyms. Barry wondered if his passion for explaining things had come from his dad, but didn't know if that was good or bad news.

He pondered his parents—his dad fully occupied with a chicken wing, while his mom daintily nibbled corn kernels off a butter-soaked cob. Barry didn't mention his job. Neither parent questioned any aspect of his life, his future, his aspirations. Never mentioned their granddaughters, whom they'd seen on Boxing Day, thanks to their mother and asshole Mervin. Nor did his parents talk about the exciting things they did in retirement, even when questioned.

Right. He was a living witness to their excitement and had been all day.

He stayed until seven and neither requested nor was offered an overnight stay, though he knew his room would have been ready, immaculately cleaned, likely that morning.

Barry never cried when he saw his folks during visits, or when he hugged them goodbye, or when he drove away. Nor did they.

* * *

Barry thought about Macondo on his way home. Two cement plugs—one deep, the other shallow—aged like fine wine for the past two months. *Patio-hard cement*, Jessica would have said.

Mixed thoughts kept him awake. The lights of Houston across the eastern horizon.

The lonely road.

Jessica.

JANUARY 2010

CHAPTER 15—Back on Track

Barry answered a late-morning phone call.

"We lost the *Marianas*," Barry's boss said. "It's finally repaired, but our contract ran out so it's not coming back. Though my rig schedule is shot to hell, the good news is I've got us a replacement rig for the Macondo well, and I want you on it."

Barry pumped his fist. "That's good, sir. Been a long time coming. When? What rig?"

"Early February. Transocean's *Deepwater Horizon*."

Barry knew the rig, had worked it before—twice. "I'll start planning now."

"Barry, this'll be a sensitive well. It was scheduled for 2009, so we're going to be spending unplanned dollars from somebody's 2010 budget."

"What's the day rate?" Barry asked.

"Just under half a million, for a total cost of about a million a day. Which means the pressure's on. You're going to get a lot of help keeping costs down, and not just from me. And just so you know, the rig's already booked for another project, which means the schedule's tight."

"Not a problem. Any program changes?"

"We submitted the rig change and got approval from the MMS.[58] The well program's the same—at least for now. Your job's to finish the well and release the rig ASAP."

"Consider it done," Barry said, neither surprised nor intimidated. He and his boss spoke the same language when it came to well costs, and especially to cost overruns, one of the evil villains of year-end performance reviews and bonuses.

After lunch, Barry pulled his notes and reminded himself about the Macondo work they'd done to date, the remaining aspects of the well program, and the people who'd been there with him. He called Jessica's Houston office number.

A staff assistant answered and told him Jessica was on a rig. She'd been back-and-forth offshore, different rigs, for the past several weeks.

Barry sent Jessica an e-mail: New rig available for Macondo early February. Let me know if you're interested and available.

An hour later, Jessica's response: *I am, but I'm not.*

---

[58] MINERALS MANAGEMENT SERVICE—MMS—The regulatory authority for offshore operations in federal waters. By October 2010, the MMS was reorganized and renamed the BOEMRE (Bureau of Ocean Energy Management, Regulation, and Enforcement), which was then replaced by the BSEE (Bureau of Safety and Environmental Enforcement) and the BOEM (Bureau of Ocean Energy management). Which means the MMS no longer exists.

CHAPTER 16—Deepwater Horizon

On 5 February 2010, three minutes after he'd taken a half dozen pictures for Bailey, Barry's chopper landed on the helideck of Transocean's *Deepwater Horizon*.[59] The rig appeared newer than the *Marianas*, because it was, and the typical word enormous seemed inadequate. Barry had studied the rig specs and reminded himself of the numbers: the behemoth was longer and wider than a football field. Though he hadn't seen the contract, he understood that the rig, built in 2001, had never worked for anybody but BP.

Unlike the *Marianas*, which had been anchored to the seafloor, the *Deepwater Horizon* was DP—dynamically positioned. The rig's GPS-guided station-keeping duties were managed by eight thrusters that collectively sent almost 60,000 horsepower to the giant screws. The colossal rig, floating directly over a wellhead for months on end, was akin to a helicopter hovering directly over a shopping cart in the middle of a parking lot—a mile below.

An internal newsletter had announced BP drilled a world-record well with the *Deepwater Horizon*—just a month before Barry had spudded Macondo with the *Marianas*. With the record-depth offshore well having been drilled to 35,050 feet in the Gulf of Mexico Tiber area, Barry had no doubt the *Deepwater Horizon*, rated for 10,000 feet of water, would be able to handle the paltry 19,650-foot-deep Macondo well in 5,000 feet of water.

A big man met Barry at the foot of the steps from the helideck. Barry's double-take led to astonishment, then, "Tanker Forster, what the hell are you doing here?"

"Well, sir," Tanker said, shaking Barry's hand, "with the *Marianas* out of commission so long, then headed into a new contract with ENI—that's the Italians—Transocean did a little rig shuffling. And since me and my crew started your well, we get to finish it."

"Tanker, I'm glad you're out here. Fact is, if I ever go to war, I want you on my side. Give me a chance to unpack my gear and I'll let you buy me a cup. Your office or the galley?"

Diagram 7

"My office," Tanker said. They walked side-by-side under the helideck, down a flight of stairs, to the BP quarters. The first room was the BP office, with Barry's room next door. "Come see me when you're changed," Tanker said. "The coffee's always hot."

Barry had camped in the room twice before. It was big—twenty-five feet deep and a third that wide. Dinky built-in bathroom with sink, toilet, shower. Two twin beds, the far one never used as Barry didn't need a roommate. Two small chests of drawers and a three-drawer file cabinet. The austere steel desk supported a hard-wired black rig phone with a built-in speakerphone and a PA—public-address—speaker. For e-mail and calls to and from the rig, a vendor-supplied second phone and fax-copy machine provided Barry with 24/7 satellite phone service to the mainland and BP's offices.

Bolts locked the furniture to the floor and walls—not to thwart thievery, but to combat storm-driven heave, pitch, roll. A porthole, its glass thick and tempered, adorned the far wall above the second bed. Barry had never bothered to look out the window, as an easy glance would tell him all he needed to know—dark, light, dim, foggy.

He pulled his laptop and accessories from his backpack and stuffed it under the desk. His travel clothes—Levi's, a blue golf shirt, and a BP-emblazoned lightweight jacket—went on a wall hook where they would stay for the duration. He emptied his duffel bag on the bed and found what he wanted. His once-white BP hardhat went to a hook on his exit door. A borrowed mantra—*don't leave home without it.* Then, three pairs of coveralls—one to wear, one in the wash, one ready to go. They were all a faded camo-green. Convenient. He climbed into a pair. Slipped on his favorite grungy-but-comfortable, steel-toed boots. Also convenient. A second wall hook received his empty duffle bag and his heavy canvas jacket for winter cold fronts, though rare.

Minutes later, laptop in hand, he entered the adjacent same-layout room. The BP office. His office. Three Spartan desks, two with satellite and rig phones. He chose the closest desk, which had its own fax-copy contraption as well as a PA phone and speaker mounted on the wall.

The PA was silent, but it would soon erupt in a barrage of announcements to mimic an airport departure lounge. Messages yet to come would include weekly safety drills, transportation departure times, the occasional football score. More often, the PA calls would be for the toolpusher, or the rig mechanic, or himself—to ring a given number. Other announcements—accidents, well control, fire—though rare if ever, would garner instant facility-wide attention and trigger overdoses of adrenaline, testosterone, fear. Only the criminally ignorant ignored PA announcements.

Barry, appreciating the silent PA, shut the door behind him as he left his office. He followed the hallway to Transocean's quarters on the far side of the rig. The well-worn path took him past the medic's office and through the back end of the media room. Conference table on his left, multiple lounge chairs on his right, facing a big screen.

The darkness seemed unnatural. Soon, the horde—drilling, ancillary, and service crews—would get into the swing of overlapping schedules. The table would be surrounded by off-duty domino and card players. An endless stream of around-the-clock movies would fill the wall-size screen, interrupted only by college and professional sports.

Barry couldn't remember the last time he'd watched a complete movie or sporting event on an offshore rig. Not because he was too good to join the masses during their hours off, but because at ten to twenty thousand dollars an hour just for the rig that would drill his well, he felt he could better justify his time in the office getting ready for and optimizing schedules, logistics, and procedures than half dozing in a deep recliner watching week-old football highlights.

Beyond the media room, he skirted the galley to get to the Transocean offices and senior-staff quarters, the mirror image of the BP side of the rig. The first room was Tanker's office. As soon as Barry entered, the big man handed him a full cup of well-creamed coffee.

"How about logistics?" Barry asked.

"We've been on location for a day," Tanker said. "Mostly general maintenance and rigging up for stump testing the BOP. There's a workboat on its way out. It's got all the pipe that was on the *Marianas*, including a full string of 18-inch casing. Shore base manager says the workboat's also got another thousand barrels of oil-based mud. We'll pump it aboard as soon the boat gets here."

Barry wasn't surprised—he'd made the arrangements himself.

"How about personnel?" he asked.

Tanker ran his finger down the IADC report. "Halliburton cementers, two M-I Swaco mud daubers, one Sperry Sun mudlogger." He leaned back in his chair. "Mudlogging unit's hooked up and ready to go. It's butted up against the rig-floor firewall, with a door for access to the drilling station."

"Jessica would've liked that," Barry said, though when he recalled her fear of heights and of deep water perhaps *like* wasn't quite the right word.

Tanker's entire face crinkled. "You two made a pretty good team for BP. Shame to break it up."

Barry shrugged away Tanker's comment, thanked the big man for the coffee, and went back to his office to check his e-mail.

But Tanker's words got him to thinking about Jessica. He appreciated her quick wit, her desire to learn and understand, her ability to apply new concepts to old problems. With a work ethic as solid as his own, she was a good listener, an eager student, took bullshit from nobody, and had been on a good team. At least according to Tanker.

What the hell, worth a try. He called his boss. Secretary answered, wanted to take a message. Barry declined, said he'd send an e-mail. Which he did. Requested Jessica.

Barry headed to the rig floor, but did a double-take as he stepped onto the pipe deck. Coming toward him was the steward, in escort mode. Barry tilted his head to shade his eyes under the brim of his hardhat. Jessica?

Yes. Street clothes. Dark slacks. Yellow blouse. Tiny little running shoes. Like *Best dressed* in a woman's magazine.

"What are you doing here?" Barry asked, knowing his boss wasn't that damn quick but grinning so hard his cheeks hurt.

"Oh, nice greeting," Jessica said. "Do you ever look at your manifest? I'm on it this time, too, probably under the Fs again."

Barry slapped the handrail. "Doesn't matter. Welcome back."

The steward shifted the bags in his hands. "I'll show Miss Jessica to her quarters, sir."

Barry gave the man a nod, then said to Jessica. "Get settled, then let's meet in the office." He checked his watch. "Fifteen minutes?"

Walking away, she tossed a glare over her shoulder, her spiky blond hair dancing in the sun.

He called behind her, "Okay, twenty."

---

[59] See **Diagram 7**—(Photo—Transocean *Deepwater Horizon*). Shown on page 103 herein. Source and photo credits on file. The rig was designed and built in 2000 to Reading and Bates Falcon RBS-8D design specifications. The dynamically positioned (DP) semisubmersible was also referred to as a MODU—Mobile Offshore drilling Unit. Water-rated to 10,000 feet, and well-depth rated to 30,000 feet. Rig 396 feet by 256 feet with 242-foot-tall derrick. Because rig was DP while hovering over a fixed location on the seafloor (the well), the Coast Guard considered it a "vessel underway," which is required to have a Master (captain). Additionally, because the facilities were massive, with accommodations for 130 personnel, federal regulations called for an Offshore Installation Manager (OIM). A toolpusher was, of course, in charge of all drilling facilities and rig personnel associated with drilling the well. Tanker handles all three jobs in the story.

CHAPTER 17—Reunion

"My room's huge," Jessica told Barry, who looked forlorn in the also-huge office. She'd unloaded her stuff onto the desk next to his. The one he'd offered her. The one with the other satellite phone. "Room's shaped like a trailer house. It has a window," she said, hiking her thumb over her shoulder, "like that one, but all I can see are sky, water, and maybe a speck of light on the horizon."

"Which room?"

She pointed toward the hallway, then to the right. "Five or six doors down, this side of the hall."

"Good it's not in the catacombs."

"What do you mean?" she asked as she booted her laptop and plugged it into the BP Sat phone.

"Main deck's above us, outside. We're on deck two. On the third deck, below us, there're more rooms. The rig sleeps a hundred and thirty."

"Then I'm glad for my room with a view," Jessica said, visualizing exits, stairs, escape routes. She pointed to the unoccupied desk. "That the toolpusher's?"

"The whole crew's out here. Tanker. All the Transocean guys."

The unexpected words sunk in. "Tanker's here?" She pumped her fists. "You said all the guys? How about my sip-spit-and-grin buddy Daylight?"

"Him too," Barry said.

"Yes! Great!" A three-step jig jumped out of her feet. "I knew I needed to get back to Macondo—let's go say hi."

"Soon. And no, that's not Tanker's desk." He stood and pointed to a layout drawing of the rig. "Our rooms are along the north side of the layout, running east and west." He tapped the drawing as he talked. "We're here, far northeast corner, under the edge of the helideck and bridge."

"Then I'm disoriented," Jessica said. "The rig wasn't facing east when we landed, not if that big ball of fire in the southern sky was the sun."

"Since the rig can face any direction, this drawing is for reference only. Up is north, but more important, it's the port side of the vessel. On level one—the main deck—the bridge and its helideck roof are cantilevered over the port-side forward corner, because that's the bow."

"The bow? Like for a ship?"

"Exactly. The bridge—nothing more than a big room loaded with controls for running the facility—has a lot of windows that face forward. They're mighty handy if you want to see where you're going when the rig's making way." Again, he pointed. "We're one floor lower, on level two. Transocean's offices and management quarters are also on level two, directly across from us, starboard side. Quarters for the rest of the guys are below us."

"Why's Tanker so far away?" Jessica asked. "I don't remember a rig where the company man and the toolpusher don't share an office. How the heck do you guys talk?"

"Other than an entry into the galley and the medic's office, there's nothing but an open hallway between our offices. I go there a lot—he comes here. You're welcome to do the same. And the intercom is easy. Pick up the handset, push the red button, and say 'Tanker Forster, call 105,' or whatever number you're calling from. Then hang up and wait for the call, which will be private."

Jessica was all too familiar with PA systems, but two zones were crucial to her work. "What about the rig floor and mudlogging unit?"

"Same," Barry said. "If you get paged to call a number, do it from wherever you are. If you page me and give me a number, I'll call you. The only place anyone's safe from the PA system is in their bathroom." He cocked his head as if shaking a memory free. "Well, not always," he added.

Jessica ignored his grin. "Have you been to the mudlogging unit?" she asked.

"Sure. Why?"

"Hope you don't mind getting your exercise," she said. Cold chills accompanied her thoughts, just imagining the long flight of steps, almost two stories up to the rig floor.

As a blossoming teenager interested in math and science she had finagled her brain to get as far as possible from her often-repeated nightmares—heights, drowning, and fire. Geology seemed the perfect answer. Rocks are on the ground. Rocks don't drown people. Rocks don't burn. She collected rocks. Studied rocks all the way through grad school. Then joined BP as a rig-site geologist. Which had dragged her inexplicably onto deep-water, high-elevation, offshore rigs.

"Great location," Barry said. "Private. The unit's back door—though the mudloggers seldom use it—will take you through the rig-floor firewall, around the drilling equipment, directly to the driller's station.

Best of all, its front door opens with a clear view of the sea. You've got your own steps down to the pipe deck, or in the opposite direction down to the shaker room. And you'll get the best breeze on the rig."

"One man's breeze is another man's gale," Jessica said, her stomach tight. "The first forty-knot wind will blow me and the mudlogging unit off the rooftop, right down to the pipe deck, which is where I'd rather be anyway."

"Wind might get *you*," Barry said, "but not the unit—it's welded down. The big base it's on is solid steel—that's the cement building, where the Halliburton guys do their magic."

Jessica waited for a flash of teeth that didn't come.

Barry stood. "Let's go see Tanker's office. Maybe he'll buy us a cup of coffee."

A buzz from Jessica's Sat phone and a blinking light. "Looks like you've already got mail," he said.

She punched in her password and read the e-mail—from her boss.

Jessica, one of our HR managers (and I) expected you to stop by the office between rigs. He says he's being bugged by an attorney trying to see you in person. We're guessing a subpoena. Anything I need to know? Please give me a heads-up before you come in so I can make arrangements.

Not good, Jessica thought as she closed the message and signed out. She'd been warned the day would come.

Barry asked if everything was okay.

She wondered if something in her face had tattled on her, since Barry wasn't normally that perceptive about feelings. "Yeah," she told him, though she was far from okay. "Family stuff."

She walked beside Barry, through the movie room and around the galley, her thoughts on how, if high winds blew her to her death from the elevated mudlogging unit, at least she wouldn't have to ever face another attorney.

And she'd damn sure haunt Barry forever.

CHAPTER 18—The Landing

Barry found the correct stairwell and headed down, Jessica behind him. At the bottom of the steps, he stomped his foot. "*This*, is the lower deck. You can't go farther down without getting wet." He reckoned she'd yank out her tally book.

"Holy moly, it's like a warehouse." She yanked out her tally book. "What is all this stuff?"

"Storage, mostly—not designed to be pretty. It's where we store the blowout preventers, plus flow lines, hydraulics, tensioners—all the stuff that connects to the top of the riser once it's installed." He walked over to a handrail that surrounded a 20-by-90-foot opening in the floor, the view down to the sea. "And this—" He spread his arms over the opening. "—is the moon pool. It's our access to the water, right down through the heart of the rig." He pointed straight up. "And that's the underside of the rig floor."

Jessica, too, looked up, though Barry noted she'd latched her hands with a death grip onto the rail. She looked down for a few seconds, then backed up, her hands blindly grabbing for security away from the rail.

"Wow," she said.

"You okay?"

"Yeah, but I wouldn't want to live down here."

"That thing you're leaning on . . ."

Jessica turned to look.

". . . is the lower-stack of the BOP."

Jessica stepped back, looked way up. "Ohmagod, it's huge."

"Thirty feet tall. Remember the drawing for the *Marianas* blowout preventers? This is a twin of that stack—same BOP."[60]

"If this is the lower part, the part with the shear rams and the variable rams, it looks nothing like the drawing."

"The rams are all in there, though they may be hard to pick out since they're surrounded by control lines, choke-and-kill connections, support braces, and hydraulic plumbing, none of which was on the drawing. Plus it's all hooked up for stump testing."

"And stump testing is . . ."

"The stump takes the place of the wellhead, so we can function test and pressure test each ram, every component, all the controls. Not much to look at, just gauges, but we've all got a vested interest in making sure everything's in good working order before we run the BOP stack."

He checked the time. "Tanker and Daylight will be here in a few minutes to run the tests, and I need to witness each step. You're welcome to stay and watch, or I can call you when we're done."

Jessica did a quick glance toward the moon pool, then re-anchored her grip to a BOP stanchion. Her face seemed the wrong color, kind of a mix of Transocean gray and a pinch of green.

"No," she said. "I think you can handle a few pressure gauges. I need to check out my mudlogging unit."

"Do it," he said, "but I recommend you join me on the rig floor when we land the stack on bottom. It'll be several hours."

"Call me," she said over her shoulder without looking back.

* * *

Barry was delighted with Tanker's leadership on the *Deepwater Horizon*: toolpusher, master, and offshore installation manager. Relative to the stump test, Barry's job was to confirm each successful step for the records.

The testing took less time than Barry expected, since all the hook-up work, for prepping the BOP to test, had already been done. Barry and Tanker independently recorded the results.

After the stump test, time to run the BOPs. Barry's mind was on the task at hand—getting the big sonofabitch from the rig, through 5,000 feet of water, and safely latched onto the connecter he and the *Marianas* had left on the seafloor three months before.

Without dropping it.

Or damaging the wellhead.

Such thoughts reminded Barry all too vividly of a post-BOP-repair incident he'd experienced in the North Sea. A blustery winter night. Big seas. Lots of heave, pitch, and roll for the anchored floating rig, but not as bad as it'd been for days. He made a tactical decision and talked to the toolpusher. Simple plan. Run the stack, 400 feet to the sea floor. Latch it up. Get back to drilling the already-12,000-foot-deep well. But with the stack a hundred feet above the seafloor, something happened. The drillpipe landing string, hanging through the rig floor, jumped, banged, and clanged for a scary few seconds, and then everything went quiet. Barry recalled even the wind stopped howling. The driller's weight indicator showed the running string had lost some weight. About 200 tons. About the weight of the BOP. Barry's first thought at the time was that dropping the BOP on a soft muddy bottom and having to hire a crane barge to pick it up wasn't nearly as important as the early-age heart attack he thought he was having. Then he'd remembered the wellhead.

The bottom half of the Macondo connecter, already attached to the wellhead and waiting for the descending blowout preventer, was located ten feet above the mud line, which was at—

Crap. Couldn't remember. He checked his tally book. Which reminded him to call Jessica. He did. Gave her a heads up that they'd be near bottom in another hour.

While waiting, he rounded up Tanker. "We need to talk about rig-floor elevation," Barry told him.

Turned out, not a problem. No records had to be changed. Tanker had already done the minimal ballasting necessary to ensure the *Deepwater Horizon*'s variable draft fixed the rig floor at plus 75 feet—the same as for the *Marianas*. Which meant the mud line was still at 5,067 feet, and the wellhead was still at 5,057 feet.

* * *

Barry saw Jessica coming. She entered the rig floor through the side door from the mudlogging unit.

"What's going on?" she asked Tanker and Daylight.

All activity stopped dead. Gloves came off and bare hands reached out to her, open for handshakes. Tanker. Daylight. The derrick man. Roughnecks. Simultaneous voices. "Hey, Miss Jessica." "Glad you're back." "Welcome aboard."

When the hoopla died down, Barry poured her a half inch of rig-floor dregs, which he considered only marginally better than the dog-house coffee on the *Marianas*. "The BOP stack you saw below," he told her, "is now hanging on the landing string just above the wellhead. We'd moved the rig off location before we lowered the stack, and we're just now getting back on location."

"Why'd you move the rig?"

"In case we dropped the stack."

"Does that ever happen?"

A second flash of the ugly memory. "Only in my worst nightmares."

Jessica sipped her brew, then licked her lips. "Now that's good coffee."

Barry pretended to sip his own. He motioned for her to follow him into the driller's station. "Let's just watch the ROV monitors—they don't need us for this."

Under Tanker's guidance, Daylight and his crew lowered the heavyweight stack, a foot at a time. Daylight called out the numbers of the closing distance. "Thirteen feet."

"Lining up the two halves of the connector," Barry said, his voice quiet, "in a mile of water from a hovering rig requires a gentle three-dimensional touch, plus eyes on the seafloor." He pointed to three screens on the long console. "Eyes in the form of video cameras, mounted aboard two ROVs. Same kind of unmanned submarines as on the *Marianas*."

"My favorite silent movie," Jessica whispered, tally book in hand.

"Ten feet," Daylight said, and the rig floor got even more quiet. Nothing moved. Daylight, without taking his eyes off his monitors, slacked off another inch and told his floor hands to take a break.

"Looks like an air-traffic-control center," Jessica said, "except the camera views are underwater." She pointed to the backs of three people seated next to Daylight. "Who are they?"

"The woman's the Transocean DPS engineer," Barry said. "DPS is the dynamic positioning system. She manages the lateral location of the rig. The guys are service hands, trained operators who manage the ROVs. They grew up playing joy-stick computer games. Now their joy sticks are wired to high-tech remote operated vehicles—submarines."

"Neat careers," Jessica said. "I have a nephew who'd fit right in."

"Seven feet," Daylight said.

Tanker rumbled, "View the lower connector."

Barry continued, his voice soft, meant only for Jessica. "They guide the submarines around the wellhead and view the connector halves from whatever direction's necessary. The ROVs have headlights, video zoom cameras, pinchers, water jets. They've already cleared debris—ocean sediments—from the top of the lower connector."

Jessica slowly shook her head, stared at the screens. "Different world to me."

"If needed," Barry said, "the ROVs can take measurements, close valves, tighten connections, or just watch and stream video."

"Four feet," Daylight said.

Tanker: "North 35 east, about three feet."

The DPS engineer leaned forward toward her video screen.

"She's inputting new GPS coordinates," Barry said, "which will direct the thrusters to move the rig. Any direction, however many feet, however many inches. We need perfect lateral alignment of the BOP and wellhead before we can latch."

Jessica held onto a stanchion.

"You won't feel it," Barry said. "The movement will be silent. Slow. Glacial."

As the ROVs, guided by the contract ROV operators, sent video of the diminishing misalignment, Tanker called out new coordinates. The DPS engineer again moved the rig—the process iterative.

Barry watched Jessica watching Daylight, who watched the slow-motion action on his own screens. Using the ROVs' cameras as his eyes and depending on Tanker and the DPS engineer for lateral alignment, all Daylight had to do was gently lower the massive BOP toward the wellhead connector. Which he did. A foot. Another foot. Then inches at a time. And again. Until the funnel guide on the bottom of the BOP swallowed the waiting wellhead and connector and aligned the mating halves.

Done. Latched. In total silence.

Barry hadn't realized he'd been holding his breath. And sweating like an NFL running back. And he wasn't alone. The driller's cabin, which normally sported the pleasant aroma of brakes, grease, and shitty coffee, had blossomed to mimic a high-school locker room.

"What'd you think?" he asked Jessica.

She hadn't moved. "I'll never go to the moon," she said, slowly shaking her head, "but I'm guessing that's as close to a Neil Armstrong landing as I'll ever witness, and it took place under a mile of seawater."

---

[60] BOP—BLOWOUT PREVENTERS—The *Deepwater Horizon* blowout preventer components were the same size and function as those on the *Marianas*. See **Diagram 5** (page 68 herein) and **Diagram 6** (page 73 herein). Manufactured by Cameron, all the rams in the 18.75-inch stack were rated to 15,000 psi. Of the two 10,000-psi annular BOPs in the lower marine riser package, one had been reconfigured to accommodate 65/8-inch drillpipe, which de-rated the unit to 5,000 psi.

CHAPTER 19—Feedback

Barry, Jessica, and Tanker took advantage of the break, hiked down steps, across the pipe deck, and down another flight into the galley.

They ate light—Tanker scallops, Barry beef, Jessica eggs, cheese, fruit, and crackers.

"I mostly sat around the office since the hurricane," Barry said, with no plan to mention his family visits. "What'd you guys do?"

"Fished the Atchafalaya every chance I got," Tanker said. "Set me a new record."

"Double-digit bass?" Barry asked, thinking ten pounds, maybe twelve.

"Nope." The big man grinned. "Fell out of my boat, twice in one day—a new record."

Barry laughed. "No way."

"Easy to fall out of a boat," Tanker said.

"That's true," Barry said, "but I'm trying to visualize you crawling back in."

Tanker didn't share Barry's laugh, just turned to Jessica and said, "How about you, ma'am?"

"Worked two other rigs," she said. "Then my boss called and told me the *Horizon* was headed to Macondo. Asked me if I wanted the job. Easy decision, and here I am—ready for a discovery."

"Sounds like exploration hype," Barry said.

Jessica pushed her tray aside as if for a fight. "Ninety million bucks says the hype has a few backers on BP's board."

"Just saying . . ." Barry said, before backing off, not wanting to discuss well costs.

Tanker stood, picked up his tray. "Daylight should have the landing string stood-back in the derrick and out of the way any minute, then we'll be running the LMRP. You two coming up?"

"Why thank you, Tanker," Jessica said. "I wouldn't miss it."

Once out on the main deck, she walked beside Tanker.

Barry followed.

* * *

While Barry and Jessica watched from the rig floor, Tanker and his drilling crews picked up and connected additional joints of riser, the repetitive process sinking the lower marine riser package ever deeper into the sea.

Jessica wrote in her tally book, maybe a drawing.

Barry wanted to peek, but didn't.

When she looked up, she said, "Tell me about the riser, then I need to go back to the mudlogging unit."

Barry pointed. "As complex as it looks, the riser's just a tool. When all's going well, it's nothing more than the path, a big pipe, for the drill bit to follow to get down into the well."

"And before we forget," Jessica said. "It's for me to finally get my drilled cuttings."

"Yep. At other times, like during well control, the riser's our life line— our connection to the BOPs."

"Okay, the riser's our friend," Jessica said. "Now what's the bad news?"

"Who says there's bad news?"

"Oh, stow it—you can't wait to tell me."

"Well, since you asked, riser problems can eat your lunch. I can think of four or five. Weather's the most common problem. Like a cartoon—the immovable seafloor and the unstoppable rig connected by nothing but the riser. Then there're major eddy currents—fast-moving looping swirls in the Gulf that last for years. They last so long they get names. Eddy Cantor. Nelson Eddy.

Jessica squinted.

He continued. "The loop currents—which often make three or four knots—can push so hard on a long riser we can neither pull nor run it. They can even keep the EDS, the emergency disconnect system, from releasing the LMRP from the BOP."

On a roll, with Jessica scribbling in her book, he added, "Another major problem for a riser is loss of rig power, especially on a rig like this with no anchors. And then, of course, the *coup de grâce* for potential riser problems occurs during well control, where all the—"

The memory exploded from hiding. "I'll be damned."

Jessica froze, then grabbed his sleeve, her eyes big. "You're freaking me out. You'll be damned—what?"

"I remember. The *irony of Piper*. Oxy's Piper Alpha platform—the North Sea fire you found on the Internet."

Jessica looked hard, as if trying to read his mind.

He saved her the trouble—the memory clear. "When you read about the *Marianas* and its previous life as the . . ."

"Tharos."

". . . there was a major explosion and fire on the Piper platform. Maintenance related, as I recall. Fortunately, somebody pressed the emergency stop button, which closed major seafloor pipeline valves and the subsurface safety valves in each well, which halted oil and gas production." Barry pictured himself in the middle of the disaster. Tried to imagine—

"I'm sorry," Jessica said, her face a bit pale, "but I missed the *irony* part."

"It's this. While all the producing wells stopped producing, which took away the fuel, the fire bloomed bigger than ever and consumed the entire facility. There were lots of deaths, even before most of the structure collapsed and sank."

"A hundred sixty-seven died," the lady with the perfect memory reminded him. "Sixty-one survived."

"They were the ones who jumped overboard, a hundred feet, maybe more, and were lucky enough to survive the fall."

Jessica shuddered. "The *irony*?"

"The guys on the platform did everything right to isolate the producing wells, which should have starved the fire. But the fire *grew*, and that's the irony. The fire was fueled not by the producing wells but by gas feeding *back* to Piper Alpha, from an adjacent platform, directly into the fire, through a miles-long, high-pressure, subsea, gas pipeline."

Talley book and pen at the ready: "Help me understand."

"The Piper lesson is this, and it's applicable to drilling risers. If there's pressured gas in a seafloor pipeline between a closed valve and the facility, a breach in the line at the facility—like from Piper's fire—will allow the gas to feed *backwards*, to the facility, until the gas pressure in the pipeline bleeds to zero."

Jessica nodded. Said nothing.

"Now instead of a pipeline and big valve on the seafloor, picture our mile-long, vertical drilling riser." He pointed to a 90-foot riser joint hanging from the blocks. "To avoid the irony of Piper, the last thing we want is gas in the riser between the closed BOP and the rig, the top end of which is wide-open, directly under the rig floor."

"How could that happen?"

"It shouldn't. Ever. That's why we check drilling breaks for kicks. If there's a kick, we've got one mandate—immediately close the BOPs, control the situation, control the gas. The only gas we ever want above the BOP is in the choke line—never in the riser."

"What about trip gas, or drilled gas?"

"That's a different story. Because of drilled and trip gas, we're always sending mud returns from the riser through the mud-gas separator. But if it's a major kick, with big gas volumes and high pressures, we have to kill the kick, using the BOP and the choke and kill lines of course, without allowing the gas to get in the riser."

"So the obvious question. What if you get a big wad of gas in the riser?"

Barry had heard war stories, a hot topic during mandatory BOP training schools. "It'd be a problem. The gas above the closed BOP would expand and start unloading the riser. We'd immediately divert flowing mud from the riser overboard, but the rate would accelerate, pushed by the expanding gas. We'd be okay as long as the flow rate didn't overcome the low-pressure packing in the diverter."

Jessica's eyes asked the question he didn't want to answer.

"The diverter's directly under the rig floor. If it failed, we'd have no way to stop the flow. The mud would go straight up, through the rig floor, followed by the gas."

Jessica finished writing notes, her knuckles white, as if she were trying to push her pen through the page. "Which tells me," she said to Barry, "that if the well kicks, you better get it shut-in damn quick. Not only to keep gas out of the riser, but because I plan to be on the rig floor to witness every step you take."

"I look forward to it," Barry lied, hoping she would never need the notes she'd just written.

* * *

Barry went below and watched riser joints disappearing into the sea through the moon pool, the process slow, careful, deliberate. His mind wandered, his thoughts on operations, money, time. Like, how long until he'd see the riser back on the rig? That evening, if something went wrong? A week, the BOP pulled to the surface for repair, like on the *Marianas*? A month, to finish the well? Six weeks? Under budget? Over budget? After a dry hole?[61] Questions for which he had no answers.

He recalled Jessica's words. A big discovery. Maybe a mega-discovery. A billion barrels. He tried to picture Jessica dancing a billion-barrel jig.

A mega-discovery—he'd been there before. What fun it would be to go down that path again. Everybody excited. Thinking about the value of their stock. Blissfully ignorant of the work to be done. Just to drill the footage would be a challenge. Gas cut mud. Intractable kicks. Lost circulation. A million-dollar-a-day operation. No doubt, he'd get a lot of help—technical and political—from the beach. Exacting operating procedures would follow. Step-by-step minutiae. Heightened concern at all levels of management for even the smallest error. Big magnifying glasses. Zero tolerance at the extreme.

Just the way he liked it.

The exploration staff would be declared heroes—well deserved. The drilling staff—off to the next well—wouldn't even look back, the rig too expensive to dawdle in the limelight.

Then would come the questions—by BP accountants and equity partners after dissecting every invoice—about how goddamn expensive the well had been. All on Barry's and his bosses' shoulders.

Yeah, right. What fun.

But worth it.

---

61 DRY HOLE—nothing economical discovered; no reserves. Note: when all the formations are wet (full of water), the well is said to be dry (no commercial oil or gas), and the well is called a dry hole.

CHAPTER 20—Road Block

With the BOP, riser, and lower marine riser package installed on the wellhead and tested, and with Barry's approval and Tanker's guidance, the *Deepwater Horizon* drilling crew reentered the Macondo well.

Daylight wasted no time drilling out the two cement plugs, then resumed drilling below the 22-inch casing.

Simple goal, Barry mused. Recover the 500 feet of hole the *Marianas* had temporarily abandoned three months before.

Jessica reported she'd collected nothing but mushy cuttings from the first hundred feet. She told Barry she saw nothing new, nothing exciting, and questioned what the hell was taking him so long just to get the drill bit to the bottom of the already-drilled hole.

The old hole was sticky, he explained, not at all like they'd left it. Wet shales had turned to gumbo, a gooey clay that could pack as hard as concrete around a drill bit. Formations had sloughed into the wellbore. The new drill bit found only remnants of the old opening.

Daylight worked the drill bit up and down, then made headway and worked the drill bit again. Using expert skills, he worried the drill bit deeper by the hour, eating into the tunnel they wanted to reclaim.

But he didn't get far. The drill bit stopped. Refused to go deeper. With Barry and Tanker watching, Daylight engaged the drawworks, pulled up on the drillpipe, then slipped the brake and banged down on the drillpipe. He twisted it to the right, and then to the left. He slowed the pump rate. Increased the pump rate. Pulled more, banged more.

"Tarnation, boss," he told Tanker, "we stuck. We got us a fishin' job."

Fishing job.[62] Sonofabitch. Nothing but downtime and dollars.

Though fishing came with the job, Barry hated having to get unstuck. Being forced to recover or abandon whatever blocked the wellbore. Having to spend the time and money. With no progress to show for the effort. But none of that mattered. He had to either get unstuck and clean out the well, or abandon the stuck drill string—the fish—and drill new hole.

Tanker and Daylight had an assortment of fishing tools for those occasions when the drill string had backed off or twisted off. Not so in this case. The drill string hadn't unscrewed, nor was it damaged, it was merely stuck.

So Barry called the experts. Directional drilling experts and fishing experts. From the beach. The directional drillers would be needed if the fishing experts failed. The fishermen brought string shots[63] —explosive devices run on wireline used to make the pipe unscrew at a selected depth. Tungsten-carbide mills, in case the drillpipe twisted off and the jagged pipe needed to be milled back to its original geometry. Large overshots used to slip over and grab onto the top of the fish in open hole. And, when all else failed, hydraulic jars, designed to latch onto and beat the shit out of the stuck pipe, hoping to jar it free.

All such attempts did exactly that—they failed.

Barry had two choices. Keep fishing, or sidetrack. Fishing, if continued, had a chance of success, or it could prove to be an endless money pit. Sidetracking called for again backing-off the drillpipe, recovering the fishing tools, and drilling a new hole around the fish-filled old hole. Hence, the need for the directional drillers.[64]

He wrote notes, compared options, guessed at probabilities, estimated hours. The lesser of evils stared back from the page. He again called his boss, which he had been doing seemingly around the clock since Daylight had first uttered the word *stuck*. Barry didn't bring up the subject of the Macondo AFE, which included the rig schedule and budgeted costs. He simply discussed the pros and cons of fishing versus sidetrack. Made the recommendation to back-off the drillpipe as deep as possible, set a cement plug, and sidetrack. Got agreement.

Which meant the fish—the original bottom-hole assembly, aka the BHA,[65] which was comprised of drillpipe, drill collars, stabilizers, and a drill bit—would never again see sunshine.

After the new cement plug hardened above the top of the abandoned fish, Daylight and the directional driller drove the drill bit along a new path a few feet away from the old wellbore. The new sidetracked wellbore, once vertical, paralleled the old hole and the abandoned fish.

Though during the fishing job Jessica had taken prolific notes, she'd done so without hiding from Barry her blatant ulterior motive—she wanted more drilled cuttings.

As did Barry. Any kind of downtime, which the Macondo AFE included as *contingency*, or *trouble time*, tore at his sense of purpose. The first delay had been the *Marianas* BOP. Then Hurricane Ida. And getting stuck. Giving up footage of hard-earned hole. Having to sidetrack. Zero progress for days. Dollars mounting—a half million a day just for the rig. And if he added casing, drill bits, mud, cement, contract personnel, helicopters, boats, fishing experts, he was spending a cool million a day. At that rate, too much trouble time could easily kill his budget. Not a pleasant thought.

A dozen hours later, he spotted Jessica on the pipe deck. "We just drilled past our previous deepest depth."

She hoorayed, which made him happy. Then said, "About damn time," which reminded him of his ex.

"Join me in the galley—we'll celebrate."

"No thanks, this gal's got work to do, evaluating new hole, new footage, cuttings and marker fossils never before seen by man." She pointed toward the rig floor, then to the white Sperry Sun mudlogging trailer. "I'll be up there. Give me a call when you break for dinner—maybe I can join you."

Barry thought about dinner throughout the slow-moving remainder of the afternoon.

Dinner.

With Jessica.

A good team, Tanker had said.

A damn good team, Barry reckoned. His well. Her rocks.

Drilling the *well from hell*, he'd heard in the galley, though the convenient rhyming moniker for every tough well worldwide had long ago reached its cliché shelf life.

---

[62] FISHING—Not for bass. On a drilling rig, a fish is any man-made stuff in the hole that's not supposed to be there. A 36-inch-long pipe wrench dropped into the well would be a fish. A drill bit stuck in the wellbore, even though attached to drillpipe, is a fish that needs to be fished. Fishing is any activity related to recovering a fish, or trying to get unstuck.

[63] STRING SHOT (and back-off)—Follow the logic: By measuring the inches of stretch in a stuck drill string when pulling, for example, an extra 50,000 pounds, the footage of "stretching" pipe can be calculated (let's say 8000 feet). That means everything below 8,000 feet is stuck solid—so solid it didn't even stretch when pulled on. The goal, therefore is to unscrew the drill string as deep as possible, where it's known to be free (not stuck); i.e., at 7,900 feet. So the driller adjusts the hook load to pick up the weight of 7900 feet of drill string, so at that depth the pipe is in neither tension nor compression (it's neutral, relaxed, "easy" to unscrew). The driller then turns the drill string "to the left" as if trying to unscrew the drill string at the neutral depth, but stops before anything actually unscrews (because such "blind back-off" is an inexact science). The fishing experts then set an explosive device (STRING SHOT) inside the drillstring at

7,900 feet, and pull the trigger. The result, more often than not, is a clean back-off (unscrewing) near 7,900 feet, which becomes the TOF (the top of the fish). Then the real fishing job starts.

64 DIRECTIONAL DRILLERS—Use 3-D surveying instruments to map the course of the wellbore, then use down-hole tools that force the drill bit to drill slightly off course along a predetermined path. Convenient for drilling around junk (fish) left in an abandoned wellbore. At the extreme, wells can be forced to make gentle curves, starting off vertical, then increasing the angle away from vertical, heading, for example, northerly away from the rig, and continuing to increase the angle until the drill bit drills a horizontal well headed due north.

65 BOTTOM-HOLE ASSEMBLY (BHA)—Consists of heavy-weight drill collars to help apply weight to the drill bit, as well as large-diameter stabilizers to keep the drill bit and the drill collars centralized in the well (so as to drill a vertical hole). As drilling progresses and the diameter of the wellbore (the hole) gets smaller, other devices can be added to the BHA, including turbine motors, directional-drilling assemblies, and pressure-while-drilling and logging-while-drilling tools.

CHAPTER 21—Tolerance

A day later, Barry considered his options as Daylight cheeked his Skoal, spit in one cup, and sipped coffee from another. The driller had pushed the drill bit to 8,969 feet, where he stopped drilling and picked-up off bottom. He'd have to pick up another stand of drillpipe from the derrick before he could drill deeper.

Barry conferred with Tanker, who then shut down Daylight. Simple instructions—make a short trip up into the 22-inch casing, then circulate bottoms-up.[66]

Barry made his way into the back door of the mudlogging unit, where Jessica was studying drilled cuttings from somewhere above bottom. He greeted the on-duty mudlogger and told him it might be a good time for him to take a break. The mudlogger took the hint and closed the airtight door behind him as he left.

Jessica didn't look up. Since getting her first sack of fresh-drilled Macondo cuttings, she had shown ferret-like interest in every chip of rock removed from the well. Her head immobile, she looked through a stereoscopic binocular microscope, which, she claimed, allowed her to view tiny objects in three dimensions. Great for logging cuttings, identifying marker fossils, and watching for shows—traces of oil and gas.

"I need to see your gas traces," Barry told her. "We may be at a casing point."

"And a pleasant hello to you too," Jessica said. "And if you'll remind me after we look at the charts, I'll show you Herkimer's diamond." She sat up and rubbed her eyes, then rolled her chair over to the appropriate analog recorder and pointed to a mark on the page. "That's where you looked at background gas this morning," she said. "Though the numbers aren't all that high, the gas is double now after an all-day increase. These little spikes are connection gas—they've been increasing too. What's going on?"

"I've been watching background gas on the rig-floor monitor," Barry said, wondering why she'd mentioned a diamond and somebody named Herkimer, "and I don't like the steady climb. I think we need a couple more points of mud weight, but we'll know for sure after we circulate bottoms-up and look at trip gas."

"Barry, I watch and record gas readings—connection, trip, background—around the clock while we're drilling, because I'm always looking for shows. I see gas, I get excited. So make me understand the flip side—why trip gas may be the deciding factor for running casing."

"All gas readings are important," Barry said, "but they mean different things. Background gas is easy. You crunch up a foot of rock that contains a little gas, you end up with a mix of rock chips and gas being pumped up the hole, regardless of your mud weight. Trip gas, on the other hand, is more related to mud weight and formation pressures."

"How so?" Jessica asked, standing to check a gadget that'd beeped.

"It has to do with swabbing," Barry said. "You pull the drill bit up the hole, it's like loading a syringe. It sucks in formation fluids. When that gas eventually reaches the surface, we call it trip gas."

Jessica nodded, took notes. "What about the opposite? Like when the drill bit's going in rather than being pulled out."

"It's called *surge*, and *opposite* is the right word. Like a plunger in a toilet, you cram the drill bit in the hole, the mud below the bit has to get out of the way. It's either going up the annulus, or if we push too hard and the formations are weak, we can cause lost circulation."

"Then back to trip gas and swabbing. Why the concern?"

"There're a couple of problems with swabbing," Barry said. "The closer the mud weight is to being balanced, the more fluid we pull in from exposed formations." He readjusted his position in the seat to get more comfortable, wondering if this was her way of announcing her engagement to some shithead named Herkimer. "A big wad of trip gas might indicate we need to increase the mud weight.[67] Second, gas in the wellbore is never good, even if it's just swabbed in. We can't ignore it, and we better be ready for it when we circulate it out of the hole."

"Why all the concern for gas?" Jessica asked. "And you haven't even mentioned oil."

"They're night and day," Barry said. "If we swab a little oil into the wellbore, it'll mix with the mud and we might never see it. Same thing with formation water."

"Oh, my," Jessica said, "how I would love to swab in a batch of oil."

"Oil would be nice," he said, "but if we're swabbing gas every time we pick up the drill bit, we may need to raise our mud weight."

"And the problem is . . ."

"The leak-off test. If we're drilling with 9.9-pound mud and the shoe can take only 10.1, how much do you want to increase the mud weight?"

Jessica gave the ceiling a thoughtful look, then turned her chair and faced him. "So even if excess trip gas tells us we need to raise the mud weight, we won't, because we might lose circulation. Something illogical about that. Like if the well can't take more mud weight, especially if it kicks, why are we drilling?"

"Good question," Barry said, "and you're not the first to ask. It's all part of the concept we call *kick tolerance*."[68]

"Which I'm sure you'll explain."

In answer, Barry pushed an intercom button to the driller's station.

"Yo, rig floor," Daylight answered, his West Texas drawl filling the mudlogging unit.

"Daylight, Barry—how much longer for bottoms-up?"

"Half hour, give or take."

"I'll be with the mudloggers—watch her close."

Jessica tapped a fingernail on a wall-mounted, pump-stroke counter. "Why call Daylight when the answer's right here?" She read the chart. "Bottoms-up in twenty-eight minutes."

Barry stared at the chart he depended on 24/7. "Makes Daylight feel good to give me an answer I need, and makes me feel good that he knows where I am."

"Good, then you've got time to tell me about kick tolerance."

"First things first," Barry said. "Tell me about the diamond."

She rolled her chair over to the binocular microscope, made a few adjustments, then rolled out of the way. She handed him a needle-pointed plastic rod. "Your turn—you can either get a chair or stand."

Barry leaned down and adjusted the binocular spacing to his eyes. Tiny cuttings, looking like boulders, filled the screen. "What am I looking for?"

"The glassy piece in the middle, pointed on both ends. It's a Herkimer diamond—a double-terminated quartz crystal. Eighteen facets—I love it."

Barry searched and found the crystal. Poked it with the needle. Six sided. Two points. Ultra tiny, but prettier than the stone he'd given his wife twenty years before. He rolled it over. A Herkimer damn diamond. He grinned.

"Never saw one before," he said, likely related to his zero interest in collecting rocks, "but I like it more than you know."

"Big ones—an inch, even bigger—come from Herkimer County in New York. The one you're looking at, about two millimeters, is typical for the Gulf. You want it?"

"Sure," Barry said. He moved aside so Jessica could roll her chair to the microscope, the Herkimer question a nonissue. While she did something with a pair of tweezers and a glass vial, he checked his watch and found the wall display he was looking for.

Jessica joined him. Side by side, they watched a blue line on the small screen. No bells or whistles, just a line on a graph that measured gas in the mud from the annulus. Same graph Barry had watched on the rig floor, the systems redundant.

A minute later he said, "There's your trip gas. Right on time. And it's not trivial."

Jessica pointed to the line on a pit-level chart.[69] "We've seen more, but it's definitely up."

They watched in silence.

"Gone now," Barry said, pointing to the blue line. "Two minutes—one belch of gas off bottom, a blip on the pit-level, and then back to normal."

"So what's the verdict?" Jessica asked.

"As predicted. With background gas creeping up and a fair amount of trip gas, we've got work to do."

Barry used the speakerphone, called the rig floor, and asked for Tanker. "Casing point," Barry told him. "We need to come up a couple points on the mud weight, from 9.9 to 10.1, then we'll log and run a liner. Give me a half hour to call my boss and I'll meet you in your office."

Tanker asked Barry about logging.[70]

"It'll be Jessica's show when we log," Barry said. "She's here with me now."

Barry hung up, and Jessica said, "I need to roust up my Schlumberger guys. How much time before you're ready for them?"

"Plenty," Barry said. "We'll sort out timing when we meet with Tanker."

"Good, because I need to know why you're increasing the mud weight to 10.1 pounds when you said the casing shoe couldn't take it."

"Good catch, but if we experience lost circulation," Barry said, "I'd expect it to be minor and fixable. Truth be told, I'd rather be running casing in the 9.9-pound mud. We were probably at zero kick tolerance fifty or a hundred feet back up the hole."

Jessica dipped her head and peered through the fringe of spiky bangs. "*Zero kick tolerance*?" she asked. "Would you have called it *Z-K-T* if you'd been talking to anybody but me?"

Barry thought of his visit to his parents, about whom Jessica had been right—they'd never really been interested in his tech talk, had never once asked him to translate an acronym, had never expressed beyond a nod even a modicum of understanding about his work. That visit and Jessica's admonishments had convinced him that *Keep it simple* wasn't so bad after all.

"Probably not," he said. "It's too serious a subject for an acronym. Heck, I'm not sure I'm even authorized to talk to a geologist about it."

She scrunched her face.

"But I will," he said, "as soon as I make a call."

Barry, the tactical guy, dialed his boss, the strategic guy. Gave him the scoop. Made his recommendation. Got agreement to run casing. Hung up.

"Okay," he said to Jessica, "kick tolerance."

Jessica washed and dried her hands. "Not in here—I need a break."

"Our office?"

"No. Come with me, I've got a better idea."

She led him down a double flight of stairs to the pipe deck, around the pipe rack, to the handrail overlooking the sea, the view westerly. "As I told you, I'm afraid of heights," she said, her hands on the upper handrail, gripping hard. "I come out here as often as I can. Therapy."

He adjusted his shades because of the glare. "I thought you were afraid of water."

"I never told you that. You been spying on me?"

"No. I picked up on it when we were talking about how deep the water is and when you looked into the moon pool."

"Perceptive," she said. "I'm not afraid of much, especially when I have choices. I don't drink or do drugs, but I drive fast, snowboard even faster. I'm not afraid of guns, and I've had just enough martial arts to put a lesser-skilled big man in a world of hurt."

Barry moved half a pace left.

"That'll do you no good," she said. "You're still in my range."

Barry laughed and moved back, though still two feet away. "Sorry, I got you off-track. You were about to say something about water and high places."

"Three things frighten me," she said. "Falling, drowning, and burning—in my nightmares. Usually, just one at a time. Sometimes, two will gang up on me. Never all three."

"Must make living offshore kind of tough."

"Being out here is self-imposed therapy," she said. "Plus sharing my phobias with others. And that's why we're standing here, looking at this beautiful view, the horizon miles away, five or six stories above water a mile deep."

Then something changed. She looked at Barry, crossed her eyes, gagged a dry heave, mumbled something about lunch, and leaned over the rail.

Barry backed up. "You okay?"

"Ease up," she said, upright, again looking toward the horizon. She squared her hardhat. "Just wanted to make sure you were listening."

Barry stared at her profile until she turned his way. "Was I listening? Yes. But I have phobias too—tears, brown diapers, fresh puke. You get sick on me, I get sick right back."

"I'll consider myself warned."

Barry checked his watch. "Eight minutes—you ready to go back to the unit?"

"Not yet. I want you to tell me about *zero kick tolerance* while I continue to squeeze the paint off this handrail. Which means you need to summarize a lot, because I get pretty testy when my arms begin to cramp."

"Well," Barry said, "the topic's critical. It was gonna take an hour, but I'll give you the three-minute version."

"Forget the preamble. Just talk."

"Yes, ma'am. Example: If we test a casing shoe to fourteen pounds per gallon and we're drilling a couple hundred feet below the shoe with twelve-pound mud, it's wrong to think that fourteen minus twelve means we can take a two-pound kick."

"It seems logical you could," Jessica said, "so why not?"

"During any kick, annulus pressures jump all over the place but generally increase throughout the process."

"Because . . ."

"Gas."

"Why?"

"Because as gas moves up the hole while we're killing a kick, its volume increases and decreases with changing casing pressure. You start out with a 500-psi kick and end up with a 1,000 or 1,500 psi on the casing annulus. No telling what's happening at the casing shoe. And if you close the BOP and wait too long, the gas will continue to rise. Since it's unable to expand, it'll increase the wellbore pressure, which will likely breakdown the shoe. Lost circulation."

"So," Jessica said, "fourteen minus twelve will get you in trouble with a two-pound kick. What's the answer?"

"Critical to every kick is the amount of underbalance, the proportion of gas, and, the only thing we have a chance of controlling, its volume. Therefore, the best we can do at the first sign of a kick is shut-in the well immediately, which restricts the volume. Restricting volume is also why we check drilling breaks,[71] trying to catch kicks early."

"I could use an example. A short one."

"Ten feet of a high-pressure sand will pump formation fluids into the wellbore five times faster than two feet of the same sand. And if you let it flow for five minutes, rather than one minute, the difference is another factor of five. Combined, those two cases differ by a multiplier of about twenty-five. Even if the kick is raunchy high-pressure gas, one barrel is a lot less bad than twenty-five barrels."

"I missed where *zero kick tolerance* fits in," Jessica said.

"Kick tolerance is nothing more than asking ourselves how big a kick we can take. Which means we have to answer a few easy questions. Like for the depth at which we're drilling, and for the mud weight we're using, and for the measured weakness of the last casing shoe, can we afford to take a two-pound kick, which might be 1,000 psi? If that's too big to handle, how about a 10-barrel, 500-psi gas kick? Probably not? How about a measly 100-psi gas kick? Still no? Can't tolerate it?"

"Let me guess—zero kick tolerance."

Barry nodded. "Kick tolerance is a serious subject. *Zero* kick tolerance means we can't afford to take a kick, even a small one, without it getting complicated. If kick tolerance is zero, there's no room to drill even one more foot, and it's time to log and run a liner."

"I got it," Jessica said. "ZKT is important. The end of the world as we know it. Now if you'll help me unlatch my hands from the handrail, I could use a potty break, dinner, and fulfillment of your promise."

"Promise?" Barry asked. "What's that?"

"It's a thing you make and aren't supposed to break."

"No," he said. "What promise?"

"Full casing string versus liner."

"Oh, that," he said. "I thought it was something important."

"All promises are important to me," Jessica said, loosening up and exercising her forearms, her wrists, her fingers, and fists. "Keeping them says something about trust. And I'm big on trust."

"Then as soon as I visit with Tanker about the casing let's meet for dinner—we can talk afterward." Barry watched her fists open, close, open, close—the movement hypnotic. "Or maybe after you work out, you know, like curl a joint of drillpipe, bench press a couple sacks of cement."

She grinned and turned away, slowly shaking her head.

Barry reckoned dinner couldn't come too soon.

---

66 CIRCULATE—First, circulate refers to pumping mud (or cement) down the inside of pipe (drill string or casing) and back up toward the surface outside the tubular (in the annulus). Second, reverse circulating is the opposite—like down a casing annulus and up the inside of the drillpipe. Third, circulate bottoms-up (CBU): example: one minute after the drill bit drills from 5,000 feet to 5,001 feet, that foot of rock, now cuttings, and the fluid that was in the rock, are mixed with mud in the annulus just above the drill bit. In that moment of time, the rock/fluid/mud mix can be defined as "bottoms." To see "bottoms," for any number of reasons, the driller stops whatever he's doing and circulates bottoms-up (CBU), which will arrive at the rig an hour or more later, depending on depth and how fast he pumps. For a short trip (as opposed to a bit trip, which brings the drill bit all the way to the surface), the driller pulls the drill bit perhaps a thousand feet off bottom, then goes back to bottom and fully circulates bottoms-up (CBU), to see what he gets, for example important cuttings, trip gas from swabbing, evidence of a minor kick, etc.

67 INCREASE MUD WEIGHT—Mud engineers add crushed barite (a heavy mineral) from massive bulk tanks in the "mud room" to increase the density (weight) of the drilling mud. Barite is 70% more dense than quartz rock, and is 4.5 times more dense than water.

68 KICK TOLERANCE (KT)—the well's ability to withstand a kick, measured in pounds per gallon (ppg). KT (simplified) is inversely proportional to the volume of the kick fluid, the pressure of the kick, the amount of gas in the kick fluid, and the weakness of the rocks already drilled. Zero kick tolerance (ZKT) means the well can't afford to take a kick of any size or pressure—it's time to quit drilling and run another (smaller diameter) string of casing (or a casing liner).

69 PIT-LEVEL—Critical to drilling is an active record of the number of barrels of drilling mud in the system—both in the wellbore and in a number of mixing and storage pits on the rig. While drilling, the system is dynamic, with mud being pumped into the well and mud returning up the annulus (the riser). At other times the system is static—nothing moving. Whether static or dynamic, sensitive gauges and meters compare pit volumes and pump rates 24/7, with a simple reality: any loss of mud in the system likely means lost circulation, and any gain of mud in the system (called a pit gain) likely means a kick. The driller watches such readings (which are audibly and visibly alarmed), so he can take immediate action as necessary. The mudloggers monitor the same charts and alarms around the clock on behalf of the company man (BP).

<u>70</u> LOGGING (WIRELINE)—the process of lowering sophisticated tools into the wellbore (hanging from a high-strength electrical cable) to take readings that allow geologists and engineers to interpret rock and fluid properties and take samples. The output is a series of graphs, on a foot-by-foot basis, that portray rock and fluid data. Hard-copy "logs" for a 20,000 foot well might be 150 feet long, folded into an accordion shape. Digital well logs take up less room, use fewer trees.

<u>71</u> DRILLING BREAK—a rapid change in drilling rate; usually indicative of change of formation; i.e., firm shale to soft sand or vice versa. Since water, oil, and gas can't flow from shale, but could flow from sand, it's wise to check drilling breaks for fluids flowing into the wellbore—a kick.

CHAPTER 22—Promise Kept

The galley was evening busy, half the tables occupied. Voices rumbled. The heavy aroma of fire-grilled red meat made Barry's jaws ache, though he and Jessica had just eaten.

He continued his story. "Ninety-one marked the beginning. We both graduated, got married that weekend, and towed a U-Haul fourteen hundred miles south to Lake Charles."

Jessica tabled her cup, her mouth and eyes wide open.

"Lake Charles," Barry clarified, "in Louisiana. Conoco's Gulf of Mexico field office. Wasn't my first choice, but it was nice to have a paycheck, and we discovered no food in the world is better than in South Louisiana. Crawfish. Roast pig. Blue crabs."

Jessica's eyes seemed to indicate she'd never heard of Louisiana. She mumbled, "Married? Conoco? I didn't—"

"Yep. Six years and three baby girls—Buie, Brandy, and Bailey."

Jessica sat back in her chair, her face red. Her hands went to her hair, as if to keep it from falling out. "Sorry, Barry, man of many secrets, I've got a brain ache. Probably something to do with me thinking you were career BP and you having three kids and me not knowing you were married until thirty seconds ago."

"I'm not. Ninety-eight was another milestone year. I came home from a week offshore and the house was empty. Moving-van empty, except for my stuff. I couldn't even find her. No sign of the kids. But the divorce papers found me. As did the wedding invitation a month after the divorce was final. She'd apparently found true love with a Conoco finance manager—asshole named Mervin—while I was offshore earning a living."

Jessica rocked in her chair. "Sorry. I didn't know."

"Not an everyday topic," Barry said. "Now enough about me and my dysfunctional family. Tell me about you and yours."

Jessica stiffened, then broke off a chunk of chocolate-chip cookie and popped it in. "Not much to tell." She chomped. Sipped coffee. "Mom and a sister. My dad's gone."

"Where are they?" he asked.

"My mom's at home, in Colorado Springs. My sister? I haven't heard for a while. Could be lost, for all I know."

"Colorado surprises me. I thought you were a southern girl."

"Atlanta, Georgia, until I was fourteen. We buried my birth mother there, when I was eight. She had breast cancer, and it was everything bad you've ever heard about it, including almost killing my dad."

"Sorry to hear that," Barry said, an uncomfortable mumble.

Jessica nodded. "When I was twelve, my dad remarried. She was a nice woman, very good to me. My dad cherished her and her four-year-old daughter, but he and I kept our special relationship, never forgetting the person we both lost. I was his crutch and he was mine. Though he never asked, he liked it when I started calling his wife Mom and her daughter Sissy. Sissy was fun to be around, but I missed a lot of her growing up after I left for school, so we're not close."

"How old is she?"

"Twenty-two."

"You think she's lost on purpose?" Barry asked.

Jessica leaned back in her chair, her left index finger massaging her cheek scar. She stared toward him, through him, but her eyes told him she wasn't home.

"Your folks divorced?" he asked.

"My dad was killed in an accident. I was driving."

Barry froze. Waited. Wanted to ask—

"Thanksgiving night, 2007." Her words were soft. "We were broadsided in an intersection, passenger side. I lived. He died."

"I'm sorry about your dad. Where were you?"

"Wyoming. We were headed south to Colorado Springs. Got to Rawlins, stopped for gas. My dad gave me the keys. I took over driving— just in time for a guy to run a red light. All I saw was the truck, on my right, no headlights, coming fast."

"Let me guess—drunk, illegal, no license, no insurance,"

She met Barry's eyes. "That would have been better. The guy was a local, sober, with his wife. He was a big name in town."

Barry nodded—said nothing.

"When I woke up, paramedics were everywhere. My dad's door was where he'd been sitting. His head . . . his eyes . . ."

"Jessica, you don't have to do this."

A trace of crinkle crossed her face. A beat of silence. Thinking. "The couple, they were both seriously hurt. The man's *I'm sorry, so sorry*, repeated like a dirge, changed a few days later when he and his wife suddenly swore their light was green—that I ran the red."

"Any other witnesses?"

"Just me and my dad." She tilted her head as if sifting memories. "Less than a minute before we got hit, I was pulling out of the gas station and got cut off by a car driving in. The last thing my dad said—his SLD, science lesson of the day—was that *Right of way is something you've got only if the other guy gives it to you.*"

Barry recalled the words, the quote, the SLD, in Jessica's tally book.

"So a quarter mile later, I'm talking to him about the traffic light. The only traffic light. The one directly ahead. The one we were both looking at. It was green. Not red. Not yellow. Just green."

"But you got the blame—the ticket?"

An almost-nothing nod. Then she sat upright, laced her fingers. "Yes. The ticket. I lost my dad, got charged with vehicular homicide though I'm free on a technicality, insurance companies are in an ongoing battle, and my stepmom and stepsister aren't big on forgiving Cinderella."

Barry had seen and witnessed a number of deaths during his years offshore—a slick-line decapitation, an overboard drowning, two dropped-load incidents, a rig-floor flash fire—but had never known anybody charged with a homicide. "I've never lost anybody close to me," he said, though his divorce had hurt mighty bad. "I can't imagine . . ." He searched for the right words. Had no clue.

Jessica broke off another chunk of cookie. "Let's change the subject."

He pulled the empty cup from her hand and kick-started his left brain. "My turn," he said, shoving back his chair, "then it's liners and long strings. As promised."

Jessica, too, stood. Stretched. "I've got an hour before we start logging, but let's keep it short. I need my personal time."

* * *

"Here's the situation," Barry said. "We've got 22-inch casing down to about eight thousand feet, plus a thousand feet of new hole that needs casing and cement."

"That'd be the 18-inch casing."

"Yep, but there's a decision to make—a liner or a full string of casing, not both."

"And the simple difference?" Jessica asked.

"Liners are shorter, cheaper, and faster to run, and are easier to pressure test."

"Then it's a no-brainer—run the liner."

"We often do. If we have a thousand feet of open hole, the easiest thing to do is run twelve hundred feet of liner, giving us a couple hundred feet of overlap into the previous casing."

"And when would you *not* run a liner?"

"If the already-installed casing isn't pressure rated to handle deeper drilling yet to come, then we'd bury it behind a new long string of casing."

"Which is . . ."

"The long string goes all the way from the wellhead to the bottom of the well. It's the one with the donut."

"Ah, the donut returns. Otherwise known as the casing hanger."

"Good memory," he said.

Jessica wrote in her tally book, yawned, then glanced at her watch, the double hint far from subtle.

Barry checked his too. "If you're good with the kept promise, I need to visit with Tanker, tell him we're running the liner."[72]

"Liner? For sure?"

"Yep. Done deal. Approved by Houston. Not negotiable."

Jessica snapped shut her notebook and slapped it against her palm. "Thanks, boss. Good lesson."

Their paths separated when they left the galley—Jessica farther down the hall to her room, Barry into the BP office, alone. She'd needed her private time. And she'd called him *boss*, whatever that meant, because he wasn't. He tried to imagine their relationship, their personal relationship, otherwise. Tried to imagine private time—together. Couldn't.

He wasn't on her radar. Never would be.

But he was pleased she was on his side, in his court. He wondered what it would be like if they ever got cross-threaded—two joints of pipe with galled threads, male and female, neither yielding.

Not a pretty picture, he guessed.

---

[72] See **Diagram 8**—Install 18-inch Liner. See page 137 herein. The drilling system, including BOP and riser, is now ready for the "drill as deep as we can before we have to run more casing" mode.

Diagram 8
Install 18-inch liner

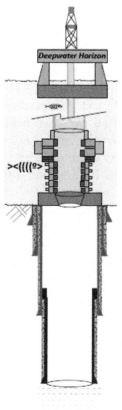

Rig Floor (RKB) at zero feet

Sea level at 75 feet RKB

Marine Riser

Blowout Preventers (BOPs) on top of
Wellhead at 5,057 feet

Seafloor at 5,067 feet

36-inch structural casing at 5,321 feet

28-inch casing at 6,217 feet

22-inch casing at 7,937 feet
with Wellhead attached

18-inch liner at 8,969 feet
(top of liner at 7,489 feet)

Diagram 8

CHAPTER 23—One More Foot

Days later, drilling below the 18-inch liner, Daylight was on high alert. He'd been in the chair since sunup and had already made 125 feet of new hole. Making hole was his job, but making hole *fast* was another matter.

The company man, Mr. Barry, had repeated once again his several-times-a-day message: "You get a drilling break, let's check her out." Tanker had told him the same thing not twenty minutes later.

Skoal and coffee—Daylight's best friends. He spit and sipped, then palmed the brake and made another few inches of hole. He was drilling shale, but he was ready for what lay ahead. Whatever, and wherever, it might be.

He called down to the mud room and talked to one of the M-I Swaco mud engineers who'd been working to increase the mud weight as ordered by the company man.

The engineer confirmed the new mud weight, 10.6 pounds per gallon, both into and out of the wellbore. He also spouted off a bunch of other mud numbers like viscosity and gel strength and fluid loss. Important numbers that kept the hole from falling in and the drill bit from getting stuck.

Important, yes, but Daylight, on high alert, cared only about the mud weight. Had to be right, he reckoned. Too high, he'd lose circulation. Too low, the well would kick. Middle ground—that's what he liked. He depended on his watchdogs—his pit-level and flow meters—to occasionally scream their warnings. Losing mud. Gaining mud. But if he was drilling and the well was neither sucking mud nor kicking, then the mud weight had to be somewhere between too high and too low. Yep, middle ground. Middle ground always worked. Always.

Always, that is, unless he got a drilling break. And a drilling break was exactly what he was looking for. Not because he wanted it—but because he knew it was coming. He drilled another foot. No break. Then he forgot about that small victory, because he had to drill the next foot too. And the one after that. One foot after another. Hour after hour.

Can't drill shale forever. Where's the break? A hundred feet deeper? Maybe only five? Two? Maybe in fifteen seconds?

He spit and sipped.

How about this next foot?

CHAPTER 24—Transition

Barry watched drilling data for each new foot of hole, his gaze on the logging-while-drilling screen.[73] He and Jessica were in the mudlogging unit, as was a trainee mudlogger. Jessica hovered over the binocular microscope, focused on cuttings.

Having a trainee onboard, working for the rig's two senior mudloggers or whichever one was on duty, was common. A number of service companies did the same as part of their on-the-job programs. Barry avoided trainees whenever he could.

"We're in a transition-zone shale,"[74] Barry said to the young mudlogger, "which means formation pressures are increasing faster than normal. The shale itself doesn't bother us because it's impermeable, but any sand layer in or below the shale could be over-pressured and permeable."

Jessica rolled her chair away from the microscope and rubbed her eyes. "So what happens when you drill into the sand?" asked Jessica, the straight man.

"Two things," Barry said, aware she knew the answer. "First, there'll likely be a drilling break—a change in the drilling rate—going from shale to sand."

"And we just love sand," Jessica said, "because that's where we find oil."

"Second," Barry said, staying on track, "permeable sands, if accompanied by high-enough pressures, are the prime ingredients of kicks."

Jessica raised her fist in a victory salute. "Yes. A kick. And we hope it's an oil kick."

The young mudlogger nodded like a bobble-head doll.

Barry enjoyed Jessica's banter. Enjoyed her. "Whether oil, gas, or water," he said, "if the kick is strong, the driller will know it's a kick at the same time he sees the drilling break. That's because horns go off, sirens scream, phones ring, and Tanker and I head to the rig floor."

Jessica held up her hand. "Me, too, because I've not been on the drill floor during a kick."

"Neither have I," the mudlogger said.

"Rig floor's not for sightseers during a kick," Barry said. "Besides, during a kick the mudloggers are busy gathering numbers for us, watching progress, keeping us informed. The excitement's not just on the rig floor."

Jessica turned in her seat. "I'm neither a sightseer nor a mudlogger," she said to Barry, "but thanks for the compliment."

No visible animosity on her part. Just a two-kiloton reminder Barry reckoned he needed to save in deep memory.

"If the kick is minor," he continued, "the driller will see the drilling break first, and then he'll raise the drill bit off bottom and check for flow. And that's when he'll see the kick. Or if it's really minor, we'll all see it when we make a short trip and circulate bottoms-up."

"Which will show up as trip gas or high chlorides off bottom, or even a pit gain," Jessica said.

"Kicks aside," Barry said, "it's more likely that you and your fellow mudloggers would review drilling data, recognize the transition-zone shale, and call me. We'd circulate bottoms-up to see what's down there. Depending on the nature of the shale cuttings and the amount of drilled gas, we'd increase the mud weight, as we're doing now."

The young mudlogger nodded, but he glanced again at the door like he wanted to either leave or be joined by one or both of his more-senior colleagues.

"If we've been successful with the mud-weight increase," Barry added, "the driller will still see the drilling break, he'll still stop drilling, and he'll still pick up and check for flow. And when he sees neither flow nor pit gain, he'll go back to drilling."

"Which, of course, would be the ideal world," Jessica said.

"And an everyday occurrence," Barry said, recalling few exceptions during his two decades of managing unique but similar wells, "if everything goes right."

Which it did, quite often.

Well, usually.

<u>73</u> LOGGING WHILE DRILLING—LWD: With LWD tools built into the bottom hole assembly, just above the drill bit, a number of data readings are sent to the surface via mud-pulse telemetry and are immediately available in the mudlogging unit. Such readings can include gamma ray, shale density, formation resistivity and pressure, equivalent mud weight in the annulus, etc.

<u>74</u> TRANSITION ZONE SHALE—Normal shales get more dense with depth. Transition-zone (TZ) shales (also called undercompacted) are those that didn't compact normally over the past few million years. TZ shale densities increase slower than normal, while their internal pressures increase faster than normal (hence, transition refers to increasing pressure). TZ shales are not permeable, but sands are, and any sand buried in a TZ shale is likely to be over-pressured. If the mud weight is not increased as drilling progresses through the TZ shale, a layer of over-pressured sand trapped within or below such shale can kick hard. See Reference (19)—"A Risk Analysis of Transition Zone Drilling."

CHAPTER 25—Kick

Noon, on the rig floor, everything had gone right for another 133 feet, which meant 258 feet of shale. Undercompacted shale. The next foot was different. Daylight's drill bit dropped into what felt like a hole.

"Tarnation," he mumbled through a slurry of Skoal and coffee. He immediately stopped drilling, told his derrick man to call Tanker, and picked up the drill string to get the drill bit off bottom even as the pit-level and flow alarms blared their warnings.

* * *

Noon, in the mudlogging unit, Jessica reacted to the audible alarms—she knew what they meant. Barry was already out the door. "Well's kicking," she told the senior mudlogger. "Do your thing." She hustled after Barry.

She and Tanker got to the rig floor at the same time.

Barry talked to Tanker over Jessica's head: "Pumps are off with 10.6-pound mud. Well's shut-in on the upper annular BOP."

"Tell us about it," Tanker said to Daylight.

Jessica, her tally book at the ready, stepped between Barry and Tanker so she could see and hear the driller.

Daylight spit in his cup. "Drilled to 11,583—took a two-foot break. Bit fell like a three-pound hammer in a bubble bath, and in no time, maybe sooner, the pit alarm howled like a cat in a mouse trap. I picked her up off bottom, shut down the pumps, and closed the bag on her. Looks like we got us a little kick." He tapped his finger on a gauge. "Maybe 175 pounds."

"Daylight, good job shutting-in so quick," Barry said.

Tanker shot Daylight a chin dip of respect, picked up two clipboards—each with a well-control worksheet—and handed one to Barry. "We got us a well to kill, men—let's get moving."

Ha. Three men. Jessica, admittedly nonessential, wondered if she were invisible too. She unfolded the blank well-control killsheet she'd stuffed into her tally book weeks before and gave herself an assignment—stay involved, miss nothing.

Tanker and Barry divided the work—Barry and Daylight would run the BOP control panel and the choke, while Tanker managed other critical rig activities.

Barry said the kick was small—0.3 pounds per gallon.

Jessica opened her calculator, confirmed the numbers with the mud-weight equation. Which meant they would need 10.6 plus 0.3—for a total of 10.9-pound mud—just to balance the kick. A little more to be overbalanced.

Barry directed Tanker to have the mud daubers increase the mud weight from 10.6 to 11.1 pounds per gallon. Barry said he wanted the new mud ready before he started the kill.

Jessica wrote: *10.6 mud + 0.3 kick + 0.2 for overbalance = 11.1 ppg for target mud weight.* She followed the note with a happy face.[75]

Tanker said, "We've got 500 barrels of 14-pound kill mud in a reserve pit. We'll use it to bump up our mud weight to 11.1."

"Pit gain?" Daylight asked, eyes on his tally book.[76]

"Twelve barrels," Barry said. "Won't be bad unless it's gas."

"Twelve's right," Jessica said. Part of the team. She wrote a note: *12 bbls—composition unknown. Gas?*

"Let's agree numbers," Barry said.

Daylight: "Initial shut-in drillpipe pressure, 175 psi. Shut-in casing pressure, 250."[77]

Barry confirmed the numbers.

Jessica filled in blanks on her killsheet—depth, pressures, volumes, the easy stuff. Dozens of other blanks on the page made no sense. Not yet.

"I've got the depth as 11,583 feet," Barry said.

Jessica read her notes. "I've got 11,585—including the two-foot drilling break."

Daylight: "Miss Jessica's right—11,585."

"Good catch," Barry said, acknowledging her presence. "Daylight, what do you have for slow-rate pump pressure?"[78]

"Three fifty psi, at forty strokes a minute."

Barry: "Got it."

Jessica: "Tell me about the slow-rate pressure."

"It's just friction," Barry said. "As long as we pump at forty strokes per minutes, which we'll do throughout the kill, we'll know 350 psi of our pump pressure is friction. Same as while we're drilling, though we pump at higher rates and have more friction when drilling."

Jessica wrote down the numbers. The pressure. How fast the pumps would pump. Understood the numbers. Had no clue what they would be good for. Memories of her first cement job. Except then, a screw-up meant there'd be hard cement where nobody wanted it, or crappy cement where it was supposed to be hard.

Shoot. People talking. She'd lost the thread on a serious subject, which wasn't good. Barry had said something about choke-line back pressure, but she'd missed it, and Tanker was already on the next topic with a question to Barry about the strokes and barrels necessary to fill the drillpipe.

Barry scribbled on his clipboard, while she scolded herself.

"I get 220 barrels," he said, "which is 2100 strokes. It'll take just less than an hour."

"Fifty-five minutes," Daylight said.

Jessica watched for Daylight's grin—he spit in his cup. Busy man. Focused.

She added a line of notes.

Tanker wrote something on his clipboard. "We're good to go," he told Barry. "I'll be in the mud room—call me."

"Reckon it's brine," Daylight said. "Maybe a little gas."

"Maybe," Barry said. "Maybe not."

Two minutes. Three. Jessica wanted to ask about the choke line, but daydreaming didn't seem like a good excuse for missing it the first time.

Barry picked up the rig-floor phone, punched two numbers. "Tanker, what do you have?" He listened, then hung up. "Mud daubers are up to 10.8 pounds. We'll go when they're ready."

Daylight spoke from the control panel. "Casing pressure's up from 250 to 290. Drillpipe pressure's up to 215."

Barry: "What's the leak-off test for the 18-inch liner?"

Daylight: "I got L-O-T 11.6 pounds, Mr. Barry."

Ha. Jessica understood leak-off tests. She flipped back a page in her notes. "Affirmative. Leak-off test equivalent to 11.6 pounds per gallon." *Affirmative*—so much cooler than yes, or right, or ten-four.

"With our 10.6-pound mud," Barry asked her, "what's the maximum allowable casing pressure?"

Damn, a real question. She punched numbers into her calculator. Barry's mud-weight equation—depth, mud weight, 0.052. "I get 466 psi." Definitely on the team.

"Daylight, how's the casing pressure doing?"

"Up to 310, Mr. Barry."

"That's only sixty-five percent," Jessica said, wanting no part of any pressure being too high for . . . whatever.

Barry shrugged. "If the kick is gas, 500 or 600 on the casing won't be a surprise."

Minutes passed, the rig floor graveyard quiet. Like an unexpected lull at a family-reunion dinner table. No humming machinery. No clanking steel. No bullshit dialog. Tanker in the mud room. Barry and Daylight seemingly intrigued by flat-line gauges, perhaps waiting for something to change.

Jessica whiffed fresh air. The breeze off the Gulf. Two white seagulls on the edge of the helideck. Life being wonderful—though standing on the rig floor in the middle of a kick, remembering the irony of Piper, she expected a panic attack at any moment.

But she was exactly where she wanted to be—in the action. She faded the chills, fought the fear, and stepped up to Barry. Casual. Firm voice. "Getting the mud ready before the kill seems safe enough. Is that what you always do?"

"Nope." Also casual. "Mixing early is safe for *this* kill, but only because the kick volume is small, the pressures are low, and we can afford to wait. The method has a name." He pointed to the top of the killsheet. "It's called *Wait and Weight*, which means we close the BOPs, but we wait to start circulating until we weight up the mud."

"And the opposite?"

"If there's a big volume of gas in the annulus and we wait too long while we're messing with the mud, the casing pressure might build high enough to cause lost circulation. Instead, we'd start pumping immediately, while the mud daubers raised the mud weight on the fly. Kind of the best of both worlds, though the kill itself would take a long time."

Jessica checked her notes. "What's next?"

Barry showed her his killsheet. "With the drillpipe full of 10.6 mud, we need to know the pump pressure necessary to circulate the well while keeping the well from flowing."

"Which is?"

"The slow-rate pressure—that's the friction—plus the original kick pressure, which is the amount of underbalance."

"Seems too simple."

"Simple enough. But when our heavy mud gets to the drill bit, the drillpipe's no longer underbalanced, so all we'll have is the slow-rate friction."

"So what's the big deal? Fill the drillpipe and the well's dead."

"Not quite. We're going to fill the drillpipe in ten small steps. With each step we'll pump a tenth of the drill-string capacity, using kill-weight mud. At the same time, we'll reduce the pumping drillpipe pressure by a tenth of the original kick pressure."

"Wow, I'll put that in my notes right next to *Keep it simple.* Why all the bother?"

"Because there're a couple bad guys we want to avoid—lost circulation and another kick."

She nodded. Said nothing.

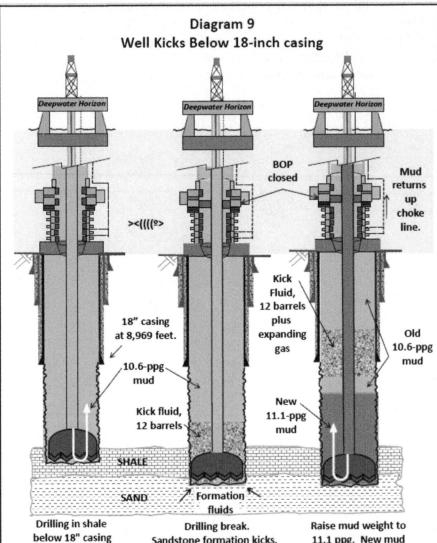

Diagram 9
Well Kicks Below 18-inch casing

BOP closed

Mud returns up choke line.

18" casing at 8,969 feet.

10.6-ppg mud

Kick fluid, 12 barrels

Kick Fluid, 12 barrels plus expanding gas

Old 10.6-ppg mud

New 11.1-ppg mud

SHALE

SAND   Formation fluids

Drilling in shale below 18" casing with 10.6-ppg mud. All is okay while drilling deeper, until the drill bit finds the high-pressure sand.

Drilling break. Sandstone formation kicks. Close BOPs. Pit gain is 12 barrels. Shut-in pressure shows underbalanced by 175-psi, equivalent to 0.3 ppg at kick depth (11,585 feet). Therefore, formation is 10.9 ppg. Need more than 10.9-ppg mud to drill ahead.

Raise mud weight to 11.1 ppg. New mud down drillpipe displaces the old mud and the kick fluid up the annulus, where gas eventually starts expanding at shallower depths.

Diagram 9

Barry, in mentor mode: "When we start pumping kill mud, the ratio of new mud to old mud in the drillpipe will change by the minute, always increasing. And that means the pumping drillpipe pressure needs to change at the same time, always decreasing. Too much pump pressure will cause lost circulation, and too little will allow another kick."

"So we divide the kick pressure—175 psi—by ten," Jessica said. "But how do you reduce the drillpipe pressure by seventeen or eighteen psi?"

"With finesse—the finesse of the kill. Since the choke line and the annulus and the drillpipe are all fluid filled, it's all one system. Remember what we said about adding surface pressure on top of the mud?"

"The pressure throughout the system increases by the same amount."

"Exactly. And we're going to manage that pressure by adjusting the choke on the casing annulus, through the choke line, while we're pumping at the slow rate. If the drillpipe pressure is a bit too high, we open the choke a smidge and let off a little casing pressure. If the drillpipe pressure is lower than called for, we pinch the choke to raise the pressure. Our goal is twofold—get rid of the kick fluid that's in the well, and prevent more from entering."

"Isn't the casing pressure, because of gas, unpredictable?"

"It is," Barry said, "but we're using it as a tool to allow us to finesse the drillpipe side, which is where we have control. The drillpipe pressure is king—it doesn't lie. Good clean mud all the way to the drill bit. As long as we stick to the drillpipe schedule, the well's under control, even if gas is screaming out the annulus."

Jessica made a note about the drillpipe pressure being king. "What happens after the tenth stage?"

The phone horned. Daylight answered. Hung up. Glanced toward Barry and Jessica. "Mud's good to go, y'all. Tanker's on his way."

Barry nodded toward Daylight. To Jessica he said, "After the tenth stage, the only pressure the drillpipe's bucking is the friction—350 psi. That means the old mud and kick fluid are on their way up and out the annulus. After that, all you need are earplugs."

Tanker arrived and they kicked in the mud pumps and started the kill before Jessica could ask her earplug question. She stood out of the way, to watch and learn, and took notes as numbers on gauges rose and fell.

But the real show was Tanker and Barry and Daylight killing the kick. Total concentration. They worked as if they were conjoined triplets. Talked in grunts. Muttered pressure readings and stroke counts. Manually manipulated the choke-panel controls. Somebody said the circulating pressure was too high, and they all leaned their butts left while somebody tweaked open the choke, and then they all leaned their butts to the right as the pressure rose. Then just the opposite. Like being on a slow-rolling cruise ship. Might as well have said lean right, lean left, left some more. No, too much left, go back right. Choreography on the rig floor. Grown men. Slow dancing.

Jessica checked her watch. An hour and a half—gone.

More cruise shipping—a long cruise.

Two hours fifteen minutes.

The ultimate low-calorie lunch. None.

Then two things: a pressure gauge on the choke panel went spastic, and a slight rumble broke the silence.

A harsh whistle from somewhere way above her head—in the crown of the derrick. Like a steam vent or a high-pressure fire hose. But it wasn't steam. Wasn't water. She knew the source—gas. Natural gas.

She raised her face. No aroma—crude or otherwise. Sea breeze. She was right where she wanted to be, watching Barry kill a kick, yet she feared fire. Her left hand, with a will of its own, gripped a stanchion—wouldn't let go—forced her to stay.

Barry, Tanker, and Daylight—still leaning, manipulating controls—didn't seem to notice the noise.

The gas got louder, screamed in deafening batches. Like baby-bear loud, then mamma-bear loud, daddy bear the loudest.

Ha. Barry's earplug warning. Right. Seagulls gone.

Then grandpa bear, beyond storytelling, double-ugly loud.

Jessica's bones vibrated. She gripped the stanchion harder, dug her nails into glossy gray paint. Seconds of fear. Minutes of terror. This was the kick she'd wanted to see—but perhaps a wee bit more than she'd imagined. She held back tears.

Then . . . silence.

She pressed her palms to her ears. Good. Not deaf.

Tanker, phone in hand, recipient unknown to Jessica, said, "Got returns?" A beat. "Call if it changes."

No returns? No mud exiting the well, as it should? Lost circulation? Jessica reckoned such info would go into her tally book in the not-so-good column.

The roar returned. More gas. Though less energetic.

The phone horned. Tanker yanked it up. Listened. Put it down. "Got full returns," he told Barry and Daylight.

"There'll be more slugs," Barry said. "We don't have bottoms-up yet." He turned to Jessica. "Make sure the mudloggers are getting fluid samples."

Yes—a real assignment, though gathering kick samples was a mandatory mudlogging responsibility. She unlatched her fingers. Called the unit. Line busy. Made her way through the tool room to the back door of the unit. Bursting in, she found the trainee mudlogger, alone, on the phone.

"Where are they?" she asked.

The young man shot back: "One's at the mud pit monitoring and sampling returns—the other's at the mud-gas separator, doing the same." He held up his phone. "I got the mud room if that'll help."

"My bad—good job. Make sure they get samples. There may not be enough rock to see, but we need to know what fluid kicked. I'll be on the rig floor—have somebody meet me there when they know." She heard the remnants of "Yes, ma'am" when she closed the door from the outside and rejoined the rest of the kill team.

"Almost over," Barry told her, as one of the senior mudloggers headed their way. "We opened the BOP, and we're circulating bottoms-up with the new mud. We'll short trip to make sure we're okay." He turned to the mudlogger. "What'd you see?"

"A few surges of gas and thirty minutes of high-chlorides off bottom.[79] Looks like a saltwater kick with a little gas."

"'Tain't no surprise," Daylight said. He engaged the drawworks and raised the drill string, not stuck—got a thumbs up from Tanker—then signaled his crew. Sign language, Jessica assumed, as two roughnecks pulled out high-pressure hoses, ready for housework.

Barry said to Jessica, "I need to call the office again," then headed to the stairs.

Jessica, hyper about the kick, needed to talk. Daylight was busy. Tanker didn't appear to be. She'd never had a one-on-one with him, decided it was time. She went to him, by the door into the doghouse. His shoulders were higher than her head, including her hardhat. She looked up. Way up. "I'm glad you were there for the kick."

He looked down, but with a gentleness that made her feel tall. "My pleasure, ma'am."

"Are they all like this one? This intense?"

Tanker gazed out the vee door and across the pipe racks, long seconds before he answered. "Nothing more important out here than the people, which means when the well's acting up, which could hurt the rig in a terrible bad way, we got us a job to do. Me, Mr. Barry, you, Daylight and his floor hands, the mud guys. Everybody has to do their part. Everybody's important. No mistakes."

Responsibility. He hadn't said the word, but Jessica felt its weight. And not just for rocks and oil shows. She nodded. Said nothing.

"So the answer's yes," Tanker said—sounding just for a moment like her dad. "Every kick you'll ever see, and for sure on my watch, will be just as intense as the one we just killed. Mr. Barry for sure wants to save the well, but when the time comes, no matter how big the kick, he'll be right beside me for the kill—like a war, intense, whatever it takes. Afterwards, we'll work our way through fixing the well. Maybe getting unstuck. Running more casing."

Jessica conjured up an echo of screaming gas, maybe would hear it forever. "I can't imagine a kick a lot worse than what I saw."

"And we don't want one either," Tanker said. "That's why, as soon as we even think we see a kick, nothing else matters. Especially offshore."

"I get the fast part, like checking drilling breaks, but why'd you say *especially offshore?*"

Tanker hesitated a beat. "Two reasons, mostly." He spread his arms, palms up. "The rig—this rig—is our home. Without it, we're all in a heap of trouble, nothing but shark bait."

Jessica shuddered. Nodded. Relived a flash of memory that used to haunt her—her dad's fatherly warning when she'd taken her first assignment in Gulf of Mexico—*Trust your life to nobody, because offshore there's nowhere to run.*

Tanker folded his massive arms, his biceps as big as her thighs. "Second reason—the people. Everybody out here."

Jessica nodded, then leaned over a bit to glance at Daylight. As if he'd been waiting, he spit in his cup and sent her a chin-up nod, which she returned.

To Tanker she said, "None better than your entire crew."

"Got that right, and there's not a one I don't trust." He paused, gathered a thought. "A few months ago I watched one of those reality TV shows, like the one about catching crabs in Alaska, but it was about drilling wells in West Texas, maybe Oklahoma." He shook his head. "The whole program was downright disgusting. Some kind of contest to drill fast, get a big contract. The hands would cuss and fight, disrespect the driller, toolpusher be damned. They'd drink in town, then show up late or not at all. A goodly number seemed to be hard workers, good hands, but the rest I'd fire on the spot and send them packing."

"I've seen the program," Jessica said. "It's definitely made for TV, because it's way over-the-top compared to the land rigs I've ever worked, whether in Alaska, Wyoming, or New Mexico. But what's that got to do with offshore?"

"Simple, from what I saw. TV program or not, those boys don't respect the well, the rig, what all can go wrong. They don't trust or care about one another. And the only difference between their well and this one is that if their rig burns down, they can walk away. Get in a car and drive to a bar. Go have a beer."

Jessica wondered just how protected her rig life had been. Even on land rigs, she'd never heard rank cussing or been aware of crews fighting, working shorthanded.

Tanker softened, hands in his pockets. "We can't go get a beer. No roads. No cars. No bars. Which means we're all in this together. The only hands you'll see out here, and on most any offshore rig anywhere, are those willing to do whatever's necessary to keep us all safe. Period."

A hug seemed appropriate, but she touched his sleeve instead and pulled out her tally book. "That's all good advice, especially about kicks. I'll make some notes if you don't mind."

"That's nice of you, Miss Jessica, but there're a lot of folks you can talk to about kicks. Mr. Barry, for one. Fact is, though, with you planning to be a company man, you'll be wanting to get certified at well-control school.80 Nothing better to teach you how important it is to understand the well and how to handle pressure problems. Me, I'm required to go every couple of years. Same goes for Daylight, Mr. Barry, engineers, supervisors—anybody involved in well control."

Jessica made a note. "Well-control school—I'll check it out."

"It's a five-day course, in either Lafayette or Houston. It's well worth your—"

"Anybody hungry?" Barry asked, back from the office.

Jessica, famished, raised her hand, but Tanker begged off, said he needed time with Daylight before dinner. She hoisted her tally book toward Tanker and thanked him—got a big grin in return.

She went with Barry but stopped along the east-side handrail. Gripped it tightly.

Barry did the same. He thanked her for her help during the kick and asked if she'd seen what she'd hoped to see.

"My goodness yes, even more." She forced herself to take deep breaths, focus on the horizon. Then at a drilling rig, miles away, and up toward a cloud that looked just like . . . a cloud. "Thank you for letting me stay."

"Were you afraid?"

"Ha, you think? I didn't expect the gas, didn't think it'd get so loud, and I wasn't keen on even the slightest chance of it catching fire."

"It was loud because we discharged it through the mud-gas separator, which vents through a gooseneck outlet at the top of the derrick. If the kick had been bigger, more gas, we would have diverted it overboard through a bigger line."

"Because it's quieter?"

"That, but more important, we wouldn't want all that gas up in the derrick. The diverter system is designed to send big volumes overboard, to whichever side of the rig is downwind."

"Wow, if that kick wasn't big enough for the diverter, what would be?"

"Nothing quantitative," Barry said, "but just so you know, if the raunchiest kick is a ten, the one we just killed was less than a two."

"And no lost circulation. Did we come close?"

"Probably. You told me the maximum allowable casing pressure was 600 psi, and we had pressure spikes over a thousand."

Jessica shuddered. "You ever see a ten?"

"Never. And never want to."

"What would it be like?"

"Ever see pictures of Spindletop?"

"The East Texas blowout? That was back when the whole world was black and white."

"For its day," Barry said, "it was a ten. Total loss of control, though they were ecstatic to have found the oil."

"I want a discovery, but not like that."

"Whether or not we make a discovery, you can bet your life there won't be a blowout.[81] A ten. Not on my watch."

Having heard similar words from Tanker, Jessica nodded her understanding. But Barry had it wrong. She wasn't about to bet her life, not even when teamed up with him, her trusted mentor.

For whom *trust* might be in question. She faced him. "Did you tell anybody *besides* Tanker that I plan to go back to school?"

Seconds passed while Barry's Adam's apple climbed his throat like mercury in a thermometer. "No."

Didn't matter anyway, but she allowed a similar short period of silence so he'd know for sure that she knew for sure that he hadn't kept his word.

"So what's next?" she asked. "Back to drilling?"

"You tell me. Pretend we have an instant replay, another kick a hundred feet deeper."

"Oh, now that wouldn't be good," she said. "Pressures would be too high. We'd lose circulation, and we'd probably—" Her jaw locked.

"What?"

"Oh, my gosh. ZKT. Zero kick tolerance. That's what we have right now. We can't drill deeper. Can't take a kick. Even a little kick." She slapped the handrail with both hands. "Wow, this is exciting. Now I understand. Really understand. ZKT. It's so clear—"

"Jessica, take a breath. Good news is, you participated in a perfect learning exercise."

Good news. Like half a sentence. She asked, "And the bad news?"

"Again—you tell me."

She weighed the goods and bads. All good. Clean kick. Efficiently handled. Good teamwork. Samples of formation fluids. Drill bit not stuck. Well, almost all good. ZKT the exception. Which meant—

She got it. "Oh, shoot."

"What?" her mentor asked.

"We can't drill deeper, so we need to run casing—the 16 inch stuff on the pipe rack."[82]

"Good job," Barry said. "Class dismissed. If you're hungry, I'm buying."

"Then you better have a bucket of money, cowboy, because this girl could eat a blossom of broccoli, a head of cauliflower, and an eight-ounce filet of acorn squash, grilled medium-rare."

Barry leaned over the rail, faked a dry heave.

"Tanker's coming," she lied.

Barry stood up so fast he almost lost his hardhat. Then laughed at himself, a rare treat.

But Jessica's own thoughts worried her. She didn't know how to handle *Barry* and *fun* in the same sentence.

Diagram 10
Install 16-inch long liner

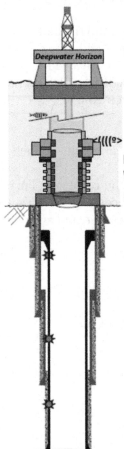

Rig Floor (RKB) at zero feet

Sea level at 75 feet RKB

Marine Riser

Blowout Preventers (BOPs) on top of
Wellhead at 5,057 feet

Seafloor at 5,067 feet

36-inch structural casing at 5,321 feet

28-inch casing at 6,217 feet

22-inch casing at 7,937 feet with
Wellhead attached

18-inch liner at 8,969 feet

16-inch "long liner" at 11,585 feet
(top of liner at 5,227 feet).
Rupture disks (✸)installed in liner at 6046,
8304, and 9560 feet as pressure-release
safety devices in case the well is a discovery
and is completed for production.

Diagram 10

[75] See **Diagram 9**—(Well Kicks Below 18-inch Casing—P. 145 herein). In this example, kick depth, volume, and pressures are hypothetical. See Reference (20), Chief Counsel's Report, P. 59, for depths and dates of actual kicks and lost-circulation events during the well.

[76] PIT GAIN—Related to, as caused by, a kick. A ten-barrel kick (ten barrels of formation fluid) pushes, or kicks, an extra ten barrels of mud into the pits; hence, pit gain.

[77] SHUT-IN PRESSURE(S)—When a kick is detected and the BOPs are activated (which shuts-in the well), drillpipe and casing pressures are measured and recorded. The shut-in drillpipe pressure (SIDPP) is key to characterizing the kick. The pressure of the kicking formation is the simple sum of the SIDPP and the pressure exerted by the drilling mud in the drillpipe at the true-vertical depth (TVD) of the kick. The equivalent mud weight of the kick is the SIDPP converted to ppg (using mud-weight equation). The shut-in casing pressure (SICP) can be high and erratic, due to the unknown volume and type of formation fluids (oil, water, and especially gas) moving up the wellbore.

[78] SLOW-RATE PUMP PRESSURE (SRPP)—A simple (pre-kick) recorded pump pressure, representing how much pressure (friction) the mud pumps see at a given slow rate. SRPP is needed during the kill of the well, so control pressures can be differentiated from friction pressures

[79] CHLORIDES—If kick fluid contains brine (salty formation water), then the mud that comes from bottom (even if the water volume is too small to see) will generally show an increase in chlorides (salt).

[80] WELL CONTROL SCHOOL—For example, the IADC WellCAP® Drilling Well Control Certification course (for company men, engineers, toolpushers, drillers, drilling managers, etc.) . . . through hands-on simulations and exercises . . . includes: regulations; theory, causes, and detection of kicks; shut-in procedures; BOP operations and equipment. Additionally, for student petroleum engineers, Louisiana State University, LSU, offers hands-on training in well control with a full-scale, well-control research and training facility, including its own well, equipment, and instrumentation for conducting training and research related to borehole technology.

[81] BLOWOUT—Formation fluids flowing up a well into the environment without control. Blowout, by definition, means lack of well control; hence the emphasis on kick-control (well-control) training. Industry wide, highly experienced rig crews (toolpushers, drillers, company men) react immediately to resolve every well-control event. When such actions fail, for whatever reason, the result can be a blowout.

[82] See **Diagram 10**—(Install 16-inch "long liner"—P. 152 herein) Size and depth data for all Macondo casing strings are from Reference 3—BP's Wellbore schematic—Page 19. Note the "long liner" covers previous casing (18" and 22") up to 5227 feet RKB. Note also the installation of rupture/burst disks at 6046', 8304', and 9560' in anticipation of possible discovery and completion of well for production. The rupture/burst disks are "safety valves" that limit how high the annular pressure can rise. This is to prevent burst of the 16-inch casing or collapse of the production casing, whether a full string or a liner with tie-back.

CHAPTER 26—The Target

Mid-March—three days after Jessica's first kick, she hosted a pity party for herself. She called Ranae Morgan, late evening, from the satellite phone in her office. Five minutes of how is so-and-so, I may have to hang up, did you get, do we have a discovery—all the typical soft words Jessica knew the two would use when one of them had an agenda.

Jessica had an agenda. "We took a kick a few days ago and ended up running a 16-inch liner. And now we're running a 13-5/8 inch liner at 13,145 feet, which is about two thousand feet shallower than planned."

"Listen to you," Ranae said, "you sound like the drilling department complaining about well costs. Now forget the lengths and widths and tell me what the problem is."

"I'm calling to vent," Jessica said, "to make me feel better. And you're not going to like the problem. In fact, you're going to end up with the same concern I have."

"Then we'll share it together—tell me about it."

"If we set too many strings of casing, the hole will be so small we can't drill deeper."

Nothing. Quiet line.

"You there?" Jessica asked.

"Yes, but I'm boiling inside. I remember telling the drilling manager that better not happen on my well."

"It's happening, Ranae, and it's real."

Another beat. "What can you do about it? Any hope?"

"Yeah, right. All we have to do is drill the next section from 13,000 to 17,000 feet, and we'll be back on schedule."

"You seem short on confidence. Which is it—a rig problem or a people problem?"

"Neither," Jessica said. "The rig is unbelievable. It just finished drilling the deepest offshore well in the world, to more than 35,000 feet. And the guys out here are great, good times and bad. They know what they're doing, they work well together, and I trust each of them as much as anybody I've ever met on a rig."

"I know you," Ranae said, "and I'm guessing you're seeing all this touchy, feely togetherness from your normal hideaway inside the mudlogging unit."

"I'm there a lot, of course, but I don't hesitate to go to the rig floor with Barry, Tanker, and the rest of the guys."

"You mentioned Barry before, but who's Tanker?"

"He's Transocean's toolpusher. Good guy. Knows his business."

"Tell me about Barry. Is it Eggerton?"

"The one and only." Jessica listened for hall noises. "I thought for a while he was getting sweet on me, but that's no longer a problem."

"Something you said?

"He's seen me at both extremes."

"And the good, as fabulous as it is, isn't good enough to offset the bad?"

"Not even close. He's good at what he does and has a strong work ethic. I trust him because he's totally focused on the well."

"Which means you're very much alike."

"Not hardly. He's a hard-core, left-brain engineer. Not a people person. Doesn't listen to music. Never reads the paper. Rarely accesses the Internet. He likes one-answer problems. No ambiguity. No interpretation. Basic premise—understand the problem, fix it. If the well's drilling slow, drill faster. Low productivity? Increase it. Downtime happens? Minimize it."

"I can see why you like him," Ranae said.

"You're funny, but not." Jessica glanced at the doorway. "He likes procedures, cookbooks—step A, then B, then C."

"Maybe he can cook too."

"Maybe, but I couldn't care less. But I do know he's a good company man. On the rig, there's no doubt he's in charge. His well, his responsibility—safety, cost, schedule, everything."

"And the problem is?"

"That's not the point. He's a loyal BP employee and a damn good mentor, but he's seriously short on tact and diplomacy, which means neither of us is willing to take bullshit from any direction."

"Two engineers," Ranae said, "one with a pretty smile. My guess is you win your debates."

"Not all. He backed me into a corner."

A beat. "Physically?"

"Figuratively. I felt obligated to tell him about my dad."

"Oh, Jess, I'm glad—if you talked to Barry, you can talk to your stepmom."

"Dammit, Ranae," Jessica said, hating the term *stepmom* for her dad's wife of almost twenty years, "my mom won't take my calls. I tried from a rig, before and after Christmas. It's like she and her daughter, Sissy, *my sister*, divorced me. And is Sissy home? Married? A porn star? How would I know? Now another lawyer's got a subpoena waiting for me when I get in. Whose insurance now? Mom's? Mine? The couple who caused the accident? Same old crap—just another attorney."

"Jess, please, start with your mom. Tell her the truth—again. All of it. And if she won't listen, tell your sister. And screw insurance—screw the money—go to jail if you have to. Then, if your mom disowns you, at least you can get on with your life. Not like now."

During that first rough year, Ranae had been Jessica's confidante. She alone had believed Jessica. Had seen through the claims about guilt and innocence Jessica could never defend. For her mom, the cross-traffic red light hadn't mattered—Jessica had been driving, therefore it was her fault. She had killed her dad—her stepmother's husband. Period.

"Seems like I've heard that before," Jessica said.

"Valid then—valid now. Is that why you called? To make me get on your case? Raise my blood pressure?"

"None of the above," Jessica said. "I guess I just needed to get your lecture one more time."

"Consider it done. Now what else did you call for?"

"The well."

"Then let's get back on track," Ranae said. "You've got a problem that's neither people nor the rig. What's left?"

"The geology. The rocks out here are typical for the Gulf, but we haven't found those nice long intervals where we get to make a lot of hole without setting more pipe. We've got weak shales, high-pressure stringers, transition zones, and more lost circulation than any well deserves. Which means we can't make a lot of hole before it's casing time."

"Make a discovery, Jess, and nobody will remember how much pipe you set."

"Maybe, maybe not," Jessica said. "We've been out here more days than originally planned, plus we've already overspent the entire Macondo budget and still have 6,500 feet to drill."

"Don't sweat it. The first part of the well may have been difficult and expensive to drill, but it's the bottom part that will deliver the goodies. You'll get there."

"Ranae, you're not hearing me. If we can't get to the target depth because the hole's too small, there won't be a discovery, and all this will have been for naught."

"Then it's simple," Ranae said, her words snappy. "You're on my prospect, spending BP's money. And I'm not about to go back to the executive committee, out of money, two thousand feet short, unable to drill deeper. It's your and Barry's job—19,650 feet—no excuses."

"Oh, thanks for that," Jessica said. "I'm looking for support and you've either had too much coffee or not enough wine. Call it an excuse, but I've learned from Barry that Mother Nature's got a bunch of physical laws that just can't be broken without consequence."

"Jessica, all I'm asking is that you keep the pressure on Barry to reach the target."

"Hah. He's got as much incentive to reach the target depth as we do, but he's got a lot of balls to balance while trying to get there. What I tell him won't make much difference."

"Tell him anyway, damn it. Don't back off, don't give up, and you'll get there."

Rarely at odds with her friend, Jessica shook her head, glad the phone was voice only.

They jointly switched to small talk. Smoothed the waters.

"Keep me posted, Jess. Call when you can."

After warm goodbyes, she hung up her phone and went to the small window at the far end of her room. The rig and others like it had yielded better times, good times that kept her coming back. But she missed Ranae, her soul mate for sharing personal issues over a glass of wine. Missed the ability to have a grubby day, at a pool, shorts and a tee, a day without a hundred eyes following her every move. Having to be the perfect professional 24/7—clad in coveralls—in charge of nothing. She looked to the stars for answers. Unfortunately, the grubby window and ocean haze ensured an empty sky. And she saw no answers for the well. Or a discovery. Or her mom. The mystery attorney. Ranae's harshness. Barry's budget and schedule anxieties. Or getting up that long flight of steps to the mudlogging unit without falling to her death.

It was good the galley never ran out of chocolate-chip cookies.

CHAPTER 27—Choices

During an early April drill bit trip, after drilling out the new 9-7/8-inch liner at 17,168 feet, [83] Barry took time to visit with Jessica. They sat at their own desks, leaning back, feet up, ankles crossed.

He thought about complimenting her pea-green-colored coveralls, but he'd never liked peas. "The only time I see you," he said, "you're in the mudlogging unit."

"Not true. Remember that day you were coming out of the galley and I was going in?"

They shared weak laughter.

"You found anything yet?" Barry asked.

"Other than Daylight's trip gas and your gassy kicks, not a sniff."

"We've got only 2,500 feet to go, so you better get busy."

Jessica's hand went to her heart. "Me? I've done all I can. Take me to a discovery and I'll dance a jig on the rig floor."

"I'll do what I can. We're drilling with 14.4-pound mud, with a 16-pound leak-off test. If we can go the distance without a kick, without more casing, we'll be there in a week."

"And if you need another liner?" Jessica asked.

"I won't be in a good mood," Barry said, imagining another couple million on top of his already-trashed well cost. "It'd probably be seven-inch, and drilling deeper would be tough. We'd have to get small drill collars and a string of spaghetti drillpipe."

Jessica's face sagged.

"I have to ask," Barry said, having seen the look before. "Are you in a lottery about the final depth of the well?"

"Nothing that easy. I'm out here for one reason, and that's to evaluate the geology that sold the well to the board, and those rocks are locked in a deep structure called Macondo at 19,650 feet. That's our goal. You've heard it before. Not negotiable."

"Actually it'll have to be negotiable, if we have a shallower discovery."

Jessica grinned. "Good reason. But could we still get to 19,650?"

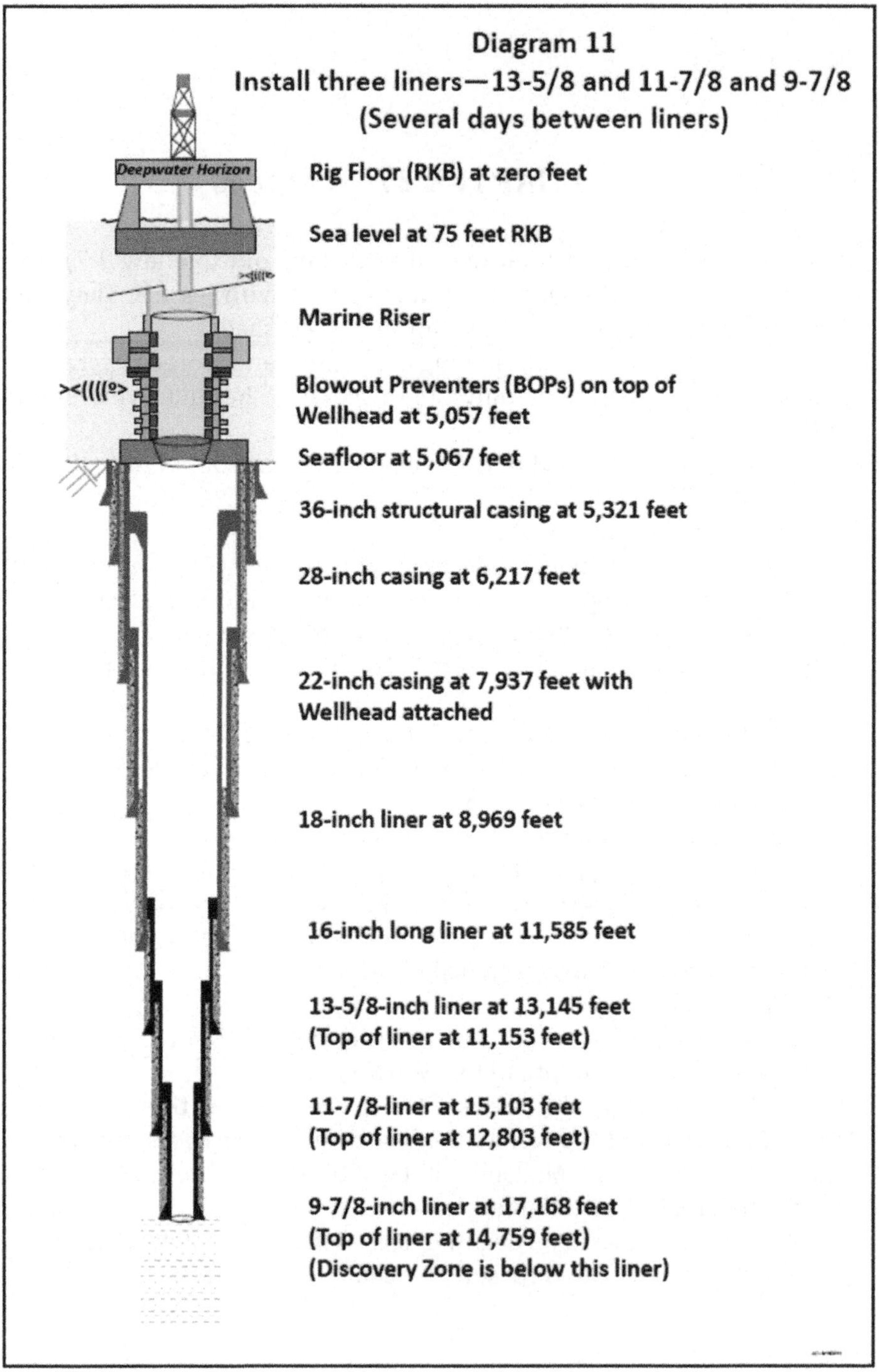
Diagram 11
Install three liners—13-5/8 and 11-7/8 and 9-7/8
(Several days between liners)
Deepwater Horizon
Rig Floor (RKB) at zero feet
Sea level at 75 feet RKB
Marine Riser
Blowout Preventers (BOPs) on top of
Wellhead at 5,057 feet
Seafloor at 5,067 feet
36-inch structural casing at 5,321 feet
28-inch casing at 6,217 feet
22-inch casing at 7,937 feet with
Wellhead attached
18-inch liner at 8,969 feet
16-inch long liner at 11,585 feet
13-5/8-inch liner at 13,145 feet
(Top of liner at 11,153 feet)
11-7/8-liner at 15,103 feet
(Top of liner at 12,803 feet)
9-7/8-inch liner at 17,168 feet
(Top of liner at 14,759 feet)
(Discovery Zone is below this liner)

Diagram 11

"Depends on what you find—depth, pressure, thickness, reserves.[84] Discoveries get a lot of attention, and there'd be a number of good reasons to cancel drilling below a big one."

"Like what?" Jessica asked.

"Anything bad you can imagine—twisting off the small drill string, getting stuck, junking the hole, losing the well."

"So, if we make a discovery, we just quit drilling?"

"Probably. Though we'd be busy evaluating and testing the pay zone, doing whatever we needed to do to get it ready for production."

"Including what?"

"We'd run a casing liner to cover the pay. Then we'd probably run a tie-back string before turning over the well to the development guys."

Jessica pulled out her tally book. "I get the liner part, but I don't understand *tie-back*?"

"The tie-back is just a way of adding casing from the top of the liner up to the wellhead."

"I didn't know you could do that," she said as she looked up from her notes. "Why would you run a liner plus a tie-back, rather than just one long string of casing?"

"If you make a discovery, the liner and tie-back is the most secure way we have for isolating hydrocarbons from the annulus. The liner allows you to cement and test both ends of the pipe, then the tie-back adds another layer of pressure security."

"So why ever run a long string?"

"There's nothing wrong with the pressure integrity of a long string given a good cement job, and there's a significant saving in rig time. Which translates to cost. The full string is so much less expensive it's hard to not consider it."

"Then why don't you like it?"

"If you run a single long string of casing," Barry said, "the entire annulus behind the casing, from the pay zone all the way up to the casing hanger, is wide open."

"What about the cement?" Jessica asked.

"A few hundred feet of hard cement makes a great barrier, if it's perfect. But if it's not good, all bets are off."

Jessica rocked back and forth in her chair. "Said he of little faith in deep, hot cement jobs."

"You got it. Which means if you make a discovery, I'll supply the liner and tie-back."

"And I'll bring my dancing shoes."

"Should I rent a tux?" Sounded dumb when he said it. Worse when she stared at him, waggled her head, and said, "Nah."

---

[83] See **Diagram 11** (page 164 herein)—Three New Liners Installed—13-5/8 & 11-7/8 & 9-7/8: Casing depths are for the last three liners above discovery zone, as per depths noted in Reference (3). Each zone was drilled and a decision made to run a new, smaller-diameter liner. Then more drilling, followed by an even smaller diameter liner, etc. Details skipped to expedite story.

[84] RESERVES (OIL)—the calculated amount (barrels) of economically producible oil in the ground. Because oil "sticks" firmly to porous rock, only about five to thirty-five percent of what is discovered can be produced (and therefore be declared as reserves). Hence, there are high incentives for sophisticated petroleum-engineering research-and-development technologies that increase the percentage of oil recovered.

CHAPTER 28—Edging Deeper

Daylight had just pushed the drill bit past 17,800 feet when everything went crazy. Pit-level and mud-return alarms shattered the heavy white noise of rig-floor drilling activity.

Wary of a kick, he reacted accordingly, picked-up off bottom, and adjusted the drillpipe tool joints to be clear of the BOP rams. Just in case. Then he realized the alarms were different than for a kick. For a reason. He shut-down the mud pumps. Left the BOPs open.

"I need Tanker," he said to his derrick hand. "Page him." Then Daylight called the M-I Swaco mud engineers.

The event was not a kick—formation fluids flowing into the wellbore. This was the opposite—lost circulation. With the mud pumps actively pushing mud down the drillpipe, no mud was coming back to the rig from the annulus. Hence, the alarms.

That loss of mud might be insignificant. Or not. If too much mud was lost, higher-pressured zones could kick. Daylight reckoned he'd rather eat his sister's sour-pickle pie than have one zone sucking and another blowing. Which called for a pinch of Skoal.

Tanker, Jessica, and Barry all got to the rig floor within two minutes of the first alarm.

Daylight greeted the threesome, then said to Tanker, "She be taking mud, boss. And she's really thirsty."

"How bad?" Tanker asked.

"No returns," Daylight said. "She be taking it all. Hundred percent, then some."

Tanker asked if he had called for lost-circulation material.[85]

"Yes, sir," Daylight said. "Done deal. Mud boys making me a 50-barrel pill of LCM in the trip tank."

"Watch for a kick," Tanker said. "I'll be in the mud room."

Barry stepped closer, and Jessica followed. "Help us out, Daylight," the company man said. "Tell us what you're doing."

"As soon as the mud boys make me my LCM pill," Daylight said, his comments directed to Jessica, "we'll pump her down the drill string. We'll lose the most part of 600 barrels, gone forever, before the pill even get there. But if we lucky, Miss Jessica, the pill will do the trick, and we'll get back to drillin' real soon."

* * *

Barry nudged Jessica away from the driller. "Lost circulation material is only part of the solution," he told her. "We need to come down on the mud weight."

"There's a formation-pressure spike of 14.1," she said, "up the hole about a hundred feet. Everything else is lower pressure. You could drop the mud weight from 14.4 to 14.2, but I wouldn't go any lower."

"Then let's go to the mud room," Barry said. "We need to talk to Tanker and the mud daubers."

Jessica thanked Daylight, who spit in his cup and tipped his hardhat in return, and then she followed Barry.

"You keep being nice to these guys," Barry said, "and they're gonna start bringing you flowers."

"Or, gosh, maybe niceness works both ways. You could try it yourself."

"Being nice to you? I thought I was."

She scrunched her face. "I'm talking about you being nice to them."

"Almost done," Tanker said as Barry and Jessica approached, saving Barry from having to figure out what Jessica was talking about.

"We also need to drop the mud weight," Barry told Tanker. "Let's go to 14.2 before we displace the pill."

Jessica punched numbers into her calculator. "That's an almost 200-psi drop on bottom, which should help a little."

Hours later—it hadn't.

"Then we pump another pill," Barry said.

Double-digit hours later—that hadn't worked either.

They cut the mud weight from 14.2 ppg to 14.1.

The measurement-while-drilling tools indicated one zone was trying to flow, while another was taking mud.

Barry, Tanker, and Daylight spiced the mud with a number of *sonofabitch*es, *goddamn it*s, and *tarnation*s before they pumped enough LCM to regain full circulation, which allowed them to get the mud weight back up to 14.2 ppg.

"Finally," Jessica said.

"Just don't stomp too hard," Barry said. "We've got a good balance, but it cost us a couple thousand barrels of mud and almost two days of rig time."

"And those barrels and days aren't cheap," Jessica added, echoing words Barry had said on a number of occasions.

He didn't tell her the bill for this latest problem would total more than $2 million, which seemed like a lot for not even a pinch of progress.

* * *

Daylight, back to sipping, spitting, and drilling, couldn't have been happier. Just past 18,000 feet, he pushed the drill bit and reamer from shale into another drilling break. A subtle bump on the weight-on-bit indicator, a slight increase in the rate of penetration. He stopped drilling and raised the blocks, which raised the drillpipe, which lifted the drill bit off the bottom of the well. He studied the pit-level and flow indicators. His job. His responsibility. He saw no extra mud. No mud losses. Middle ground. Just the way he liked it.

He kept drilling. The drill bit acted as if it were back to drilling shale. Then another little sand. He checked for flow. Checked for losses. Back to drilling shale. Then more sand. Again checked for flow and losses. Still good. Kept drilling sand. Easy drilling. Good sand.

Mr. Barry charged onto the rig floor and flashed the timeout sign. "Let's circulate it out, Daylight. Jessica needs to see what we've got. I'll talk to Tanker."

Didn't bother Daylight—anything for Miss Jessica. But he was drilling so deep, down through an upper wellbore so big—mostly 21-inch riser and 16-inch casing—the cuttings from the drill bit would take more than two hours to get to the surface.

"Tell Miss Jessica," he told the company man, "we got some nice clean sand on the way up. Special delivery, just for her."

* * *

Jessica had also been pleased to see Daylight back at work. She watched the pressure-while-drilling charts and saw nothing in the way of increased pressures. And that was good, considering the fragile lost-circulation zone they'd left behind, now a hundred feet up the hole.

Daylight drilled for two hours before Jessica sent a mudlogger to collect the first samples. Lots of shale. Very little sand.

Two hundred and fifty feet deeper. Still lots of shale. Very little sand. A few thin stringers.

Jessica feared a kick. Not because of anything scary, but because the LCM-plugged lost-circulation zone would require another liner. She wondered if she'd have to eventually tell Ranae Morgan that her 19,650-foot-deep well was out of the question. Wow, what a fun conversation that'd be.

Jessica spotted another probable sand, evidenced by its increased rate of penetration on a simple depth chart. Sure enough, Daylight picked up the drill bit and checked for flow—all on the charts. Then, back to drilling. But the sand, if that's what it was, wasn't a two-foot stringer. When Daylight drilled his tenth foot of the interval, Jessica paged Barry.

"Shut down Daylight," she told him, "and let's circulate bottoms up. I need to see what's in this sand."

Jessica normally had little stroke on the rig relative to what Barry did with the well. Not so when she was focused on cuttings, checking out a potential show. For that, she was the indisputable boss, with the entire executive-committee backing her up.

So Barry had gone straight to the driller, shut him down, then found and told Tanker. No question by the big man about chain of command.

A long two hours later, one of the Sperry Sun mudloggers carried another sample into the mudlogging unit. Jessica back-calculated the depth of origin, confirmed by the current depth of the drill bit and the number of barrels of mud pumped in the interim. She then prepared the samples and put them under the microscope.

She stopped cold. But only because she feared she'd wake up and it would be gone.

Fluorescence. Oil stains. A rainbow of colors.

A show. The first hint of a possible discovery.

---

[85] LCM—Lost circulation materials. Anything designed to plug leaking, fractured, mud-swallowing formations. Shredded cotton and ground-up walnut hulls have been replaced over the years with sophisticated lab-manufactured plugging agents. A slug or "pill" of the material, pumped down to drillpipe and into the annulus, is quite often successful. Or not.

APRIL 2010

CHAPTER 29—Showtime

Jessica paged Barry. Snatched up the rig phone when he called back. Told him what she'd seen—a strong show, oil and gas, a possible discovery. Told him he needed to tell Tanker. Reminded Barry, as she'd reminded the mudloggers, that *tight hole* means *confidential.*

"It's a need-to-know-basis only," she added, "as in nobody else needs to know."

"Jessica, I'm on your side, but I've been down this path before. Make sure you're right before you call the beach."

"I've got a few more tests to make, and I'll let you know." She hung up the phone and went back to picking and choosing the best samples for her analysis, yearning to share the joy, deciding when to call her boss, anxious to call Ranae.

She heated a small portion of the sample, then gave it the nose test. The aroma of her dreams, a sweet aroma, though ever so faint, overpowered her senses, stopped all thought, coaxed her to inhale again. Then again, the tiniest intake, nostrils spread, and again, until her lungs filled and she gently exhaled, only to repeat the cycle.

Then a bit of wet sample. Out the door and onto the landing—sunlight. A faint sheen—but the prettiest little flashes of rainbow colors she could ever hope for. Cramps in her cheeks from smiling.

She put the same sample under the stereo binocular microscope and turned on the ultraviolet-light option. When bathed in black light, the sample fluoresced yellows and blues, with faint glows of red and orange.

The drilled-gas meter had spiked high and stayed high. The composition of the gas mattered—a job for the gas chromatograph. Drilled gas and trip gas from shallower depths had registered as methane and ethane—the smallest natural-gas molecules. But the gas from the deep zone with the show was comprised of the heavies: propane, butane, pentane. The kind of larger gas molecules often associated with an oil reservoir.[86]

She scanned her notes, listing what she knew. A drilling break. Sand. Aroma. Sheen. Fluorescence. Drilled gas. Heavies. What else for confirmation?

Nothing. She'd found hydrocarbons. Oil and gas. A pay zone.

But it was all for naught if there were only five or ten feet of pay. She needed more footage for it to be commercial. Fifty feet, much better. A hundred, wishful thinking.

Barry entered the mudlogging unit.

She flashed him thumbs up, grinning so hard her ears hurt.

* * *

Hours later, though nobody talked about what was happening on the rig floor—tight hole and all—Jessica noticed a lot of whispering in the galley. She wondered if her glowing red ears were a source of concern. Yet, nobody seemed to have realized Daylight had drilled only ten more feet before she'd given the order for him to stop drilling and again circulate cuttings to the surface. And nobody had been watching the mudloggers diligently collecting drilled cuttings and hustling the little sacks up a flight of stairs to the mudlogging unit, where she'd poured over every grain, every cutting, every whiff of crude, her endorphins in overdrive.

Analysis done, she added ten additional feet to the gross discovery interval. Barry had watched over her shoulder. They high-fived each other. She told him to have Daylight drill ten more feet. "Call me when he's done," she told Barry. "If it's more sand, we'll want to look at it." Barry went to the rig floor, and she headed to the office, where she could hardly wait to call her boss.

He yahooed. Put her on speakerphone. Yelled to his assistant, named names, gathered a crowd. He told Jessica his door was closed and had her repeat her story. Followed by more kudos, like she'd brewed the oil and put it there to be discovered. Her boss agreed to keep the phone lines open, which meant Jessica was to call him with any new information.

* * *

Drilling deeper through the discovery zone, Jessica had one goal— more, more, more. She appreciated Barry keeping her grounded.

For each foot of oil-saturated rock drilled, cuttings, oil, and gas were carried up the annulus.

Jessica focused on the cuttings. Rock chips generated by the drill bit didn't change much during their trip to the surface. A chunk of barren shale was still a chunk of barren shale. A grain of sand was nothing but a grain of sand. But a *chunk* of sand—uncrushed, porous, permeable—was the jewel Jessica yearned for.

She told Barry, "Each unbroken piece carries a tiny bit of oil. I can see it, smell it, touch it. And BP's looking for oil." Ranae Morgan's oil, she didn't say.

In spite of Jessica's joy, Barry paced, studied charts. "You drill a foot of rock with oil and gas in it," he lectured, "and it's all coming up the hole. Including gas, lots of gas." He pointed to Jessica's trip-gas charts like she might not know what they meant. "You get so much gas expanding and foamy mud it looks like the well's flowing."

"It's not flowing," Jessica said. "It's drilled gas."

"Can't be sure," Barry said. "No way to know without stopping drilling, checking for flow, circulating bottoms-up, taking our time." Which he did. More than once.

Jessica, too, paced. Without more drilled footage, without more cuttings to analyze, she might as well be on the beach. She checked her watch, bottoms-up imminent, headed to the rig floor, looking for Barry.

Just in time to hear Daylight say, "Tarnation, Mr. Barry, we losing mud again."

Barry called for another LCM pill.

Daylight eased the drill bit deeper, one slow foot at a time.

The mud room again reduced the mud weight.

Jessica recorded increases in connection and drilled gas.

But all wasn't lost. Barry, Tanker, and Daylight found a delicate balance where they were able to operate without either swabbing in the well or causing lost circulation.

Barry had one goal, he assured Jessica—drill a little deeper, expose more pay, repeat the process. And he was making progress.

Jessica, thrilled to be adding tens of feet to the footage of pay she'd already observed, reminded Barry they were still more than a thousand feet from the proposed depth of the well.

As if he might not know.

---

86 RESERVES (NATURAL GAS)—including methane, ethane, propane, etc. Gas coexists with oil in the Gulf of Mexico, though it can be found without oil. From an O&G reservoir, the operator produces both, separates the two products, and sells them individually. Relative to energy efficiency—and dollar value—6,000 cubic feet (6 MCF) of natural gas is considered equal to 1 BOE (one barrel of oil equivalent). A 600-billion-cubic-foot (600 BCF) gas discovery equates to 100 million barrels of oil reserves, designated as a 100 MMBOE reserve addition. Note: in U.S. oil-field units: M = one thousand; MM = one million; B = one billion

CHAPTER 30—The Pay

The gross Macondo pay zone started near 18,000 feet and bottomed out near 18,200.[87] Jessica confirmed to her bosses she'd seen no additional pay between 18,200 and 18,360 feet, where Barry had stopped drilling in hydrocarbon-free sandy shale.

Jessica had ranted about the deeper target objective, then considered kick tolerance. With the wellbore full of high-pressure zones and fragile lost-circulation intervals, they couldn't afford to take a kick. Zero kick tolerance. Continuing to drill was not an option.

But that was good news for Jessica. Her time to shine. She and Schlumberger took over the rig, fully in charge of evaluating the 200-foot-thick discovery interval. Their goal was simple—learn everything they could about the rocks, the oil and gas in the rocks, and the ability of the reservoir to produce oil and gas into a pipeline. Wireline logs—sophisticated tools run on electrical cables—would allow measurement and estimation of porosity, permeability, resistivity, background radiation, flow capabilities, rock types.[88] Other tools were designed to capture oil and gas and water samples from the discovery interval and bring the high-pressure samples to the surface. Collected samples would be hustled to Schlumberger's laboratories in town to ensure in-situ fluid compositions were not compromised by loss of pressure, for example.

With these and dozens of other tests, which wouldn't start until Daylight got all the drilling tools out of the well, Jessica estimated she would need no less than two full days, maybe three, or perhaps more if any given test justified a new test.

Installation of the casing liner, Barry informed her, would follow immediately after completion of her logging work. The sooner the better, he reminded her.

The deliberate and careful—aka slow—drilling operation had garnered the attention of the managers on the beach. Questions from Jessica's bosses to Jessica. Questions from Barry's bosses to Barry. They wanted to know how much had been discovered, and they wanted to know *now*. Was the gross pay interval fifty feet thick? they wanted to know. A hundred feet thick? Had the hundred-million-dollar well been worth it?

The 200-foot-thick interval had quashed their concerns.

A number of phones in Houston had been manned by anxious managers day and night. One such manager was geophysicist Ranae Morgan.

Jessica called Ranae between logging runs. Ranae's pre-well subsurface maps, being adjusted based on Jessica's new data, showed the rough lateral extent of the discovery zone. Not a mega-discovery, but a decent one, perhaps in the range of a hundred million barrels. A billion would have been more exciting, they agreed, but that's the nature of the business, they lamented.

The good friends were thrilled by their collaborative success. They agreed to share a bottle of fine wine when they next saw each other.

"I'm too excited with the discovery," Ranae said, "to be concerned you're not yet at 19,650 feet."

Jessica picked up on the word *yet.* "We've got a lot to do before we drill any deeper," she told her friend, fearing the disappointment that would accompany the truth—that drilling deeper might be off the table.

"Two hundred feet is wonderful," Ranae said, "but just remember the goal. I dreamed last night we'll see a lot more oil from this well than we ever imagined."

"I hope so," Jessica said, though her own quiet words seemed to echo in her quarters.

---

[87] From Reference (3)—Depth-of-pay numbers and depths of lost circulation zones above and below the pay are from BP's post-mortem Internal Investigation—Page 52

[88] WELL LOGGING— (both wireline and logging-while-drilling). Logging tools can measure a number of parameters, including background radiation and electrical resistivity. For example, shale (like an old-fashion blackboard) contains more organic materials than does sand, hence exhibits higher natural background radiation. Sandstone full of water (brine) conducts electricity easier than sandstone full of oil (the oil is resistive to the flow of an induced electrical current). Therefore, if we're looking for oil-saturated sand in thousands of feet of hole, we look for low background radiation and high resistivity to electrical current.

CHAPTER 31—Options

Three days later—felt like a month—Barry walked fast to catch up with Jessica as she crossed the pipe deck. She and Schlumberger's wireline loggers were still busy with the well, running tests, taking pressured samples. Barry, too, had been busy. Adding up daily costs. For the well that had already consumed every budgeted dollar.

"How much longer?" he asked over Jessica's shoulder.

She stopped at a handrail and gripped it hard. "We're done tomorrow, though I could use another full day."

"What's stopping you?"

"There's pressure in the office to finish the job and release the rig as soon as possible. Another project's been waiting a long time for the *Horizon*. I think it's the Nile." [89]

The Nile—another BP code name used for a deep geological structure and its first well—was no surprise to Barry. He'd heard for weeks about BP's obligation to get the *Deepwater Horizon* off Macondo so it could get to the Nile project for ongoing work. But that wouldn't happen until Barry ran casing, secured the well, and released the rig. Though he had little control over the rig's release time and date, other than minimizing nonproductive time while executing mandatory tasks, the responsibility weighed heavy on his shoulders.

"Temporary abandonment of Macondo will take time," he told her, holding open the door, "so the sooner we get started, the sooner we're done."

Jessica scrunched her eyes. "Abandon? This well? No way."

"Take it easy. There's a big difference between abandonment and temporary abandonment. If this had been a dry hole, we'd fill it with cement and never look back. But with a good discovery, we need to leave it safe so we can come back to it when we're ready. In the meantime, your successor will be out here over the next year or two drilling delineation wells." [90]

"Just so the emphasis is on *temporary*," Jessica said. "What's your best guess for first oil? [91] Two years? Three?"

"More like five, plus or minus two. You did the exciting part, making the discovery, and now you can leave and be happy. But the tougher part, the bigger investment, will depend on the amount of oil and gas ultimately confirmed. It'll take the project-management guys a number of years and a billion dollars before any of this oil ever sees a pipeline."

"I'll be out of school long before that."

"Then you could be the engineer on the rig, maybe even the company man, when BP reenters Macondo in 2015 and completes it for production."

"That'd be nice, but this one's yours. Your well, as you say. I want my own—top to bottom." She released the handrail and popped two thumbs up. "A mega discovery if it's in the cards," her words fading as she bounded up the stairs to the logging unit.

Barry projected his voice to keep up with her departure. "When you get your own well, I'll buy BP stock, but in the meantime this one's still mine and I want it back. Soon."

She flipped something over her shoulder without even slowing down. He couldn't tell if it'd been a strand of blond hair or a well-manicured bird.

He did an about-face and walked beside the pipe rack, dragging his hand along the 7-inch casing that would soon be run as a liner, buried forever miles below the rig. On the far pipe rack, an even-higher stack of casing, 9-7/8-inch diameter, that would follow the liner as the tie-back string. [22]

Inside the quarters building, Barry joined Tanker in the Transocean office and pulled out pen and paper.

"Another list?" Tanker asked.

"Options," Barry said. "Full string versus a liner plus tie-back. Hang around; I may have some questions for you."

"Might as well run the full string," Tanker said. "Easier than running a liner and tie-back. Same result either way."

"Can't go there," Barry said, recalling all the reasons he'd given Jessica. "I need a liner and tie-back to better isolate the pay."

"Okay by me, but you'll need a couple extra rig days to do it that way."

Barry wrote more notes. He needed to be prepared, get his thoughts in order. His bosses, rightfully, never missed an opportunity to eliminate a poorly spent hour of rig time. Barry reckoned that's one of the reasons they were the bosses.

BP's formal written procedure for temporarily abandoning the well included step-by-step activities, though without a commitment to full string versus liner plus tie-back. Even though the issue was open, he'd been copied on an internal BP document in which engineering spoke against the long string.[93] No surprise.

Yet, Barry's bosses had been quiet on the casing subject, though he suspected they already knew what they wanted. The senior drilling managers were good men, and they expected Barry to propose options, with pros and cons for each, before making recommendations he was willing to stand by and was ready to execute.

He made a list for the full string of casing. Pros: faster, cheaper. Cons: must confirm good cement job, no easy option to remediate poor cement, open annulus from top of cement to casing hanger.

His list for the liner and tie-back option. Pros: better pay-zone isolation, extra barrier at top of liner, closed annulus to wellhead. Cons: two days of rig time, higher cost.

Barry reread his notes on the tie-back option. No way to escape the release-the-rig-ASAP words Jessica had heard from Houston, which were fully complemented by Tanker's couple-extra-days observation. Nonetheless, Barry had no lingering doubt the liner with tie-back was the right way to go.

He made his recommendation via e-mail to the beach.

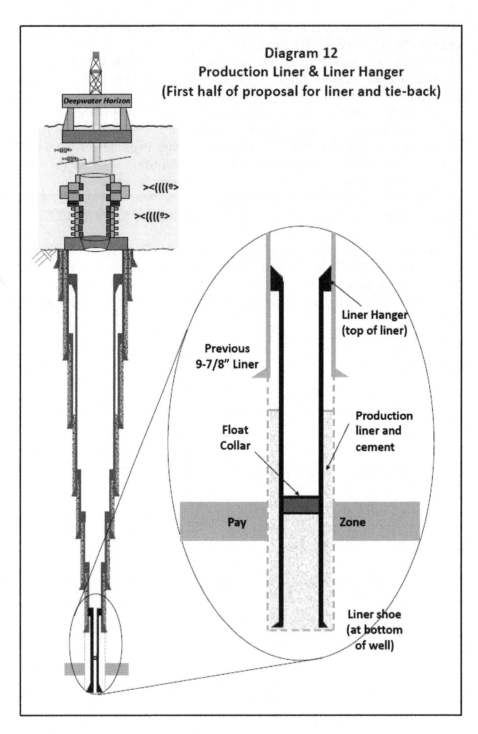

Diagram 12
Production Liner & Liner Hanger
(First half of proposal for liner and tie-back)

Liner Hanger
(top of liner)

Previous
9-7/8" Liner

Production
liner and
cement

Float
Collar

Pay   Zone

Liner shoe
(at bottom
of well)

Diagram 12

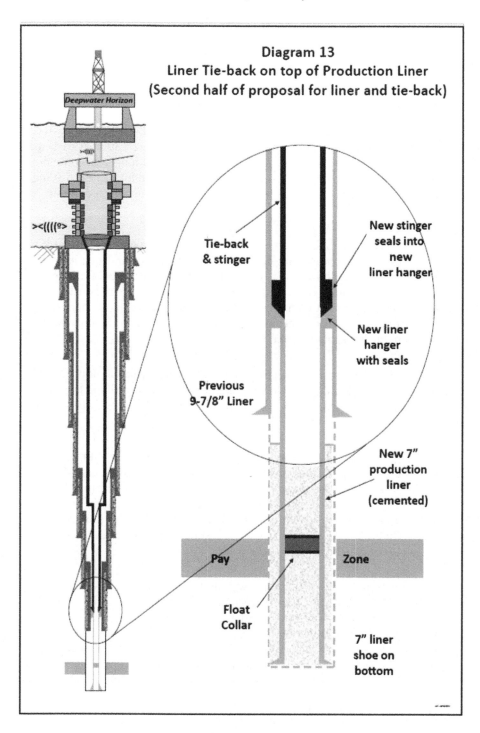

Diagram 13
Liner Tie-back on top of Production Liner
(Second half of proposal for liner and tie-back)

Tie-back & stinger

New stinger seals into new liner hanger

New liner hanger with seals

Previous 9-7/8" Liner

New 7" production liner (cemented)

Pay   Zone

Float Collar

7" liner shoe on bottom

Diagram 13

---

[89] Reference (1)—House Subcommittee Letter to BP's CEO—14 June 2010. "The *Deepwater Horizon* was supposed to be drilling at a new location as early as March 8, 2010. In fact, the Macondo well took considerably longer to complete than planned . . . By April 20, 2010 . . . the rig was 43 days late for its next drilling location, which may have cost BP as much as $21 million in leasing fees alone. . . ."

[90] DELINEATION WELL—After a successful discovery well, the operator's next well(s) in the area would be drilled to help delineate and quantify the discovery zone (the geometry of the accumulation and amount of oil and gas). These answers are needed before the operator can proceed with the design and economics of a development plan; hence, the normal three years or more from discovery, through delineation, through development wells, and the installation of development facilities, before first oil.

[91] FIRST OIL—Term used to signify the first for-sale barrel of oil produced into a pipeline or vessel. For big developments with millions-billions in sunk costs over a number of years, first oil marks the long-awaited beginning of the income stream.

[92] See diagrams 12 and 13. **Diagram 12** (page 182 herein) is the Production Liner and its Liner Hanger. **Diagram 13** (page 183 herein) is the second half of the liner/tie-back option. The diagram shows the Liner tie-back on top of (stung into) the top of the Production Liner. Though the production liner (either without or with its tie-back) was a viable option, BP elected to run the full string of production casing, also a viable option. Both options required testing to ensure the security of (or the need to repair) the well.

[93] See Reference (7)—Macondo—Rig and Well Specs—BP-Production Casing—BP-HZN-CEC022118, which recommends the liner and tie-back. Document also references pay-zone depths and pressures, plus depths of lost-circulation zones above and below the pay. See also Reference (8)—Macondo— Rig and Well Specs—BP-Production Casing— BP-HZN-CRC022145, which states the full Casing string is again the choice, and the liner is now an option.

CHAPTER 32—Decisions

The buzz of the satellite phone in Transocean's office pulled Barry back to earth. Earlier, he and Tanker had been talking about March Madness—how nice it would have been for Butler to have beaten Duke. Barry had tried to picture his dad watching the game—*Two pointer. Score's thirty-three thirty-two. Duke's ahead. By one. Halftime.* Tried to picture his seemingly deaf-and-blind mom nodding.

Tanker, eyes on his computer screen, said, "You've got mail."

Barry punched in his password and opened a short message from his boss. The note included a 21-page attachment—the operating procedures for the production casing.[24] The message advised the single string of casing, comprised of both 7-inch and 9-7/8-inch casing, was *again* the primary option. Barry read it twice. No mention of running a liner and tying it back other than as a contingency option, which would be justified only if the hole proved to be unstable. The MMS—Mineral Management Service—had approved the amended drilling permit for the full string.[25]

Kiss my ass, Barry thought. Or maybe mumbled.

Tanker ignored Barry's maybe mumble and stayed busy at his own desk.

Furthermore, per the note, Barry was to do an inventory check, send details of shortages, and supply a timeline for abandonment and demobilization. As expected. He closed the message, preferring to print the attachment before studying it in detail.

Tanker's guest chair creaked when Barry leaned back, arms crossed over his chest. It wasn't the first time his bosses had declined his proposal. He glanced at Tanker, who glanced back, a measure of understanding between them.

"So we're gonna run the long string," Tanker said.

Barry nodded, said nothing.

"Then the good news is," Tanker said, "your list won't be so long." He stood up and stretched. "I'll stop by the galley and have 'em send you a cup of coffee, then I'll make my rounds. Page me if I can help."

Barry hiked back to the BP office, printed the casing-cement procedure, and glanced through it. No surprises. Nothing extra. Nothing left out. But he had a list to make, so put the procedure aside.

He wasn't worried about the inventory of casing and tools, which he'd been through twice. Though he preferred the liner and tie-back option, his boss's decision had just uncomplicated the casing job and cut two days off Barry's schedule. Worthy of a high-five if he hadn't been alone. Determined to do the job right, his list would include everything necessary to maximize the integrity of the long-string casing and cement job. The hard part was estimating the duration necessary for each activity.

Before running casing, he planned to run the drill bit to bottom, condition the mud, and check out the wellbore. Now was no time for the kind of wellbore instability problems they'd experienced on the *Marianas*. He estimated the required hours and—

Jessica bounced into the office and handed Barry a cup whispered by steam. "I brought you dark roast with lots of cream. Tanker said you might need it."

Barry stood and thanked her, hoped she noticed his good manners.

She sipped her own coffee and collapsed into her desk chair. "What're you doing?" she asked as she picked up the casing-cement procedure off the edge of his desk and started flipping pages.

"Laying out the game plan for the rest of the well," he said, "and estimating times for each step. You on a break?"

"For a few minutes. You can fill me in as you write."

"Well, first we'll have to run the casing slowly, to minimize surge, so we don't disturb the lost-circulation zones."

"So how long will that take?" she asked, while adjusting her ear-bud wire and scanning another page.

"Eighteen hours, plus or minus."

More iPod adjustments. One glance at him, eyes intense. Then back to the clipped pages.

"After we run casing," he said, "we'll circulate bottoms-up. That'll take all day."

Her eyes asked why.

"Because our three-and-a-half-mile-deep well holds a lot of mud, and we need to pump slow. BP's standard procedure calls for the casing volume plus fifty percent. For bottoms up, it's the volume of the entire annulus. That's for—"

"Wow," Jessica said, flipping pages, "this is more detailed than a Julia Child's cookbook. You could give it to Daylight and me and we could finish the well. Heck, you could give it to me—"

"It's detailed on purpose."

"No, it's crazy." She pointed to a line on a page. "Do not need to set 16.5 mud in rat hole as volume is only four barrels." She flipped pages, ran her finger, stopped on another line. "Do not circulate liner greater than 5 barrels per minute, unless required, as it will convert the auto-fill equipment, whatever that is."

"It's part of the float collar, which you need to get familiar with."

She threw him a hard stare and then turned another page and faux scanned it, her head movements exaggerated. "What happens if something goes wrong? Or are there instructions in here for trouble-shooting? Maybe an 800 number to get a technician with a Swahili twang?" She slapped shut the document, sent the paperclip flying.

"Why slow?" she added, her words snappy, segue free.

"For what?" Barry asked, thankful she hadn't picked up on, and he didn't have to defend, the long-string casing option.

"Bottoms-up. You said you had to pump slow."

He wished he could see inside her head, take a look at the synapse-driven spreadsheet that allowed her to enjoy music, listen to him, read, recall, ask penetrating questions. "About twelve hours," he said. "Just circulating."

"That's twice as long as normal. Why so long?"

"We're trying to avoid lost circulation, and we can't afford a mistake, especially once we start mixing cement. Which means we need trip gas cleared from the annulus before Halliburton mixes the first sack of cement. And no junk in the casing."

"Junk? Like . . ."

"Anything—gloves, dentures, drilled cuttings that enter the casing shoe. Any kind of junk, anywhere in the casing, could screw up the cement job."

"So you'll circulate bottoms up," she said, "then run another exciting cement job." She checked her watch, sprang from her chair, and went to the door. "Sorry I can't stay and play, but maybe we can meet at breakfast for an update."

"And you'll give back my well?"

"As soon as I'm done with it." Soft southern words. *I'm* said like *am*. Soft smile. Then she was gone—Barry with no clue as to what was on her mind.

He had often mused about being an attorney, prosecuting or defending a tough case in court—a piece of failed equipment, a chopper crash on the helideck, a heavy load dropped through the deck of a workboat. Now, up late, he imagined Jessica working with him in court—

a good team, Tanker would have said. That vision faded and he saw her on the other side of the aisle, her hair longer and tied back, wearing a black pants suit over a bright-red neck-high blouse. The interrogator. The defender. The prosecutor. Hammering away with question after question, dug up from who knows where.

Barry the attorney? No way. Damned good decision.

---

[24] Reference (9)—BP document—MC 252 #1STOOBP01 - Macondo Prospect—7" x 9⅞" Interval. Dated 15 April 2010. "Macondo.Well.Casing.Production.Operations—Long String Specs and Ops."

[25] Reference (1)— House Subcommittee Letter to BP's CEO—14 June 2010. ". . . BP chose to install the single string of casing instead of a liner and tieback, applying (to the MMS) for an amended permit on April 15 . . . approved on the same day."

CHAPTER 33—Last Steps

Ah, Barry thought, cement. And Halliburton. Working on a half-inch-thick ham steak under three soft-fried eggs, he continued drafting his notes. All he needed was—

"Private meeting?" Jessica asked, sliding her tray onto the table across from his.

"Always room," he said. For you, he thought.

"Writing your folks?"

"Nope. Putting together my time estimate for the rest of the job."

Jessica, working on coffee, a soft poached egg, and a toasted English muffin, chewed a tiny bite and swallowed. "To finish the well? How long?"

"That, my curious friend, will get an answer only after I finish my estimate."

"What's the last thing you'll do?" she asked, crunching toast. "You know, the thing that says we're done and ready to move the rig."

"Several things, including—"

Jessica held up her hand. "I would've been disappointed had you given me a direct answer without a preamble." She looked at her watch. "So the *last* thing is?"

"We'll place an abandonment cap over the bottom half of the BOP connector," he said. "That's to keep the hole clean and make sure the connector's not damaged when we come back. In five years."

Jessica stabbed her fork into a tidbit of toast and sopped up a glob of yellow from her plate. "Okay, you win. Tell me what you'll do *before* the last step."

"Well, since you asked. After we hang the casing in the wellhead and cement it with nitrified cement, we'll install the lockdown sleeve. That's the safety device." He waited.

"Got it." She looked at her watch.

"Then we'll test the casing for leaks, from inside out and from outside in."

"What if there's a leak?"

"We fix it, then repeat the pressure tests. When everything tests okay, we'll install a cement plug in the well."

She stared at his eyes. Moved not a muscle but for the hint of a nod.

"Then the big step," he said. "We displace the heavy mud from the riser with seawater. As soon as that's done, we'll pull the riser and LMRP—the lower marine riser package."

She casually covered her mouth as she chewed. "Where's the connector fit in?"

"When it's time to pull the BOP, we'll unlatch the wellhead connector. Half of the connector will come up with the BOP. The other half—" He paused, waited.

"Stays on the wellhead and gets the protective cap. For maybe five years." She wiped her mouth on a paper napkin. "Thanks for breakfast, but I have to go." She stood and gathered up her tray, clean plate, and empty mug, then leaned toward Barry. "Some guy keeps bugging me that he's got a well to abandon, and I don't want to slow him down."

Barry applauded—a silent tap of his hands.

Jessica nodded—a single dip of her chin.

Six seconds flat, like a human dragster, she was gone. He'd wanted to tell her about the cement job, which they'd discussed only briefly, and its complexity and how much thought had gone into its design. She would've been impressed.

Then he reconsidered. Would his parents have been impressed? They would have nodded, that was for sure. And Jessica? Impressed about a cement job yet to come, while totally focused on and responsible for evaluating the discovery?

Not likely.

With no time to dwell, he refilled his mug and resumed drafting his note, cement on his mind. Ever since the discovery zone, there'd been a barrage of e-mails among players of the triad—himself on the rig, his bosses in Houston, and Halliburton's cementing gurus. The e-mails were well intended. Macondo's cement would cater for the deep, high-pressure, high-temperature well—18,300 feet, oil and gas at 12,400 psi, bottom-hole temperature near 260°F.[26] And severe lost circulation zones—above, and near the top and bottom of pay zone. It would be a complicated job—not quite like a local cement company delivering pre-mixed concrete for a patio, a big job being the whole nine yards.

Yet, Macondo was no different than hundreds of other deep wells in the Gulf, a great number of which were deeper, hotter, higher pressured, and associated with equally raunchy lost-circulation zones. And every well in the Gulf, bar none, depended on cement to isolate oil and gas zones from the rest of the annulus.

But Macondo was Barry's well. And the cement job needed to be perfect, he mused, recalling what he had told Jessica. Coming from the bottom up, the annular cement had to push the mud out of the way, totally cover the pay zone, adhere to the walls of rock and casing, and prevent the oil and gas from ever entering the casing shoe or the upper annulus.

Jessica had asked, "What can you do to increase the chances of perfect cement?"

"A wider annulus would help," he told her, "but our production engineers want the biggest casing they can get for the high production rates they expect from the well."

Jessica finished a note. "Anything else?"

"Plenty more. Pumping the cement faster would help but would cause increased back-pressure on the annulus, which our lost-circulation zones couldn't stand. We could also use a larger volume of cement, but heavy cement pumped too high up the annulus could also break down our lost-circulation zones. We've agreed with Halliburton to use nitrogen-infused cement.[27] Like folding whipped cream into Jell-O," he explained. "The bubbles of nitrogen will make the cement weigh less for the lost-circulation zones, yet allow the cement to retain its strength."

Barry pulled a printed e-mail from his pocket and noted Halliburton's estimated volumes and times for the nitrogen-infused cement job. Comparing their numbers with his own, he could almost hear the "Check," "Check," "Check."

The e-mail also reaffirmed Halliburton's recommendation for centralizers. The mechanical devices would keep the casing centered in the wellbore so the cement could surround it uniformly. The cementers wanted to use twenty-one centralizers to minimize the chance of severe gas migration up the annulus. Barry's on-rig inventory showed six.

He rubbed his temples and added a line-item message for his boss: Short 15 centralizers. Deliver Horizon tomorrow, latest. Also need Schlumberger loggers for CBL tomorrow noon.

He wanted the loggers on board to run the cement bond log in case there were problems during cementing—like lost circulation. The CBL tool wasn't perfect, but it targeted cement quality. Perfect results would show credible bonded cement up and down the annulus and across the pay. At the other extreme, it might show poor bonding, as if there were no cement. Poor bonding, combined with the open annulus all the way up to the wellhead, would require Barry to drill-out or perforate holes in the casing and execute one or more cement squeeze jobs, thus sealing the annulus.

He estimated and noted the time it would take to run the cement bond log, but, ever optimistic, not the time it would take to remediate poorly bonded cement.

He stood and stretched, reached for his toes, discovered that his legs had lengthened. Probably because he'd eaten big and sat on his ass for three days while Jessica had control of his well. He forced his butt back in his seat to finish the note. He damn sure didn't want anybody waiting on him once he got his well back.

He mused post-cement-job topics: Hang-off casing. Pressure test seals. Install the casing-hanger lockdown sleeve. Run high-pressure and negative-pressure tests. Set cement plug.

The high-pressure test would be simple. Like blowing air into a balloon. Close the BOPs. Pump mud into the wellbore to increase the pressure. Make sure nothing leaks from the wellbore into the annulus. If okay, release the pressure. Get the same amount of mud back as had been pumped into the well. In the past, he'd found junk-induced casing-hanger leaks, but never a casing leak using a high-pressure test. Like Jessica watched for shows, he watched for leaks.

The negative-pressure test—designed to reveal leaks from the annulus into the wellbore—was more complicated, would need more time. Any such leak, though rare, would need remediation and proof of repair before Barry took the critical step of displacing the riser's heavy mud with seawater.

He estimated and added time requirements to his notes.

Cement plug. Barry would set a 300-foot cement plug inside the casing a couple hundred feet below the seafloor. Reminded him of a similar plug he'd set from the *Marianas* before he'd pulled the BOPs. Just before Hurricane Ida. He'd wait for the Macondo cement plug to harden, then have Daylight bang on it with a drill bit to make sure it really was a plug. He added to his note the minimum time estimate for setting and testing the plug—assuming all went well.

Then bad memories. He'd been on wells when it seemed impossible to set a simple cement plug. He'd done everything right, waited for the cement to set, tried to find it, found zip. He pictured the heavy cement oozing down the wellbore, lighter-weight mud oozing up to replace it. Hence, his mantra—if you can't beat on the plug, it's not there.

Bored with not the favorite part of his job, he was almost done. *Displace riser, then pull LMRP, riser, and BOP.* He'd be sure to have Jessica watch as Tanker and crew installed the abandonment cap. The *last* step. Then Barry and company would be ready to haul ass to the next well.

The Nile.

Jessica's grapevine had been right—the *Deepwater Horizon* would go immediately from Macondo to the Nile project. Barry's printout for the casing procedure included a section on cleaning out the synthetic *oil-based mud* from the *Deepwater Horizon*'s riser and mud pits *before* releasing the rig to the Nile project. The Nile needed *water-base* mud. But his brain got stuck on the word *before. Before* releasing the rig. Which meant simultaneous activities—cleaning the pits while displacing the riser. He predicted a very busy time during the abandonment process.

He added the hours necessary for the concurrent operations, including the use of a workboat alongside the rig to salvage the discharged, very-expensive, oil-based mud.[28]

He looked over his note. Job complete.

Back in his office, he summarized his time estimate for temporarily abandoning the well and demobilizing the rig, then prepared and sent his e-mail.

Seconds later, the satellite phone buzzed and indicated an e-mail to Barry from his boss. Short note, unrelated to the message Barry had just sent. Took less than a minute to read.

Less than another minute to haul tail to Tanker's office.

––––––––––––––––

[26] Reference (2)—USCG—24 Aug—Halliburton depositions—pp 246— references bottom-hole temperature in Halliburton's cement-design files.

[27] Reference (16)—USCG Deposition 24 August 2010, P. 250–257, testimony by Halliburton about cement modeling results and the use of nitrified cement.

[28] WORKBOAT—large vessel used to move heavy loads, pipe, equipment, and bulk products, including liquid mud, to and from the rig. Rarely used to transport personnel. The *Deepwater Horizon*'s workboat was the MV *Damon B Bankston*.

CHAPTER 34—Visitors

"Visitors," Barry declared, waving a printed e-mail like a surrender flag.

Tanker spun in his chair. "Who? When?"

"Transocean and BP. A bunch of VIPs. Big guns. Next Tuesday. You know anything about it?"

Tanker shrugged. "Not yet. What's going on?"

"As you've seen before, the BP guys, managers from operations and engineering, visit rigs on a rotating basis. Monthly when they can, though it's the first time for Macondo. They'll do pep talks, tour the rig, look at our paperwork." He handed the message to Tanker. "But this time's different—they won't be alone. The note says your guys are coming because the *Horizon*'s gone seven years without a lost-time accident."[99]

"I know about the LTA record," Tanker said, "just not the visit."

"I'm sure it's a coincidence they're coming out the same day as the BP guys. They'll all want to tour the rig. Have safety meetings."

"Just what we need," Tanker said.

"Tarnation," Barry said, mimicking Daylight, "y'all might want to shave."

Tanker rubbed his scraggly chin and returned Barry's note. "I'll see what I can find out."

Barry returned to his office, not unhappy with the upcoming rig leadership tour. Good time to mingle with the brass, though all parties respected the priority of rig activities over visitor meetings and tours. And this one, combined with Transocean, was good timing. A joint photo op. BP's big discovery. Transocean's commendable safety record. He checked the calendar. Next Tuesday, April 20—three days. Three days to spit-shine the rig, get it ready for the tour.

Barry's satellite phone buzzed. He read a new note from his boss. Time-saving suggestions to Barry's list of final abandonment activities, including key items like installation of the lockdown sleeve and preparing the rig for the next well. Instead of running A *then* B, look at A *while* B. Instead of three trips with drillpipe to do X *then* Y, look at doing X *and* Y on the same trip. Suggestions that saved an hour here, a million dollars there. Not trivial suggestions. All good stuff. Nothing Barry couldn't handle with ease. And with the benefit of a shorter schedule.

Barry considered his options. He had already overspent the original budget by more than $50 million, and he was almost six weeks behind schedule.[100] Probably a coincidence that if he achieved every time-saving suggestion, the well would be complete and the rig released the morning after the VIP visit.

Good timing, he thought. Yeah, right.

* * *

Barry took the news like a man. Jessica would be done with her work in about an hour, she had said. It'd been three days, almost four, since the drill bit had been in the hole. And bad things can happen to an uncirculated wellbore in three days—like tight hole, sloughing shale, a wad of gas. With the entire company on Macondo alert, the last thing he needed was to stick the drill bit in the middle of a massive pay zone and have to call out the fishing experts.

Barry talked to Tanker, gave instructions for the bit trip. Tanker and Daylight made it happen. The hole proved to be in good shape—no drag, nor sloughing shale. High trip gas, after so much time logging, was no surprise.

"The spike of gas—ten times higher than we normally see—came right on time," Jessica told Barry, the duo side-by-side in the mudlogging unit. "But I was surprised to see the leading edge of the gas strung-out several hundred feet up the hole. Will you raise the mud weight?"

"No. Mud weight's good—just where we want it for avoiding lost circulation while maintaining overbalance. And the gas up the wellbore shouldn't be a surprise, since gas rises in heavy mud."

He'd made his statement as a matter of fact, but Jessica's face told him she would replay his words and turn them into notes. Always the notes. *To remind my brain*, she would have said had he asked.

* * *

As Daylight began pulling the drill bit from the hole before dinner Saturday night the 17th, Barry watched the clock tick away expensive minutes, each hour an important part of the schedule he was determined to keep. A schedule he would shorten even more, if possible.

He was further reminded of his tight schedule when the rig's logistics coordinator notified him that Schlumberger's crew for the cement bond log was scheduled to arrive on the first flight the next morning—Sunday.

Second-guessing his own decision to call them out, he knew he wouldn't need them until after the cement job.

And more than likely, not at all.

---

[99] Reference (17)—WSJ—Monday, May 10, 2010. "Rig Owner Had Rising Tally of Accidents." Article by Ben Casselman. " . . . on the day of the disaster, BP and Transocean managers were on board to celebrate seven years without a lost-time accident." The Visitors were more formally referenced by themselves and by offshore staff during a number of USCG depositions.

[100] Reference (18A) The Bureau of Ocean Energy Management, Regulation, and Enforcement (BOEM): Regarding the Causes of the April 20, 2010, Macondo Well Blowout. Issued 14 September 2011. BOEM's FINAL report on the BP Blowout. Section VI B, P. 78–79, Scheduling Conflicts and Cost Overruns.

CHAPTER 35—Production Casing

Dark-thirty—the black of night—hours after Saturday dinner, Barry asked Jessica if she wanted to join him on the rig floor for the casing job.

She laughed, but it was a nice laugh. "More casing? I don't think so. Though you can show me the float collar you've been bragging about."

"It's on the pipe rack, but there's not much to see." He led the way.

Jessica yanked out her tally book. "What am I looking at?"

"It's a dual-flapper float collar with an auto-fill tube."[101]

"Barry," Jessica said as she slapped her hardhat bare handed, "listen to yourself—you sound like a used-car salesman."

Barry thought about his parents, nodding, clueless as to what he ever talked about. "It's a check valve," he said. "In fact, two check valves, designed to make sure—after conversion of course—the heavy annular cement doesn't U-tube back into the casing after the cement job."

Jessica patted the device. "I understand check valves. Tell me about cement getting back into the casing."

"Backflow's a big problem," he said. "Since the cement's heavier than the mud, once the slurry is in the annulus, we need a way to keep it there long enough for the cement to set up."

"Hence, the check valves."

"Simple as that."

Jessica opened tally book to a dog-eared page. "So, after all that build-up, the float collar goes in the *no-brainer* column."

"Not even close," Barry said, aware that no-brainer assumptions were those destined to bite him in the ass when least expected. "There's another important aspect—the automatic fill-up system."

"And this is something I need to know?"

"Only if you want the flappers to work."

"Then I want to know." She hugged her torso. "Can we go inside and talk about it?"

"Not yet." Barry pointed inside the float collar to a stubby piece of pipe. "This is the top end of the auto-fill tube. Right now it's locked in place, holding open the check valves. As long as it's there, the flappers don't flap."

"So you'll pull it out before you run the liner."

The liner. "No," he said, thinking he'd tell her about the liner later. "We want the tube in-place while we run the pipe. With the flappers forced open, the casing will fill from the bottom up as it's being run."

"And that's important because . . .

"If the flappers were closed, all the mud being pushed out of the way by the casing would surge up the small casing annulus, which could cause lost circulation."

"Surge causing lost circulation doesn't tell me how you get the tube out of the flappers."

"The tube stays in place until we increase our pump rate to five or more barrels per minute. The higher rate will push the tube out of the flappers and down into the shoe joint. We call that *converting* the float collar—the flappers become check valves."

"The casing procedure had a reference to *converting*," Jessica said, "something about premature conversion."

"Good memory. That was to remind me to stay below five barrels per minute or I'd prematurely convert the float collar."

She folded her arms. "I get it, but what if the tube and the flappers don't work?"

He folded his. "Where do you get questions like that?"

"Hey, if you tell me the equipment's important, then I believe you. But if you tell me we'll all die if it fails, you get my attention in a way I'll never forget."

"You won't die if it fails."

She gazed at him from under the brim of her hardhat. "I didn't mean just me." She turned and headed to the handrail.

He joined her, looking west into hazy darkness. A triad of light specks in the distance—perhaps a southbound freighter. "If the flappers don't convert," he said, "we'll have to keep pressure on the casing to force the cement to stay in the annulus."

"How long?" she asked, gripping hard, leaning over the rail, looking down.

"Depends on the cement thickening time and compressive-strength schedule. Six hours, twelve hours, maybe even a day. We call it WOC—waiting on cement.

"Tell me you're kidding."

"Not at all. Green cement—cement slurry—behaves like drilling mud. Pump the slurry and it pushes the mud ahead of it, especially if the casing is well centralized. When the slurry hardens into cement, there's no more mud, and we've got a chance for good cement integrity."

"But twelve hours, maybe more? We drilled hard cement on the *Marianas* as soon as we could get a drill bit in the hole."

"We need more time when we're this hot and deep," Barry said. "If green cement is halfway down the casing and we have pump or plumbing problems, or rig power fails, we don't want the cement to set up in the casing. So we give ourselves plenty of time by adding chemical retarders. Besides, we've got lots of other work to do and the cement's got nowhere to go once it's in the annulus."

"Twelve hours, maybe a day, holding pressure on green cement. Bet that'd be a fun morning report. 'Hey, boss, we're shutting down the rig for a day or so because the flappers didn't flap and the cement's not behaving.'"

Barry started to answer, Jessica not knowing how right she was, but she rapped her tally book on the handrail. "I'm done," she said. "Too much information. Can I buy you an early breakfast?"

Barry checked the time—almost midnight. Tanker had estimated the casing crew would start rigging up about three in the morning—Sunday— and pick up the first joint of 7-inch casing and its reamer shoe a half hour later. "Yep, a light bite, then I need some shut-eye. Long day tomorrow running casing."

"Not me," Jessica said as they headed to the galley. "I think I'll sleep in, take an extra-long shower, maybe watch a couple of movies."

"Right," Barry said, enjoying her soft drawl. "Knock yourself out." He couldn't imagine even teasing about the luxury of free time, his own mind unable to escape the constant search for ways to cut another hour off the abandonment schedule.

* * *

The crew ran 185 feet of 7-inch casing before they picked up and installed the float collar. Barry didn't bother calling Jessica to witness the event. Additionally, as the crew ran the first 800 feet of 7-inch casing they attached in predetermined locations six solid-blade casing centralizers.[102]

The Halliburton cement supervisor, dressed in bright-red coveralls, stood beside Barry during installation of the centralizers. "All our modeling work," he again told Barry, "says we need twenty-one centralizers."[103]

"There're fifteen more on the pipe rack," Barry said after a sip of bad coffee, "but they're *bow* centralizers, which are slow to install. Worse, they're flimsy and if they fall apart they can stick my casing. I'd rather have to remediate the cement job than stick the casing with the hanger fifty feet above the wellhead."

"I'll remind you of that after we see the cement bond log," the cementer said. "Without the centralizers, the modeling work defines an increased chance of severe gas migration . . ."

A flash of bad memories reminded Barry about a casing job gone bad. The long-string casing had tagged bottom thirty feet before the casing hanger reached the subsea wellhead—like a skydiver reaching the ground before his chute opens. He'd blamed the depth, the pipe tally, the number of joints picked up, junk in the hole—didn't matter, hell of a mess. Days lost pulling and replacing casing. But it hadn't happened since. Even without the 24-page casing-cement procedure, which instructed him to leave the casing 50–60 feet off bottom, he targeted his long strings to be no less than a full joint—at least 40 feet—off bottom, leaving plenty of rat hole. And he didn't run flimsy cement centralizers.

The cementer had stopped talking, arms folded across his chest, perhaps waiting for Barry to change his mind.

Fat chance. Barry turned and went to the doghouse. There, he refilled his paper cup with a shot of black brew and gave attention to rig-floor activities. Daylight and the casing crew were focused on running pipe. Nice and slow, as they'd been told to do. Had to be slow, though every minute, every hour, cost Barry a bundle.

Barry took a power nap in the early morning hours, then called in his morning report. He ate breakfast, alone, while the casing crew continued picking up joints of 7-inch casing off the pipe racks. Coffee and a bacon-biscuit sandwich kept him at the table while he thought about the cement supervisor's dig about the centralizers. It wasn't like Barry hadn't supervised a couple *hundred* cement jobs himself. He had no doubt the centralizer confab would never be more than a point of contention on an invoice: mathematical model versus operations reality. And the cement would never know the difference.

Cement. Jessica's patio-hard cement, he mused. Too bad she wasn't with him so he could share his thoughts with her. He got another biscuit and refilled his coffee mug, hoping she might walk in and join him for a bite to eat.

She didn't. He pictured her sleeping in. Enjoying a lazy day. A well-deserved break after her days of running the logging job. And taking over his well. And the whole damn rig.

He took his tray to the counter.

Instant replay for lunch. Crews still running 7-inch casing. No Jessica.

* * *

Four hours later, as the last of the 7-inch casing—about 5,800 feet—disappeared below the rig floor, Barry checked fourteen hours off his schedule. The 9-7/8-inch casing—about 6,600 feet—would take even more time.

Numbers filled his thoughts. A million dollars a day. Forty thousand dollars an hour. Seven hundred dollars a minute. About a dollar every tenth of a second.

He glanced at Daylight and the casing crew, then walked casually across the rig floor to the exit stairs. He needed a pit stop, but he damned sure didn't want to interrupt anybody's flow of work. Not even for a few seconds.

* * *

Barry followed without deviation the slow-and-easy procedure for running the 9-7/8-inch casing—attached to the top end of the 7-inch casing.

Jessica had appreciated his toilet-plunger example for swab and surg. This was no different, except lost circulation while running casing could easily become a bottomless toilet.[104] Guaranteed avoidance of the problem wasn't possible, because the large-diameter casing made a very efficient plunger. It pushed mud efficiently. It sucked mud efficiently. Therefore more surge, more swab, than with a drill bit in the hole.

The casing/cement procedure called for running the casing at 30 to 40 feet a minute. Barry told Tanker his target speed was one minute per 40-foot joint, which Tanker passed on to Daylight. The running speed was not only slow, it didn't count the time it took for other necessities. Like to set the slips on the joint of casing just run. Or for the casing crew to pick up the next joint through the vee door and lift it into the derrick. Or to apply a thin layer of special pipe dope—a sealing lubricant—to the threads, and to lower the joint carefully and stab the threaded pin into the box protruding up through the drill floor. Or to tighten the connection with the massive casing tongs.

Barry pushed the stopwatch button on his Casio and counted the seconds it took to run the next joint of 9-7/8-inch casing. Sixty-two seconds—just less than 40 feet per minute.

He thought about asking Tanker to have Daylight speed up the next joint by a couple of seconds, but reckoned he might be pushing his luck with the big man.

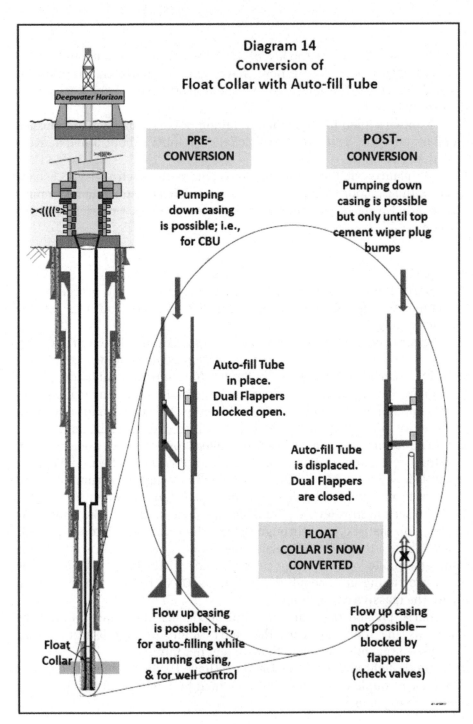

Diagram 14

---

101 See **Diagram 14**—Conversion of Float Collar (page 202 herein). The float collar is a cylindrical device (looks like just another piece of casing from the outside) installed near the bottom of casing, which contains one-way check valves (flappers) that prevent backflow of heavy cement from the annulus into the casing. BP chose a 7-inch Weatherford double-valve float collar with an auto-fill tube. The auto-fill tube blocked the flappers and allowed the casing to fill with mud as the casing was lowered into the wellbore. Then at a manufacturer-designated minimum pump rate of five to seven barrels per minute (specified as six bpm minimum for the Macondo mud), the auto-fill tube would dislodge from the check valves. This action, which must take place before the top wiper plug lands, converts the float collar, allowing the flappers to close against annular back pressure to more than 3,000 psi, whether from heavy cement or formation fluids.

102 CENTRALIZERS—mechanical devices for helping to ensure the casing is centered in the wellbore so the cement has a better chance of pushing the mud out of the way. If the mud is not fully displaced, it can form channels through the cement, through which gas and fluids can flow.

103 Reference (16)—USCG Deposition 24 August 2010, P. 250–251, testimony by Halliburton about cement modeling results, the need for 21 centralizes to minimize annular cement-mud channeling, and the well's potential for severe gas flow if fewer centralizers were used.

104 Reference (9)—From: BP—Production Casing Operations—7 X 9⅞" Interval—April 15 2010. Company man instructed to calculate swab / surge pressures for various running speeds and to select an acceptable running speed to ensure formation breakdown pressure was not exceeded.

CHAPTER 36—Doubt

Jessica finished dinner, alone, then climbed the steps to the rig floor. She sent a casual head nod to Daylight, who spit, grinned, and returned the gesture. Loud clanging noises came from the contract casing crew, heaving massive tongs, making up 45-foot-long joints, ever increasing the length of the casing string.

She nudged Barry's elbow. "Glad to see everybody's busy, but if I'd known you needed my help with a dinky seven-inch liner job, I would have come out to play earlier." She pointed her thumb over her shoulder and dropped her voice to a low alto. "You want me to chew out the casing crew? Kick their butts?"

"It's 9-7/8-inch casing. We're running the full string."

Jessica studied his face for the hint of a smile. Nothing. "Tell me you're not."

"Yes. I am."

"Barry," she said, voice calm, nails digging into her palms, "I believe this calls for an explanation. Care to enlighten me?"

"The full string is my choice. That's just the way it is."

"Actually, Barry, that's not good enough." She yanked out her tally book, slapped her open palm, forced slow and easy words. "You gave me a dozen good reasons for running a liner-tie-back combination and never once favored running a full string of casing.[105] What's going on?"

"It's faster. Two days faster. And that's two days less rig time. Which is a lot of money."

"Whoa, time and money—that's not you talking, it's your bosses. What happened to *best for the well*? *Security first*? You're supposed to be the guy in charge."

"You're right, Jessica. My bosses, like your bosses, make strategic decisions. Good strategic decisions that occasionally don't match my recommendations. Me, I'm tactical, which means I get to execute whatever big decisions are made, including what's best for the well, best for well security. And I'm damn good at my job, which is why I'm out here. So, yes, I'm in charge. My well. My responsibility. And I'm running the long string."

"Nice speech, but I have a brain and it tells me that *faster* and *cheaper* don't justify going against all you taught me, including standing up to your bosses when you think they're full of crap. Did you argue with them? Give them all the reasons you gave me?"

He stared. Said nothing.

"I guess not." She thought of her dad, what he'd say, how pissed off he would've been.

"Anything else while you're here?" Barry asked.

"Just this. As company man you owe me nothing. As my volunteer mentor, which I appreciate, you owe me nothing. But my nose tells me I shouldn't trust you when you say one thing and do another, especially without the professional courtesy of keeping me in the loop on an issue this important to the well."

He started to say something, no telling what, but she wasn't interested. She turned hard and bounded down the rig-floor steps two at a time. She was going. Going anywhere. Didn't damn matter.

A trust had been broken, and she was big on trust.

And for the first time, an ugly word—*doubt*—doubt about Barry, doubt about the well, doubt about the security of the discovery, filled her thoughts.

She didn't look back.

---

[105] See **Diagram 15** (page 206 herein): Install 9⅞ X 7-inch Production Casing. Note the casing string is installed in two sizes. The deepest portion is 7-inch diameter, which is connected to the upper portion, made up of 9⅞ casing. The entire "long string" of production casing hangs from the casing hanger (the donut), inside the wellhead. See also—Reference (12)—Macondo Casing Specifications

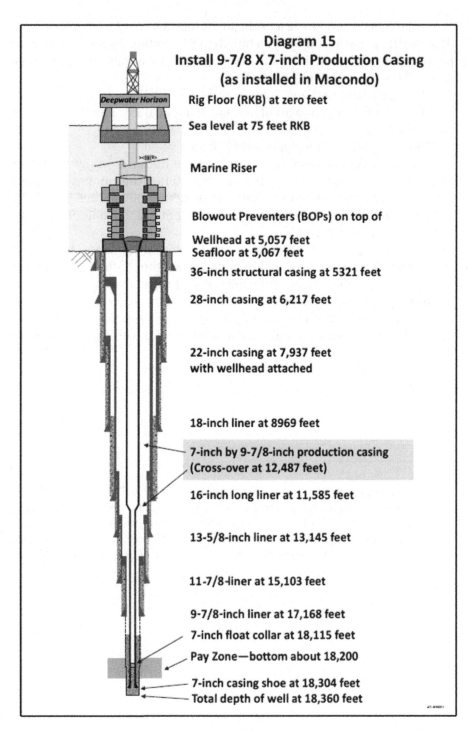

Diagram 15
Install 9-7/8 X 7-inch Production Casing
(as installed in Macondo)

Rig Floor (RKB) at zero feet

Sea level at 75 feet RKB

Marine Riser

Blowout Preventers (BOPs) on top of

Wellhead at 5,057 feet
Seafloor at 5,067 feet

36-inch structural casing at 5321 feet

28-inch casing at 6,217 feet

22-inch casing at 7,937 feet
with wellhead attached

18-inch liner at 8969 feet

7-inch by 9-7/8-inch production casing
(Cross-over at 12,487 feet)

16-inch long liner at 11,585 feet

13-5/8-inch liner at 13,145 feet

11-7/8-liner at 15,103 feet

9-7/8-inch liner at 17,168 feet

7-inch float collar at 18,115 feet

Pay Zone—bottom about 18,200

7-inch casing shoe at 18,304 feet

Total depth of well at 18,360 feet

Diagram 15

CHAPTER 37—Puzzle Pieces

Barry paced the rig floor throughout the night, except for a three-hour power nap and his morning report to the beach. He skipped breakfast. He'd pissed off Jessica, that was for sure. Couldn't be helped, though he probably should have told her earlier about the change of plans. He wondered if she had thick enough skin to ever make it as a company man.

For every day-to-day, hour-to-hour, minute-to minute decision he was expected to make entirely on his own, and for which he accepted full responsibility, there were others, from the beach, he just had to accept. Drilling directives based on future production needs—like a full casing string versus a liner and tie-back—had often produced tough pills to swallow. This was just one example. Didn't matter. Thick skin. He had a job to do, and he'd do it well.

After a number of quick stops in the galley for delectable coffee— Jessica in neither the galley nor the office—he returned to the rig floor and stayed until the casing crew picked up the last joint of 9-7/8-inch casing, about 9:15 Monday morning. From the doghouse, he paged Jessica, guessing she'd want to see the casing hanger—her donut—being lowered through the rig floor.

She called back immediately. Thanked Barry for the offer. Had just showered. Needed three minutes to get dressed.

"If you're here in one minute max, you'll see it."

"Then I'll have to miss it."

Barry hung up and checked the coffee-can coffee. Horrid. Didn't need a cup that bad. He joined Tanker and Daylight, who had already started lowering 13,000 feet of production casing, which hung from what would eventually be a mile of landing string, [106] comprised of 6-5/8-inch drillpipe. The casing hanger was already out of sight, as was the cement-wiper-plug diverter sub[107] that was needed for the cement job to follow.

"Forty feet a minute," he told Tanker, so he'd refresh Daylight's mind, as if Daylight might have forgotten. "Not a foot faster. And I want to know if we lose even a barrel of mud."

Tanker spent thirty seconds with Daylight, who spit in his cup, grinned, and gave Barry a thumbs up.

"Late breakfast?" Barry said to Tanker.

"I already ate," the big man said, "but I hate to see a grown man eat alone." He headed for the steps, Barry behind him.

Seated in the galley, Barry had sopped up a last glob of egg yolk, when a third tray slid onto the table. Jessica's fare included a single waffle and a bowl of grits. She greeted the twosome—got a "Morning, Miss Jessica" from Tanker—then left and came back with a cup of coffee.

"Light breakfast," Barry told her, testing the thickness of the ice.

"It's a repeat of what I had at midnight. It was good then and I suspect it'll be good now. What's on tap for today?"

"Three Cs," Barry said, glad to see she was no longer pouting, her skin perhaps one layer of cells thicker than he'd imagined. "Casing, circulate, and cement."

"This would be the fancy nitrogen cement you mentioned," Jessica said.

"One and the same," Barry said. "Nitrogen infused."

Tanker leaned away from Barry and turned in his seat to face him. "Nitrogen what?" he asked.

Jessica straightened in her seat, shoulders back. "Please, Barry, tell us about it."

"It's simple, guys," Barry said, wary of Jessica's smile. "Because we have so little room to play with mud weight in the open hole, we're making our cement heavy enough to keep the well from kicking, yet light enough we don't cause lost circulation."

"Now that's a neat trick," Jessica said. She scarfed a bite of waffle, then retrieved her tally book. "You talk. I'll write."

Tanker waited politely.

"We're going to do that," Barry said, "by injecting nitrogen into the cement as we mix it. The slurry will exit the cement unit all fluffed up with nitrogen bubbles. But the deeper we pump the slurry, the smaller the bubbles will get, and the denser the cement will become. By the time it leaves the casing shoe, we'll have the right weight of cement to isolate the pay zone."

"Is that a BP specialty?" Jessica asked.

"No," Barry said. "It's a Halliburton specialty. It just happens to be the right cement for our job, so they recommended it."

"Seems funny," Jessica said, "that we spend time and effort making sure we get all the trip gas and background gas out of the wellbore, then we purposefully pump a wad of nitrogen into the annulus."

"Nitrogen locked up inside hard cement," Barry said, "is no bigger risk than natural gas in a pay zone."

"But I've heard you say otherwise. Natural gas in a pay zone is considered good news. But if the same natural gas comes out of a pay zone and enters the wellbore, it's considered bad news. Likewise, I'd assume that nitrogen gas coming out of the cement, for whatever reason, would be considered bad news."

Barry sat back in his chair. "I love the way you think, but it's not mandatory for you to be a pessimist twenty-four seven."

Jessica sipped her coffee.

Barry had seen the look before, like her brain was stirring thoughts.

Her fork beat a silent tattoo against her lower lip. "I've had three good teachers for the past few months," she said, playing eyeball tennis with her tablemates. "One of the most important things you two and Daylight taught me is that everything out here is interconnected—the geology, the drilling, the wellbore. Like in a giant puzzle."

A tiny bite of waffle, then she wiped her mouth with the corner of her napkin.

Barry waited, knew more would come.

Tanker, of course, silent.

"I understand the geology pieces," she said, "but if there's a drilling piece missing, I want to know what it is, how it works, and how it fits into the big picture. So I ask simple questions and expect truthful answers that'll help me fill in gaps." She looked directly at Barry, added a big smile. "Gaps, once filled, you know I'll never forget."

Barry nodded his thanks, though he had the feeling a fuse had been lit.

---

106 LANDING STRING—Transocean's work string for Macondo was 6-5/8-inch diameter—both 32 and 40 pounds per foot. This high-strength, heavy-weight drillpipe was used as the landing string for running all casing jobs (the casing would hang below the landing string and be left behind, in the wellbore, when at the proper depth and cemented in place). The landing string, with proper connectors, was also used to run and pull the lower blowout preventers.

107 DIVERTER SUB—The diverter sub is pre-loaded with bottom and top cement wipers plugs, which will come into play during the cement job. After the cement job, when the casing hanger is landed in the wellhead and the running tool is released, the diverter sub will be pulled along with the landing string.

CHAPTER 38—A Simple Step

By about 2:00 that afternoon, the casing hanger hung securely in the wellhead, with the casing shoe suspended above bottom.[108] Still miffed at Barry, Jessica had Daylight fill her in. Open ports in the casing hanger, he told her, would allow mud to be circulated before and during the cement job. After the cement job, he added, they would install a seal assembly in the wellhead to block the ports.

Back in the unit, she made a note about Barry's plan to circulate bottoms up: *Commence CBU at __*, but left the time blank. Her laptop, plugged into the vendor phone system, gave her Internet access in the mudlogging unit. Good for music and news on the long, boring day yet to come.

She tuned in to NPR—National Public Radio—out of New Orleans, and then watched gauges and pump-stroke counters for Halliburton's and the rig's mud pumps. Nothing. No action. No pumping. No CBU. She wanted to see the pre-cement-job trip gas and quantify it, but that process wouldn't even begin until Barry told Tanker to tell Daylight to kick in the mud pumps. Barry, king of the mountain. A mountain she'd one day climb.

She saw a number of pressure spikes on the casing, but no volume pumped. Like a valve was closed. Or maybe one of the rogue gloves Barry had feared. Whatever the problem, they didn't need her help.

But two hours was too much. She again called the rig floor. Daylight told her Barry, Tanker, and Halliburton were having difficulty initiating circulation down the casing.

At 4:20 P.M., the pressure spiked above 3,000 psi, then dropped to 350 psi.[109] The mud-pump gauge, previously stuck on zero, showed 3.1 barrels per minute. Jessica noted the time and numbers in her tally book. The pump rate stayed constant—CBU in progress. Finally.

The CBU pump rate was slow, less than a third of the rate while drilling. She did the math. Annular volume—almost 3,000 barrels—at 3 bpm. Crap. Seventeen hours. Twelve hours at 4 bpm. She could read two Lee Child novels in twelve hours.

Surprised by the numbers, she converted them to clock time. Done circulating, the next morning, Tuesday, about five. And that meant she might be eating breakfast a little late.

She paged Barry and hung up.

Her phone rang. Barry. Like magic.

"Glad to see you're finally circulating," she said. "Should've called—I would've helped."

He didn't laugh.

"I'm off to a late lunch," she said, "or early dinner, if you're interested. I'll probably be up all night watching for bottoms-up. After that, I want to watch Halliburton mix the nitrogen."

"I've eaten," he said, "but I'll track you down when we've got bottoms up."

She ate alone—her plate a bouquet of green, orange, red, and white. Except the chocolate-chip cookies.

Back to the mudlogging unit. Still early evening. One of the mudloggers droned his way through an endless joke—something about a goat, a jar of olives, and a bowling ball—when Jessica noticed the barrel counter was static and the pump pressure was at zero. It'd been that way for nine minutes.

She checked the rate during the active period. Just over 100 barrels at 1 bpm, followed by 250 barrels at 4 bpm. Two and a half hours—350 barrels.[110] Hours short of bottoms-up. Hole problem? Mechanical problem?

She paged Barry. Waited. Nothing. Men's room? Didn't want to page a second time.

She called his phone in the BP office. No answer. Waited a minute. Pumps still not pumping. Called again. A dozen rings. Nothing.

Getting pissed.

Tanker would know.

"I'll be on the rig floor," she told the mudlogging supervisor. "If Barry calls or shows up here in the next few minutes, tell him I'm looking for him."

Jessica found Tanker and Daylight in the dog house working on the IADC report. "We're a long way from bottoms-up," she said to Tanker. "Pump problems?"

"No, ma'am," Tanker snapped. "Mr. Barry says we're done circulating, so I guess we're done.[111] That's all I know."

"That can't be," Jessica said. "A couple hours circulating is not nearly enough, and I know Barry. Something's wrong. Is he in the office?"

"The cement unit," Tanker said. "We just finished our safety meeting with Halliburton, and they're rigging up to cement."

Jessica flew down the rig-floor steps before she realized she had neither thanked Tanker nor told him he was dead-wrong about the cement job.

Or she *hoped* he was wrong.

---

[108] Reference (21)—Halliburton—9.875" X 7" Foamed Production Casing Post Job Report—Page 4—20 April 2010

[109] Reference (3)—BP's Internal Investigation—Final Casing Run, timeline, 19 April, P. 23.

[110] Reference (18A)—The Bureau of Ocean Energy Management, Regulation, and Enforcement (BOEM): Regarding the Causes of the April 20, 2010, Macondo Well Blowout. Issued 14 September 2011. Article V, Section C—Page 71— "Reducing (CBU) from 2760 bbls (barrels) to approximately 350 bbls (perhaps as few as 261 bbls, see Reference (1) below) . . . prevented rig personnel from examining . . . mud for . . . hydrocarbons." (Such reduction also means trip gas had been abandoned in the annulus.)

[111] Reference (1)—Congress of the United States, House of Representatives, Subcommittee on Oversight and Investigations. "BP did not fully circulate the mud (i.e., did not CBU). Instead, BP chose an alternate procedure that had been 'written on the rig . . .' BP's final procedure included circulating just 261 barrels of mud, a small fraction of the mud in the Macondo well casing and annulus."

CHAPTER 39—Production Cement

Barry conferred with four Halliburton service hands—the two regular crew, plus two experts who would manage the nitrogen-infusion process. The foursome wore matching, bright-red coveralls—four beards short of Christmas jobs at Wal-Mart.

Fast-moving boot heels rattled the steel grating.

Barry checked the source—Jessica.

Her face reflected the glow from the red cement unit. Well, maybe.

Breathing hard, she said to Barry, "My barrel counters say we didn't come close to bottoms-up. What's the problem?"

"I don't have time," he said.

"For me?" she asked. "Or for circulating bottoms-up?"

"For circulating," he said. "For you, I've got time."

Jessica's neck stiffened. "Cut the crap, Barry. What's going on with the CBU?"

He started to razz her about using an acronym but thought better of it. "We had a hell of a time establishing circulation, trying to convert the float valves. Thought for a while we might've damaged the casing. More likely it was junk in the float collar, or a formation problem, and I don't want to wash out the hole or lose circulation."

"You're leaving trip gas in the wellbore," Jessica said. "Lots of gas by my count."

Barry stepped away from the unit and turned so he could watch the cementers—the covey of bright-red coveralls—over her shoulder.

"Jessica, you logged the well, then we made a bit trip and circulated bottoms-up. That's enough. End of story."

"Wrong, wrong, wrong," she said. "We got trip gas off bottom after logging. Lots of gas, more than a thousand units. Extended duration. So, now, when you pulled the drill bit to run casing, a good engineer might assume—as you taught me—the same amount of gas got swabbed into the wellbore. It's not only still there, you just pumped a small fraction of the annular volume, which did nothing but move the gas up the hole."[112]

"And the problem?" he asked quietly, hoping she'd whack a few decibels off her attack.

She slapped her tally-book pocket. "Your words," she said, fortissimo, having missed the hint. "They're written in ink—*Never leave gas in the annulus.*[113] Have you calculated where the gas is? At what depth?"

"No. And I don't need to. The hole's clean. The well's not kicking. And we circulated with normal, even less-than-normal, pressures." He checked his watch. "Now I'm getting ready to mix cement, which you'll be happy to know will clear the wellbore of residual gas. You can stay here with me if you—"

Jessica scooted around him, stormed away, rattled the grating. She held her fists at her side, straight-armed, speed walking, like she had somewhere to go.

Barry could only watch, even after she'd disappeared around a corner. Reminded him of the last time he'd seen his ex, who'd also been really pissed.

"You ready, Barry?" The cementer's voice, deep and hollow, dragged him back into the bowels of the rig, suddenly more harsh and stark than normal.

Barry pulled out his own tally book and stepped to the cement unit, having memorized every step of BP's formal operating procedure for cementing the long string. Until that time, all had gone well—other than the pressure necessary to initiate circulation, which had been higher than normal. Other than time wasted, the event had been anticlimactic, though he looked forward to bumping the top cement wiper plug, which would confirm the casing hadn't been damaged. Because if it had, he'd spend the next couple of days having to repair the problem.

"Let's finish the numbers," Barry said to the Halliburton cementers, all too aware that for the rest of the procedure to go as planned, they had to get it right too.

"Check," said a man in red.

A dozen *checks* later and Halliburton started pumping—just before 8:00 P.M., Monday evening. They started with a slug of base oil, followed by 14.3-ppg spacer—designed to flush mud from the annulus ahead of the cement.

Barry called Tanker. "We've mixed and pumped just enough cement to clear the lines. You can launch the bottom dart.[114] I'll call again when we're ready for the top dart."

"Miss Jessica's here," Tanker said. "You want to talk to her?"

"Can't now. Call me back when you confirm the bottom dart's launched."

The entire cement job depended on the darts and their corresponding cement wiper plugs, which would chaperone the cement down the wellbore, to and through the float collar located just below 18,000 feet.

Barry waited. Paced. The minutes expensive.

Tanker called back. "Got a clean launch, Mr. Barry."

As soon as Barry told the Halliburton team, they resumed mixing. The lead slurry was mixed to 16.7-ppg. Heavy stuff—the kind that would make a good hard patio.

Then, while one team mixed additional cement, the nitrogen-infusion experts did their thing—adding tiny volumes of pressurized nitrogen into the mixed cement slurry, producing the foamed tail cement. Barry took notes. One digital meter confirmed uniform slurry densities. Another, barrels of slurry. He checked pressure gauges and nitrogen gauges, then went back and checked densities, then back to the stroke counter. Wanted everything right.

As soon as the nitrified cement had been mixed and pumped, the team switched back to making heavy cement—16.7-ppg—designed to fill the casing between the float collar and the shoe. Shoe cement—more of the good, hard stuff, with plenty of retarder so it wouldn't set up too soon.

With all the cement mixed, Barry instructed Tanker to launch the top dart.[115]

Seconds later, Tanker confirmed a good release.

Halliburton did the finesse pumping. More spacer. Then just enough mud—at four barrels per minute max, Barry advised—to push the top dart and the cement down the 5,000 feet of landing string. The dart arrived at the top cement wiper plug on schedule. Halliburton pressured up just enough to free the wiper plug from its container, which meant the cementers were done pumping.

Barry called Tanker. "It's all yours," Barry said, the potential for lost circulation still heavy on his mind. "I'll come join you, but in the meantime don't displace at more than four barrels per minute. I don't want to blow the bottom out of the well, and we've already converted the float collar."

Barry and the cementers compared notes—volumes, rates, pressures—for the record. On the way to the rig floor, Barry visualized the cement. Reminded him of a small train on a journey—base oil, spacer, bottom wiper plug, lead cement, nitrified cement, shoe cement, top plug, more spacer—chugging along, down through a 3-1/2-mile-deep vertical tunnel, getting ready to go around a corner and pull into a station. The annulus. [116]

Barry watched for lost circulation. Glancing at gauges. Ensuring no pump ever strayed above what he, and his bosses per the casing/cement program, considered to be the maximum acceptable rate—four barrels per minute.

Even if the well didn't kick, loss of circulation could trash the cement job. In that case he'd have to run a cement bond log to locate and quantify potential cement problems in the annulus, followed by a cement repair job—two or three days minimum—which he hoped wouldn't be needed. He checked with Tanker, who checked with Daylight, who checked with the mudloggers a number of times while the slurry was being slowly displaced down the casing. All parties reported mud returns throughout the period—no losses.

Near the end of the five-hour job, with the cement approaching the float collar, Barry stepped closer to the pressure gauge on the casing, watched for the expected spike in pressure. He'd kept track of the number of barrels of mud pumped. Counted down.

"Bottom plug bumped," Tanker said.

The pressure spike and its quick return to normal pump pressure confirmed the cement was passing through the bottom plug and through the float collar, on the way to the annulus. Now, one goal—don't over-displace. He'd seen wells where all the displacement barrels had been pumped, then more had been pumped, then—

"Top plug's down," Tanker said, his voice over Barry's shoulder, into a phone, on line with Halliburton.

The top plug—the one that had pushed all the cement down the casing—had bumped the float collar on schedule. Even as Daylight stopped the mud pumps, the casing pressure jumped to almost 1,200 psi, as expected. The pressure held steady—no leak past the top plug. No leak in the casing.

Barry took the phone. "You ready?" he asked.

"Yes," the cement supervisor told him. "And Jessica's down here—want to talk to her?"

"No," he said. No time, he thought. "Bleed down the casing pressure. Get a barrel count, and we'll monitor ballooning and backflow."

Barry wanted no backflow. Backflow would mean the pressure of the cement slurry in the annulus was somehow leaking past the flapper valves and reentering the casing. Not a likely scenario. Especially since the pressure outside the casing, even with the heavy cement in the annulus, was calculated to be no more than 50 psi higher than the pressure inside the casing.

The cementer reported by phone he'd bled the pressure from 1,200 psi to zero and got five barrels of mud back—ballooning—as expected.

Then a finger-size trickle. Then nothing. Five barrels—typical ballooning of the casing under pressure.[117]

The flapper valves held. No backflow. Cement job complete. To be left undisturbed for two or three years, maybe more.

Barry checked his watch, made a note in his tally book: *12:36 A.M., Tuesday, 20 April.*

Perfect, he reckoned. The start of a good day.

Visitors' day.

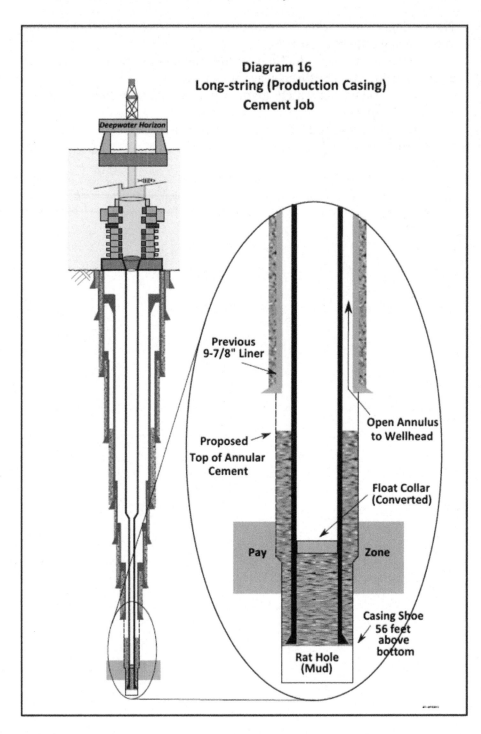

Diagram 16
Long-string (Production Casing)
Cement Job

Previous
9-7/8" Liner

Open Annulus
to Wellhead

Proposed
Top of Annular
Cement

Float Collar
(Converted)

Pay   Zone

Casing Shoe
56 feet
above
bottom

Rat Hole
(Mud)

Diagram 16

---

[112] Reference (11) BP's *Deepwater Horizon*— Accident Investigation—Static Presentation—Slide 6 of 39. BP statement: "No evidence that hydrocarbons entered the wellbore prior to the cementing operation."

[113] Reference (9)— Macondo Well Casing—Long String—Cement—Ops—BP-HZN-CEC017621. Requires (company man to) circulate volume of work string plus casing capacity prior to cement (this to clear the casing of debris). CBU, by definition, requires the entire annular volume to be circulated to the surface to clear gas from annulus. See also Reference (12) Macondo casing specifications.

[114] BOTTOM and TOP DARTS—A Cement Head installed on top of the landing string (at rig-floor elevation) contains two *darts*. Likewise, near the seafloor, the bottom end of the landing string is attached to the top of the casing hanger with a *running tool* and a Diverter Sub. The Diverter Sub contains two pre-loaded cement wiper plugs. Before cement is mixed, the bottom dart is launched from the Cement Head and travels down the landing string ahead of the cement. When the bottom dart reaches the Diverter Sub it picks up the Bottom Cement Wiper Plug and they travel down to the float collar (as bottom cement wiper plug wipes the casing clean). When the last of the cement is mixed, the top dart is launched and picks up the top wiper plug from the Diverter Sub, and together they follow the cement down to the float collar. When the bottom cement plug reaches the float collar, increased pump pressure of about 1000 psi ruptures a "burst tube" that allows the cement to pass through the float collar. The top cement plug is solid—it does not rupture when it reaches the float collar and therefore holds pressure from above (enabling the casing to be pressure tested from the inside).

[115] Ref (21)—Halliburton—9.875" X 7" Foamed Production Casing Post Job Report—Page 2. April 20, 2010. Started with 7 barrels of base oil, followed by 72 barrels of 14.3 ppg spacer. Then the lead slurry—5 barrels of 16.7 ppg Class-H cement—followed by 39 barrels of foamed tail cement. The shoe cement—7 barrels of 16.7-ppg Class-H to fill the 189 feet of casing between the float collar and the shoe. Twenty more barrels of spacer followed.

[116] See **Diagram 16** (page 218 herein). Long-string (Production Casing)—Cement job . With production-casing cement in place, note open annulus from the top of the cement to the wellhead, and rat hole under the bottom of the casing.

[117] BALLOONING—Expansion of a closed vessel when an internal pressure is applied—like for a child's balloon. When the pressure inside a cemented string of casing is increased above zero by injecting, for example, a small volume of mud, the ballooning effect should return the same small volume of mud when the pressure on the casing is bled back to zero.

CHAPTER 40—Positive Test

20 April 2010—12:36 A.M.
21h:13m:00 to Zero Hour

Immediately after testing for backflow, Barry, Tanker, and Daylight followed another exacting procedure. With the casing already hanging in the wellhead, they slacked-off the final few inches on landing string and installed the seal assembly, which blocked the circulating ports between the casing hanger and the wellhead. The same wellhead installed by the *Transocean Marianas* five months before, even as Mother Nature had birthed Hurricane Ida.

The weight of the cemented full string of casing—Barry had noted 450,000 pounds of hook load—bore down on the two metal-to-metal seals built into the hanger devices.

With another milestone behind him, Barry took a deep breath. So much more to do.

A quiet figure—Jessica—stood by the rig-floor exit to the mudlogging unit. She had been there, arms crossed, twenty minutes, but she had said nothing. At least not to Barry.

He joined her. "What we just did," he said, "was set the casing hanger in the wellhead. If you recall, you actually touched the seal profiles when the wellhead was on the pipe rack. We called the casing hanger the *donut*."
118

"I do remember," she said. "It's been a long time coming—thanks for showing me."

He started to raise his hand for a high-five, but something in her voice—a determined quietness—warned him off.

Didn't matter. He had work to do.

Barry, Tanker, and Daylight closed the BOP, then the Halliburton crew applied pressure to the annulus on top of the casing-hanger running tool to further seat the metal-to-metal seals.

Still off to the side, Jessica wrote notes in her tally book. He wondered if what she'd seen would be considered an SLD—science lesson of the day. He wasn't about to ask.

Daylight worked with contract specialists and released the casing running tools at 2:50 A.M. A positive pressure test on the hanger proved satisfactory.

Thereafter, Tanker and Daylight, in control, directed the drilling crew to commence pulling the 5,000-foot-long landing string, a four-hour job.

Barry planned ahead. Time for a nap. Up at 6:30 A.M. Call the office.

Then breakfast with Jessica. Well, maybe.

––––––––––––

[118] See **Diagram 17** (page 222 herein)—(Subsea Wellhead, Casing hanger, & Lockdown Seal Ring). Though the casing-hanger lockdown seal ring (LDSR) had not yet been installed, there was an expectation that the casing-by-casing annulus was now sealed, forever.

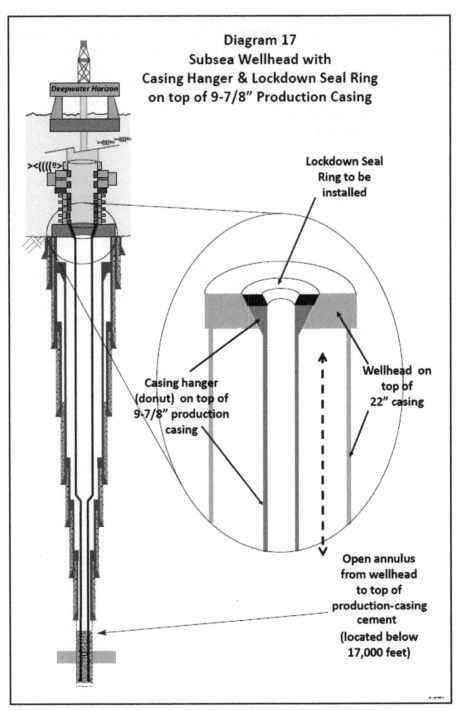

Diagram 17
Subsea Wellhead with
Casing Hanger & Lockdown Seal Ring
on top of 9-7/8" Production Casing

Lockdown Seal
Ring to be
installed

Casing hanger
(donut) on top of
9-7/8" production
casing

Wellhead on
top of
22" casing

Open annulus
from wellhead
to top of
production-casing
cement
(located below
17,000 feet)

Diagram 17

CHAPTER 41—The Itch

20 April 2010—3:30 A.M.
18h:19m:00 to Zero Hour

Jessica took a couldn't-sleep shower. Lots on her mind. Barry's out-of-character behavior. Her first day back in the office fast approaching. The attorney—whatever the hell he wanted. And her mom. Always her mom, dammit.

She sat on the bed, toweled her hair, and dialed the number she'd called a hundred times before—during normal hours. The number that never answered—during normal hours.

A middle-of-the-night feeble voice answered. It'd been seven months since anybody had answered a call from Jessica.

Jessica identified herself, twice.

"What do you want?" Hacking phlegm. "What time is it?"

"Mom," Jessica said, "it's three thirty and I'm desperate. I've got an appointment with an attorney next week, so you and I need to talk. We need to talk about Daddy."

"Why are you bothering me?" her mother snarled. "Always calling. You've done all the bad to this family you'll ever do, and I've heard all I need to hear about appointments and attorneys."

"Mom, it's almost two years—we need to resolve this. The same thing would've happened regardless of who was behind the wheel. Even you."

"Me? I wasn't even there. What are you saying?"

Jessica had faced her stepmother's defensive wall before. Debates with deaf ears. Phone calls unanswered. E-mails ignored. Visits out of the question. "You and Sissy and I can mourn Daddy forever, but I need you too."

"Oh, of course—it's all about you. My husband's dead, even after I forbid you to drive, and I'm supposed to feel *your* pain? What about *mine*? He's gone, and I'm alone. You think I'm having fun? Traveling? Playing cards? Hell, no. I can't eat or read a book, I'm sick to death of TV, and my bed is as cold as death. How's that for pain?"

"I hurt too," Jessica said. "Every day. And I understand your loss as much as anybody in the world. I'll come home this weekend, if you promise to be there. Just to talk."

Silence. Then, "What are you up to, calling in the middle of the night. If it's about the insurance money, you're not getting a penny."

"No, Mom, it's not about the money, or the house—any of that stuff. It's about losing Daddy. We all lost him. But now you're working hard, shutting me out, ensuring we lose each other."

"No. You did that all by yourself."

Jessica held back the words she so desperately wanted to scream. "Please tell Sissy what I told you, that I want to come home. Have her call me, send me an e-mail."

A beat of silence.

"If you want me out of your lives, I'll understand. But I won't quit loving you. Not ever."

Her mom said nothing.

Jessica waited. Checked her phone.

Line disconnected.

Seated on the edge of her bed, she thought not of her father, or her mother, or her sister, but about Ranae Morgan, who during those terrible post-death months had held Jessica and told her it was okay to cry.

She'd cried then, but not since. Not even as she put away her phone and pulled back the covers. Though her eyes felt moist when she rubbed them, while planning her trip. To Colorado Springs. Saturday morning. Back Sunday night.

* * *

Jessica spent Tuesday morning in the mudlogging unit. While two mudloggers cleaned up their files and equipment, she gathered her loupe, pens, pencils, an extra ear bud. Her wireline logging reports—the proprietary well logs—were already in town, in the Exploration Department. All other mudlogging and rig-floor drilling data continued to be sent wirelessly and automatically, 24/7, from the rig to the BP office, as it had been since the beginning of the well.[119] She expected little critical data to be generated before releasing the rig, since all they had to do—for the record—was the negative-pressure test, whatever that was.

She typed and sent an e-mail to her boss: As requested, I'll be in the office Monday April 26 for meeting with HR. Prefer afternoon. Please advise.

"You going home today?" the senior mudlogger asked.

"No," Jessica said. "I want to see them pull the BOP stack, which I'm guessing will be sometime tomorrow."

"We'll go this evening," the young mudlogger said, "if we get a ride."

"Don't even think about it. The well's not secure." She pointed to a pit-level monitor. "They're running pipe now, and you guys are supposed to be monitoring returns."

"We are, but they're fixin' to do some pressure tests and set a cement plug. As soon as that's done and they bang on the cement with a drill bit, we're out of here."

"Okay by me," Jessica said, "but in the meantime, you still work for BP, for Barry, which means your job doesn't change. You monitor everything."

The mudlogger pointed to a multi-colored live display of gauges on the wall. "Roustabouts and the mud guys are already messing with the pits, cleaning 'em for the next well. Plus they're getting ready to pump all the mud down to the workboat. In a couple of hours there won't be anything to monitor."

His words sunk in. The end of the well. It'd been a long time coming.

On her way to her quarters, Jessica stopped and visited with two Schlumberger engineers headed to the helideck. They were the cement-bond-log specialists.

"The company man, Mr. Eggerton," one said to Jessica, "told us the cement job went well. Full returns, therefore no CBL. Which means we're on the way to the house."[120]

"That's too bad," Jessica said. "Whether or not there's good bond, I wanted to see the log showing nitrified cement across the pay." Across *Ranae Morgan's* pay, she didn't say. The log itself—looking like an EKG for a really sick patient—would've made a nice souvenir for Blasé Ranae. Jessica could think of nobody else who'd want a copy.

"You may get to someday," the other CBL expert said. "There's little correlation between full returns and good cement jobs in deep, hot wells."

The other logger added, "And even less with lost returns."

Jessica shook hands with the departing team, then pulled her tally book from her pocket. She jotted a short note about cement and the bond log, then wrote *Barry* and circled the word.

In her quarters, she packed. She couldn't nap, because she was pissed at Barry for not including her in either the casing decision or the CBU decision. He'd been forthright that he could live with the less-desired long string of casing—he just had to make sure the cement job was perfect.

But leaving gas in the annulus? *That* was a different story. Early on, months ago, he'd been adamant it was a major dumb-ass thing to do, but hadn't really explained why. And then he'd done it himself. There had to be reasons she didn't understand. Two sets of reasons. One, why it was bad to do it. And two, why it was okay to do it when he did.

She popped off the bed and stepped to the mirror. "Get over it, girlfriend—you've got work to do."

Her last well before graduate school. She needed to finish the puzzle while she had a chance. A chance that might not recur for a number of years.

She opened her tally book and flipped through three months of notes and numbers she'd collected for selfish but important reasons. But with the well almost done, she'd made no recent entries about the stuff Barry was doing. Nothing about nitrogen cement. Or cement equipment. Or the abandonment procedure.

She checked her watch. Just after noon. She pocketed her tally book, fluffed the hardhat crinkles out of her hair, and took a deep breath. Didn't matter that she had no balls, she was damn sure ready for whatever the day would bring.

* * *

She found predictable Barry in the galley, getting ready to wrap himself around a predictable burger with an inch of red meat bleeding between the buns. His face lit up when she joined him, but the glow browned out in seconds.

"What's happening on this ominous day?" Jessica asked, ready to keep the conversation light, just mentor and student.

Barry swallowed. "Ominous? How so?"

"NPR trivia—the twentieth day of April. Hitler's birthday. Columbine High School. To lots of people, the day's a black cloud. Maybe that's why the VIPs are coming out here, psychobabble reassurance, making sure we're okay."

Barry said nothing.

"Sorry I got you off track," she said. "You were probably getting ready to tell me what's happening."

Barry wiped pink grease off his chin. "I'm sure I was. We just finished testing the casing. In addition to bumping the plug with 1,200 psi, we did an official test to 250 psi, to make sure nothing fell apart, then to 2,700 psi. Thirty minutes—held like a jug."[121]

"The casing?" She retrieved her tally book and a pen, ready to mine gold in the galley. "Isn't it rated for like a gazillion psi? Twenty-seven hundred seems a pittance."

Barry assumed his tech-mode lecture position—hands folded, on the table. "The burst rating of the 9-7/8-inch production casing is more than 13,000 psi, but we're testing other things too."

"Like?"

"Well, since we tested on the underside of the blind-shear rams, we know they're working."

"Might they not?" she asked, surprised the BOPs would be on the test list.

"Always good to test the stack," Barry said. "Never know when you might need it. The test also confirmed the integrity of the casing hanger and seal assembly. If the test pressure had leaked into the long-string annulus, even a small pressure increase could have knocked the bottom out of the well."

"I'm truly happy the hanger tested."

Barry paused, then a simple nod. He seemed . . . wary.

Jessica scolded herself. Behave. Just take notes.

Barry said, "We also tested the top wiper plug from above."

Jessica wrote *Wiper plug* in her notes. "I don't believe I ever met one," she said. "What's it wipe?"

"We used wiper plugs on every job out here, starting with the 18-inch casing. Sorry you missed them. Very educational."

Jessica accepted the well-deserved dig, gave half a nod.

Barry continued. "The wiper plug's a short cylinder that fits inside the casing. It has hard rubber flaps that expand and contract depending on the diameter of the pipe. We use a dart to release the top wiper plug into the casing immediately above the last of the cement we just pumped. With additional pumping of mud, the plug and the cement go down the casing together. The plug keeps the cement from being contaminated by the mud that's pushing it down the hole."

Jessica had Barry where she wanted him—eating red meat and answering questions. She wrote *Cement contamination*, then drew a stick picture. "Where's the stuff go that the wiper wipes?"

Barry leaned forward, scrunched his face, chewed.

Jessica answered before he could ask. "If the wiper wipes stuff off the casing walls," she said, "seems like the *stuff*, whatever it is, ends up below the wiper."

Barry fidgeted in his seat. "There are two wiper plugs. One before the cement, and one that follows the cement. The bottom wiper plug clears the casing walls of mud and debris so the cement behind it doesn't get contaminated. The debris ends up in the annulus, buried in the cement.

The top wiper plug has a single job, and that's to push the cement down the hole ahead of the mud."

"Ah, but it wipes too," Jessica said.

"There's nothing to wipe. The lower wiper plug got rid of all the mud and junk."

Jessica added a few notes to her stick drawing, then looked Barry square in his face. "I like it when we study together. Thanks for being so patient."

He nodded. A guarded look she'd seen before. Smart man.

"So the old layer of mud and dirt inside the casing," she said, pointing to the drawing, "gets replaced by a new layer of cement. Like cake batter in a clean bowl. Really sticks. Wouldn't the upper wiper plug pick up that small layer and push it ahead? Picking up a little bit more each foot of the way? Through the landing string, then through 13,300 feet of casing?"

"Sure," Barry said, "but it's still cement."

Jessica flipped back a page in her tally book. "After the foam cement, the last cement you pumped—about seven barrels according to Halliburton—was for the casing shoe. They said it was heavy cement, no foam, so it'd set hard as a rock."

Barry nodded. Said nothing.

"But if that's the case, then any foam cement on the walls of the casing got picked up by the upper wiper plug and mixed into the seven barrels of shoe cement."

She hesitated and gave him a hard stare because he needed it.

"In fact," she added, having learned a lot from the Halliburton guys, "the entire last seven barrels could have ended up being a mix of shoe cement and foam cement, or even all foam cement."

Barry's head bobbled, a mix of shake and nod. "Maybe a trace, but not enough to worry about."

Trace. A weasel word. But she had too many other good questions that needed answers. She wrote: *Contamination—later.*

"By the way," she added, "how did the cement get past the first wiper plug—the bottom one?"

A beat, while he chomped another bite of medium-rare comfort food. "By design," he said, "the lower plug did its job wiping, but it had a short lifespan. When it reached the float collar, which it couldn't go past, an internal disk in the plug ruptured and the cement continued its way *through* the plug and down the hole. And before you ask, the upper plug has a solid center—it won't rupture."

"Halliburton told me the same thing," she said, "and that's how you got the 1,200-psi test when the plug landed."

"Which was great news, because it proved we hadn't damaged our casing when we initiated circulation. All our internal pressure tests on the casing will be on top of the upper plug, which will be there until it gets drilled out—in a few years."

He again checked his watch.

"Before you go," she said, "what's next? It's only hours before we're done, and I don't want to miss anything."

"We did the high-pressure test," Barry said, "and now we're running drillpipe to about 8,300 feet.[122] We'll fill the drillpipe with seawater and do a negative-pressure test on the casing and casing hanger. As soon as that's done, we'll start displacing—"

"Wait," Jessica said. "Tell me about the negative test, and what it's for."

"After tomorrow, the BOPs will be gone, the riser will be gone, the heavy mud will be gone, the rig will be gone, and seawater will fill the wellbore down to the cement plug we haven't yet placed, at about 8,300 feet."

"Wow, it's coming quick," Jessica said. Then she noticed they were nodding in unison, synchronized, like parakeets in a pet store. She broke the connection.

He did too. "The negative-pressure test is mandated by the Minerals Management Service to simulate abandonment.[123] We have to—"

Jessica held up her hand. "Gotta stop you. Negative pressure? Like the opposite of positive pressure? Which means it sucks?"

Barry leaned forward, all serious, like sharing a secret. "Here's the deal. Three scenarios. First, we currently have 18,300 feet of heavy mud in the hole, which makes the well secure."

She nodded to keep him going.

"Now let's skip number two for a minute and go to three. When we leave tomorrow, the abandoned well will have a cement plug near 8,300 feet. Above the plug—8,300 feet of seawater. Below the plug—10,000 feet of heavy mud. If you use your mud-weight equation, you'll see that replacing 8,300 feet of heavy mud with seawater will drop the wellbore pressure by about 2,400 psi. We need to know if that's going to be a problem before we actually dump mud and pull the riser and BOPs."

Jessica pulled out her calculator. "Give me a sec, I need to do this." She punched numbers. "That'll reduce the equivalent mud weight at 18,300 feet from 14.2 ppg to 11.7 ppg." She let the numbers sink in. "Good it's only a test and that you've got the pay zone isolated."

"And that's exactly why scenario number two is the test itself. The negative-pressure test is designed to *simulate* the abandoned condition of the well. To do the test, we need 8,300 feet of open-ended drillpipe full of seawater. The drillpipe will have pressure on it because the heavy mud outside the pipe will U-tube, or push, against the light-weight seawater inside the pipe. As soon as we close the BOPs around the drillpipe, we'll bleed that U-tube pressure to zero, which will put a 2,400-psi suck on the entire wellbore below the BOPs. If the drillpipe pressure holds at zero for a half hour, it'll hold forever, and the simulation's done. Then, as our last steps, we set our cement plug at 8,300 feet and install the lockdown sleeve. The well's secure."

Jessica continued writing, then looked up. "Since the negative-pressure test is a *test*, it's looking for *what*?"

"Leaks. With that much pressure drop, even the smallest leak will be obvious. Like a high-pressure kick."

"And if something leaks during the test?"

"We fix it. We get rid of the drillpipe and the seawater, make sure the wellbore's still full of heavy mud, and fix the leak. Whatever we have to do. Can't leave a leak behind. Period."

Jessica finished her note. "You mentioned a cement plug at 8,300 feet.[124] But when you pulled the BOPs on the *Marianas*, you said Minerals Management required the cement plug just below the seafloor. Why are you going so deep this time?"

Barry drained his coffee mug with a flourish, which meant he was frustrated, or running out of time, or patience. She wondered if he'd again check his watch.

He checked his watch. "It has to do with the casing-hanger lockdown sleeve," he said. "I need about 100,000 pounds of weight hanging below the tool in order to lock it into the casing-head profile. I'll get the weight by hanging about 3,000 feet of heavy drillpipe below the sleeve, which defines the clearance needed between the wellhead and the top of the cement plug."

Jessica didn't want to miss out on Barry's remaining decisions, all seemingly important. Like his don't-tell-Jessica decisions for the long-string casing and not circulating gas out of the annulus. "If we install the lockdown sleeve before the negative test," she said, voicing her willingness to participate, "we'd have the room we need for the hang-down pipe, and we could do the cement plug later, near the mud line."

Barry wiped his chin, hard, with vigor, like it itched.

She reckoned she might be the itch, making suggestions when he and his boss had already agreed the steps to be taken.

Barry fisted his burger-stained napkin and dropped it on his empty plate, then leaned forward in his chair and lightly rubbed his temples.

"Jessica, I enjoy answering your questions more than you know, but it's about to get really busy on the rig floor, testing the casing, displacing the mud, all the other little steps. Let's plan to get together tomorrow after we pull the riser, riser package, and BOP stack."

Like *go out and play while I do important stuff.* "Or, here's another idea," she said, hoping her face didn't look as hot as it felt. "You do your work, and I'll follow you and take notes. I'll even take notes when you're in the galley visiting with the show-and-tell VIPs. That way, when we get together tomorrow, I'll know what questions I need to ask you." She paused. "And vice versa."

Barry squirmed like a kid in a dentist's chair, then stood. "That'll work, but you'll need to run to keep up with me."

He returned his tray to the service counter and didn't look back.

Jessica refilled her coffee mug and picked up a chocolate-chip cookie. She looked toward the door he'd already exited.

Keep up with you, Barry Eggerton? Like a thistle in your sock.

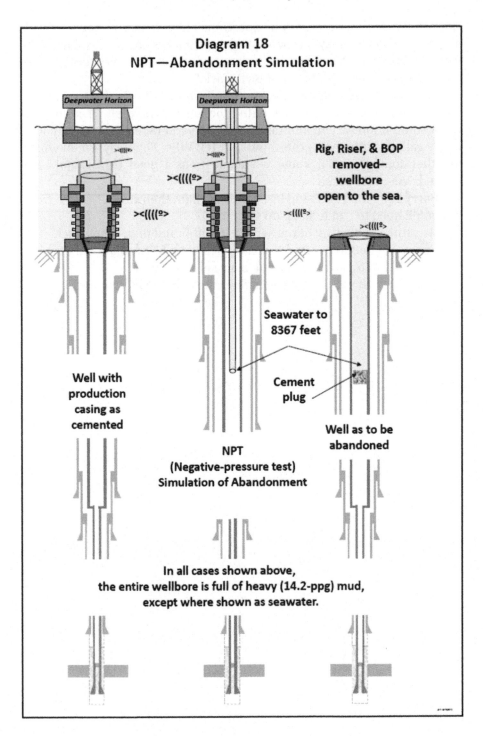

Diagram 18

[119] Reference (18A)—BOEM, 14 Sept 2011—BP and its operating partners (in offices) had access to the real-time (offshore) data through Insite Anywhere, BP's electronic data system owned by Halliburton. The system provides (and records) real-time flow-in and flow-out data, gas analysis data, ... pressures, and other drilling data. P. 96.

[120] Reference (10)— Schlumberger.MC 252. Timeline. "BP contracted with Schlumberger for a crew to be available to perform a cement bond log ... should BP request those services." At about 7:00 A.M on the morning of April 20, BP advised the Schlumberger crew their services would not be required for a CBL.

[121] Reference 3: BP internal investigation, pp 82–83: 20 April, 11:17–11:52 A.M.

[122] Reference (9)— Macondo Well Casing—Long String—Cement—Ops—BP-HZN-CEC017621—P. 8— Drillpipe for the test was to be 5-1/2 inch, with 1000 feet of 3-1/2-inch stinger, open ended, run to 8367 feet for the NPT. See also Reference (18C) Appendix B, John Rogers Smith, PC LLC, submitted to the MMS 1 July 2010, P. 7–8. " ... based on BP presentation ... " ran 821 feet of 3-1/2, 3696 feet of 5-1/2, and 3850 feet of 6-5/8-inch drillpipe to 8367 feet.

[123] See **Diagram 18**—(NPT—Abandonment Simulation), page 232 herein. Simulation of abandonment is necessary to prove integrity of the production casing and wellhead assembly. Or, conversely, to disprove casing integrity, which would lead to repair of whatever's leaking. Reference 18C—P. 17—cites Code of Federal Regulations 30CFR250.442(k)—(requirements for a subsea BOP system)—"before removing the marine riser, you must displace the riser with seawater. You must maintain sufficient hydrostatic pressure or take other suitable precautions to compensate for the reduction in pressure and to maintain a safe and controlled well condition."

[124] Reference (1)—House Subcommittee Letter to BP's CEO—14 June 2010. . . . seeking permission from MMS to install the final cement plug on the well at a lower depth (near 8300 feet) than previously approved. If permission was granted, BP's plan was to displace the drilling mud in the riser with seawater and install the cement plug prior to installation of the casing hanger lockdown sleeve. BP's alternative plan, if MMS did not approve the proposed depth of the final cement plug, was to run the lockdown sleeve first, before displacing and installing the cement plug at a shallower depth (near 5300 feet). On April 16, a BP e-mail: "We are still waiting for approval of the departure to set our surface plug. ... If we do not get this approved, the displacement/plug will be completed shallower after running the LDS." (LDS stands for the lockdown sleeve.)

CHAPTER 42—The Plan

20 April 2010—12:45 P.M.
9h:05m:00 to Zero Hour

After a short nap and a shower, Barry found Jessica on the rig floor, huddled with Daylight, chatting like old buddies. He watched their backs while he bothered with a cup of ugly coffee. She'd ask Daylight a question, then he'd spit, talk, sip, laugh, talk some more, then go back to spitting, then yak a little more. Then she'd ask another question.

Barry stayed out of the way until Jessica spotted him.

She thanked Daylight, lightly touched him on his shoulder, and joined Barry at the coffee pot. She hoisted her empty cup. "Brew's great today," she said. "Two shades better than wonderful."

"Get what you need?" he asked.

"Coffee?"

"No. From Daylight."

"Of course. If all other sources of credible information fail, I ask my buddy."

"And . . . what?"

"He's running drillpipe. It's open ended, nothing on the bottom. We'll get the depth as soon as he's done. Also, one of the Halliburton guys called a few minutes ago. They're ready for the NPT—the negative-pressure test."[125]

"Pardon me if I'm not surprised by my own schedule."

She slapped her tally book in her open palm. "Wasn't trying for surprise. Just wanted you to know your well-mentored student is up to speed on what's going on. How'd I do?"

Daylight—leaning and listening—was too obvious for Barry not to notice.

Barry moved and turned so his back was to the driller. "You did okay, but the negative-pressure test is just the beginning of all we have to do."

Jessica scanned a page in her book. "Looks like you and Daylight agree—lockdown seal ring, NPT, cement plug, displace riser. Then we start pulling the big stuff. The riser and—"

Barry held up one finger, waggled it. "That's the trouble with rumors," he said. "You've got it all backwards."

She grinned. "You're going to pull the BOPs first?"

He tossed his coffee cup in the trash can. "I've got no time for games. No time for bullshit. You want to hear my schedule, ask. Otherwise, I'm busy."

Jessica's lips tightened. "I'm asking."

"Then I'll cut to the chase. I've got a better chance of a good cement plug if I set it in seawater, rather than mud. *Then* comes the lockdown seal ring. As soon as we pull the stack, we're out of here. Tomorrow—maybe midday."

She stared at him, wordless.

He leaned an inch toward her. "If you need to take notes, I'll say it again."

"I got it, but—"

Tanker joined them. "Private party?" he rumbled.

"Not at all," Barry said, turning away from Jessica. "If you'll check on the workboat, I'll make sure Daylight and Halliburton are lined up. We'll start pumping as soon as you're ready."

"Either of you need me on the rig floor?" Jessica asked.

"Nothing to see here," Barry said, answering for Tanker, too, as the big man headed down the rig-floor steps. "We'll be pumping fluids and watching pressures for the next few hours."

"That's good to know," she said to Barry, "because I'll be around, too, watching pressures and counting barrels. The mudloggers and me, you can be sure of that."

Barry tossed her a nod, then picked up the rig-floor phone. Speed dial—Halliburton. By the time he hung up, Tanker was on the main deck, looking up and chatting with the crane operator perched thirty feet above his head.

Barry waved, hailed Tanker. Then, "You ready?"

Tanker pointed over the side of the rig, down toward the water, and gave Barry a thumbs up. Tanker's unspoken message: The MV *Damon B Bankston*, BP's massive contract workboat, was ready to commence receiving the first of 4,500 barrels of oil-based mud directly from the mud pits.[126]

"Let's do it," Barry said to Daylight.

Driller Daylight kicked in the rig's mud pumps.

Barry pulled out his tally book and noted the time they'd started pumping mud to the workboat—1:28 P.M.—as Tanker arrived back on the rig floor.

"Captain on the *Bankston* will watch his gauges," Tanker told Barry. "We call on the marine radio, and he'll let us know how he's doing, how much mud he's taken on."

Within minutes, Daylight pointed. The first of two scheduled VIP choppers was on approach to the helideck. Tanker excused himself to meet the visitors—executives from Transocean and BP—and to welcome them to the *Deepwater Horizon*.

When Tanker returned, Barry found and joined the BP visitors—one of his bosses and one of his bosses' bosses. Barry, as had Tanker, spent a quarter hour with them before he went back to work.

Barry expected a low-key evening. It wasn't like the VIPs would set up a giant banquet table in the galley and have a podium for speeches. Instead, they'd go through mandatory safety orientation, eat, pose for pictures, and tour the rig. Then they'd hold safety and operations meetings with senior rig personnel, himself included.

The BP executives would not use the leadership visit to make a big deal about the discovery, since it was a need-to-know event.

But the Transocean executives would congratulate all the Transocean employees for the seven notable years without a lost time accident aboard the *Deepwater Horizon*. Tanker Forster and his crews would deservedly be listed among the non-LTAs, though they represented only a small fraction of the employees who had safely worked the rig during the record-breaking years. Seven years. No broken bones. No amputations. No deaths.

Barry wondered how many of the same guys would still be working the rig after eight years. Or ten. Twenty years without an LTA? Possible. But not likely.

* * *

Barry had hosted such leadership visits before, all congenial, but none on the last day of a well. Maybe his bosses had planned it that way—a busy time for him, but not much for them to see, giving them freedom to roam the rig on a casual basis.

He hadn't been told if the VIPs would depart after dinner, or after the seven o'clock meeting, or if they'd tell war stories late and spend the night. The headcount of 126 souls on board, including the visiting entourage, quartered among berths for 130, meant they'd each have a bed if they decided to stay.

Barry had work to do. A well to manage. An abandonment schedule to keep. Though the VIPs would not be involved and would stay out of his way, he wasn't about to let anything fall through a crack, whether menial or critical.

Two activities in progress—offloading mud to the *Bankston* and the VIPs going through safety orientation—allowed him to get ready for the negative-pressure test. Three important steps would start the process.

He had Tanker make sure Daylight lined up all the choke and kill valves so Halliburton could flush the lines with seawater and make sure there were no blockages.

Second, Barry needed an entire cocktail of displacement fluids to be pumped down the drillpipe and into the bottom of the riser. He'd done his volumetric "Check," "Check," "Check," with Halliburton and Tanker, and then had done it again. As a start, they would pump just over 400 barrels of 16-pound spacer down the drillpipe. The viscous spacer would help the lighter-weight seawater lift heavy mud up the large-diameter riser. He'd had the M-I SWACO mud daubers use two batches of leftover lost-circulation material to mix the spacer; otherwise, he'd have to transport the 400 barrels of gunk to the shore base for contractual hazardous-waste disposal.[127]

Third, they would pump just enough seawater down the drillpipe to push the heavy spacer up into the lower riser, a dozen feet above the BOPs. That would leave the 8,300 feet of drillpipe full of seawater—necessary for the negative-pressure test. Which was designed to show the feds, the MMS, the casing was secure. Though the test was a time-consuming no-brainer, it was one step closer to releasing the rig.

Barry checked his watch. Counted minutes. Reckoned he'd be off the rig in less than 24 hours.

───────────────

[125] NEGATIVE-PRESSURE TEST (NPT)—For deep-water wells being temporarily abandoned it is necessary to pull the BOPs when the well is done. But this means all the mud in the riser (above the seafloor), which had been necessary to control the well, will be replaced by seawater. To ensure the casing is mechanically secure before the BOPs are pulled and the heavy mud is lost, an NPT is designed to simulate the replacement of riser mud with seawater. If a casing leak is detected via the NPT, the fix is mandatory and may call for repairing and securing the wellhead, or drilling out and perforating and squeezing cement into the area of the leak (at a casing connection, or more commonly at the shoe).

[126] Reference (2A) USCG/MMS depositions—May 11, 2010—Deposition—Master M/V Damon B Bankston—Page 96

127 Reference (20) P. 150–151. "BP had asked M-I SWACO (during drilling) to make up at least two different batches, or "pills," of lost circulation material . . . one commercially known as Form-A-Set and the other as Form-A-Squeeze. BP decided to combine these (leftover) materials for use as a spacer during displacement. (The combined material weighed) 16 pounds per gallon, much denser than 8.6 ppg seawater. (The combined material) created a risk of clogging flow paths that could be critical to proper negative-pressure testing. . . . BP chose to use the (combined) lost circulation pills as a spacer in order to avoid having to dispose of the material (onshore) as hazardous waste. . . ."

CHAPTER 43—Juggling Barrels

20 April 2010—1:45 P.M.
8h:04m:00 to Zero Hour

Jessica learned that the mud being pumped to the workboat was unrelated to the barrels of fluid in the well. The mud was from the rig's massive pits, from which the excess needed to be offloaded before the rig departed the location. But as with every barrel of mud on the rig and in the well, each barrel to the boat had to be counted.

Barry had drummed the reasons into her head and into her tally book, where they'd made special SLD status. Her dad would have appreciated the science-lesson simplicity of her words.

Simply, the *loss* of mud from anywhere within the mud system could mean lost circulation. Conversely, a *gain* of mud anywhere in the system could mean the well was kicking.

The mudloggers were focused on the job—watching for mud losses and mud gains. They knew what they were doing. They would keep a running count of the barrels going to the boat. Everything else, every pit-level reading *unrelated* to the boat, had to be related to the well. Losses—unwanted. Gains—unwanted. Simple arithmetic.

And a good time for Jessica to take a break.

Refreshed by her third hot shower in less than a day, she kept her hair dry. Slipped into lime-green coveralls. Paused at her family picture. Then she spent sixty seconds in front of the mirror, not so much to look, but to calm her insides, say hi to her dad, hug her mom and sister.

Lunch followed. Asparagus. Grilled mushrooms. Acorn squash puddled with butter. A chunk of cheese. Elapsed time, thirty minutes. Plus another minute for a chocolate-chip cookie.

She waved to Daylight on her way across the rig floor, then made her way to the mudlogging unit. She would have used the external stairs from the pipe deck all the way up, but they seemed impersonal, cold, and a hell of a lot higher.

Daylight, as usual, spit in his cup before he gave her a thumbs up. Skoal manners.

The mudlogging unit was empty. No mudlogger. Took her three seconds to boil over. She screamed, "Hello," in case the guy was under a desk. She looked out the door. Waited. Steamed. Back inside. Checked gauges—saw pressures, flow-line readings, pump strokes, pit-level charts that mimicked yo-yos. Nothing like the readings she saw every day while drilling.

She called the driller. "Daylight, did the mudlogger tell you where he went?"

"Yes, ma'am. He went to dinner a bit ago, should be fixin' to mosey back right shortly, maybe sooner."

Jessica thanked her rig-floor buddy and called the galley. She got the steward, controlled her voice. "This is Jessica. I'm looking for Sperry Sun, whoever's on call."

"Sorry, Miss Jessica. He ate just before you did, then left. Want me to put out a page?"

"Please. Have whoever's on duty meet me in the mudlogging unit."

Sperry Sun's more-senior mudlogger entered through the main door simultaneous to the steward's PA announcement.

"What's going on?" Jessica asked.

He pulled a toothpick from his cheek and dropped it in the can. "Nothing, ma'am," he said. "Just grabbed a little dinner before a long night."

"Not what I meant. Tell me about the pit monitoring system."

He leaned over and glanced at side-by-side monitors. "Not too good, ma'am. Way too much going on down there for my comfort. Transocean's pumping mud from the pits to the workboat. At the same time, there's a crew of roustabouts working with the mud guys to clean the mud pits and the trip tank. They're moving mud from one tank to another, cleaning as they go, getting ready for BP's next well." [128]

"Can you monitor barrels exiting the well?"

"Not much better than we can monitor what's going in. If they pump a measured volume of heavy spacer, for example, and we get a pit-level reading after the fact, we can tell exactly what went into the hole. But when we displace the heavy plug with seawater, which has no volumetric tank to measure, we have to go on pump strokes, and that—"

Timeout sign, which Jessica reckoned was better than a snarl. "Let me repeat the question. Can you monitor barrels exiting the well?"

"Monitor, yes, because we've got the same flow meter we've been using, but it's not close to being as accurate as a pit-level totalizer, which we don't have."

"So you *estimate* the barrels into the well," she said, "and use a flow meter to *guess* what's coming out. How do you know if we're losing or gaining mud?"

A shrug. "Subtract the estimate from the guess?"

She gave him her best don't-mess-with-me squint. "Not good enough."

The mudlogger held up his hands, fingers spread. "Easy, Miss Jessica, the only things we have are drillpipe pressure, the driller's pump strokes, the flow-line meter, and the number of barrels the cementers pump." He waved his open hand toward a multi-track digital screen. "What you see is everything we can measure."

Jessica could barely breathe. "Are you telling me there's no way we can isolate a single working pit, one with a working pit-level gauge, to accommodate the test—the NPT? And what about the displacement process?"

"It'll stay status quo, just like we're hooked up now, unless they start dumping the mud from the riser overboard."

"Overboard? Will that make measuring volumes better, or worse?"

"If they dump overboard, we lose the flow-line meter, which means the only way to even see what's being dumped overboard is through channel 14."

"What's that?"

"A video camera. We'll see mud going overboard, but we won't be able to measure it."

She replayed words, some old, some new. Everything Barry had taught her about gaining and losing mud, each bad, each requiring immediate attention, down to a video camera. Light-headed, she knew the thumping in her chest had to be a coronary because it damn sure wasn't love.

"I'm sorry, Jessica," the mudlogger said, giving her a *c'est la vie* shrug. "The roustabouts and mud guys are doing only what they've been told to do. By the toolpusher. Company man's orders."

She controlled her voice. "What's the most important thing you do in the mudlogging unit?"

"Monitor the well."

"Good generic answer," she said, remembering Barry's lectures. "But here's a hint. Can you think of anybody on the rig paid twenty-four seven to monitor the well for mud losses and gains?"

"Me, tonight, but Sperry Sun in general."

"Then get your partner to help. Figure out a way to count barrels into and out of the well on a minute-by-minute basis. I don't care if you have to steal a trip tank away from the roustabouts, or if you use flow-line gauges to meter volumes, but when I come back, I want answers. Barrels in and barrels out."

"Ma'am, with respect, what's Mr. Barry say?"

"Good question, and I'm about to find out." She slammed the unit's air-tight back door on her way to the rig floor.

----

128 MUD PITS—the physical containers on the rig that accommodate drilling fluid. Sensitive level-monitoring devices allow measurement of mud in each tank and the total system. A loss of mud anywhere in the system could mean lost circulation. A gain in volume likely indicates a kick. The trip tank is used to monitor gains and losses while running and pulling the drill string and when running casing. Any unexplained pit gain (or loss) requires immediate attention.

CHAPTER 44—Clean Count

Barry and Tanker worked together to make sure the system was ready for the negative-pressure test.

Jessica stuck her head around Barry's shoulder. "How we doing?" she asked.

Barry needed time with Tanker. He wanted time with Jessica but didn't need it. "I'll catch you later," he told her.

Though Jessica's feet didn't move, her body seemed to back up a foot. "One question before you run me off," she said, looking back and forth, Barry and Tanker. "How do you intend to count barrels exiting the wellbore?"

Barry pointed toward the handrail. "If you go over there and look down, you'll see a workboat being fed by a large-diameter flexible hose. The crane's holding the hose in place. We're pumping mud from the pits directly into the boat's storage tanks. We'll do the same when we start displacing the riser."

"And is there, or will there be, a time log for the volume received? Some kind of printout?"

"Not a chance. But if at any time you or the other mudloggers need to know the number of barrels displaced, you can get hold of the captain either through Daylight or the crane operator."

"Convenient," she said. "I'll tell the mudloggers, which I'll remind you, I am not."

Barry plowed ahead. "The captain gauges the number of barrels he's received. The crane operator—" He pointed to the top of the crane tower. "—is on a radio with the deck hands, so he can get you an answer any time you need it."

"Once a minute will work," she said. "Sixty times an hour. Then I'll get the mudloggers to plot you a little graph, which you can use to stay out of trouble. Think we can arrange that?"

"You done?"

"Get me a mud pit with good in-and-out measurements, and I won't bug you."

"If you want data, talk to the crane operator. It's the best I can do for now." He turned his back and left her standing, then walked to the driller's cabin and caught up with Tanker.

Tanker. Business as usual.

"You rigged up for displacing the drillpipe?" Barry asked.

"We're there," Tanker said. "Kill and choke lines are full of seawater. We trapped 1,200 psi on the kill line. It's steady. Drillpipe's full of 14.2-pound mud. BOPs are open."[129]

"Then let's get Halliburton up here to go over the numbers," Barry said.

Tanker stepped back from the rig-floor phone.

Barry made the call.

The Halliburton supervisor brought his clipboard.

Barry summarized: "We start with sixteen-pound spacer, follow with fresh water, and finish with seawater." He shared his hand-written numbers for how many feet of casing and riser the volumes would occupy. "The bottom of the spacer will end up twelve feet above the BOPs, and the drillpipe will be full of seawater.[130] Then we close the annular BOP on the drillpipe for the negative-pressure test."

The Halliburton supervisor confirmed each of Barry's numbers with a "Check," then told Barry he'd go straight to the cement unit.

Barry liked agreement. No bullshit. No inane questions. No second guessing. He checked his watch—almost four. Noticed Jessica talking to Daylight. Didn't have time to join them. Instead, he picked up the phone, connected with the Halliburton supervisor at the cement unit and gave the word.

From Barry's rig-floor perspective, nothing dramatic happened. The only clues that 14.2-ppg mud was being displaced from the drillpipe with the heavy spacer were in subtle changes on two live gauges. One read drillpipe pressure. The other was a pump-stroke counter, which could be used to calculate the number of barrels of spacer pumped into the drillpipe.

The entire process, he reckoned, including the workboat and negative-pressure test of the wellbore, followed by displacement of the riser, would take him deep into the evening. Too late to allow him much time with the visitors, though he planned to have dinner with them.

But only if it was obvious he had nothing better to do.

---

129 Reference (3)—BP's Internal Investigation—Pages 86–89

130 Reference (3)—BP's Internal Investigation—Page 24. "A total of 424 bbls (barrels) of 16 ppg spacer followed by 30 bbls of freshwater pumped into the well. Displacement completed with 352 bbls of seawater, placing the spacer 12 ft. above the BOP."

CHAPTER 45—Bubbles

20 April 2010—4:45 P.M.
5h:04m:00 to Zero Hour

Jessica entered the galley alone. Earlier than normal. A group of visitors, two in street clothing, two others in fresh blue coveralls, were gathered in the TV room. Barry had told her they were coming. The VIPs weren't there to see her. Good thing, because she might have told them exactly what she thought. Long string versus liner and tie-back. Abandoned gas in the annulus. Cleaning the mud pits.

She'd asked Daylight for his two cents about the pits getting cleaned during testing and displacing the well. Not a problem, he'd told her. Said as long as he had good flow meters and pit levels, same as for drilling, he'd know for sure, and maybe even sooner, if there was lost circulation or a kick. She asked him what he'd do *without* the instrumentation.

"Tarnation, Miss Jessica," he'd said. "I'd be blind, probably in both eyes, then some. Might as well be on the beach, for all the good I'd be."

Standing in the serving line, two old friends—nervous energy and growling stomach—teamed up to dictate Jessica's fare. She made herself a huge chef's salad. Slathered butter on a slab of sourdough bread. Fixed herself a fountain 7-Up in a clear-plastic glass. Seated alone and bashing Barry, it took only one bite for the endorphins to kick in. The salad, crunchy bits, and tangy dressing gave her goose bumps. Eating, thinking, munching, debating both sides of issues, she killed the bread, which called for another piece, and more soda to wash it down. Bread and drink in hand, she mysteriously strayed all too close to the dessert table, where her left hand snagged a Big Mac-size chocolate-chip cookie, as if her not looking at it meant it didn't count. And, by damn, it didn't count anyway, because Barry was on her mind. Barry—not quite so perfect after all. And herself—damn sure better than he thought.

She chewed a crunchy bite of heaven and stared at the bubbles in her refilled glass. Carbon dioxide bubbles—forming and rising. Silent. Shaking not needed. She thought about natural gas bubbles rising in drilling fluid. Like the trip gas Barry had abandoned in the annulus.

She finished off the cookie. Eyed the residual stack only twenty feet away.

Onerous gas, Barry had emphasized. When gas rises, he'd told her, it expands and can blow drilling mud out of the well—just like a kick. But when gas bubbles rise and *can't* expand, they stay the same pressure—Barry's words. These things she'd learned in school, too, when important only to a grade in physics. Boyle's Law—the constant relationship between pressure and volume.[131] Dry, academic concepts, now important in real life. In the form of trip gas. Not circulated out. Abandoned. Trapped under the casing hanger. So why *onerous*?

Ask Barry why? Hell no—*tell* him why.

She immersed herself in what she called her tech muse. Learned it from her dad. He'd often get away by himself, he'd told her, to contemplate issues—investments, vacation plans, adding on a room. Assess all he knew, all the issues. Pros and cons. Connect the dots. Make an informed decision. Act. The same had worked for her in school. Complex engineering courses. Understanding theory and practicality. Not just a number of equations to memorize, but, the big picture. How one technical concept fit with another, how courses linked with each other. Like Barry's mentoring. Company-man responsibilities. Jibs and jabs of seemingly unrelated issues from interesting to critical. A combination of academic- and experience-based theory, equations, rules of thumb. Pieces of equipment—tiny, massive, major—welded into a miles-long puzzle. Her good memory and a pocket full of notes had kept her busy. Fueled her tech muses while humped over a microscope, day after day, picking through cuttings, her body atrophying but her brain in full gear. Connecting the dots. Applying theory. Replaying Barry's war stories. Things that could go wrong—preventative measures. What her dad called the *What ifs*. What if the mud weight is too high? What if the BOP leaks? What if the rig loses power while killing a kick? What if a cement job fails? What if?

She pushed her plate across the table, pulled the glass closer, and stared at . . . bubbles. And what? What else? *What if?*

She needed . . .

Not a cookie. Not Barry.

Herself. A clone of herself.

She escaped the galley. Found the BP office empty—no surprise. Dialed a number from memory. Direct line. One ring. Two rings. Three—

"Morgan."

Ranae Morgan—business voice, professional, PhD, technical whiz, best friend. Jessica wished they were face-to-face.

"Ranae, I need your help. You alone?"

Book 1

"Oh, Jess, I almost didn't answer. The caller ID didn't—"

"Ranae, please. Just listen."

"My door's shut." All business. "Go ahead."

Jessica convinced herself she was far enough away from the hallway door she wouldn't be overheard. "Barry, the company man, has drilled into me the importance of managing gas in the wellbore. That it can be bad stuff. That you can never ignore it. And now he's done exactly that."

"Are you guys okay?"

"Yes. But he can't have it both ways. I used to trust him—not anymore. He went against what he taught me, and he went against BP's formal procedure that says to circulate out the gas. If I'd done what he did, he'd be able to show me on paper all the reasons I'm dumber than a box, then he'd have me fired. So I need to use his lessons to help me understand exactly what's going on."

"What's your guess?" Ranae asked.

"Hey, if his only bad action is that he didn't follow a line item in the approved procedure, then slap his wrist. But what if it's more than that? What if you can't afford to ignore the gas because the consequences are horrendous?"

"Jess, he's got a lot of experience. Have you asked him why he did it? Maybe he has reasons you're not aware of."

"I tried. Got the boot." She glanced toward the door. "It's either stupidity or arrogance, maybe his, maybe mine—or there's something technical I've missed about the topic. He said abandoned gas is bad, and BP says abandoned gas is bad, and then he abandoned it. There has to be a downside that makes it dangerous, and that's what I need to understand before I go back to him."

"Then how can I help?"

"Brainstorm with me. Here's the set up. Gas in the annulus. Heavy mud. The gas rises. Why's that important?"

"What are you doing?" Ranae asked.

"Dammit, help me. Be smart. Get technical. And all we know is the gas rises, and I need to figure out why that's so bad."

A beat. "So if the gas bubbles rise, they expand."

"Good. The expanding gas bubbles push mud out of the already-full wellbore."

"So the gas rises until it gets to the surface. End of story."

"Not quite," Jessica said. "Right after we trapped the gas in the annulus, we sealed it in place with a casing hanger. Might as well have closed the BOPs."

"But the gas keeps rising as long as it can."

"Ah, yes, but as it rises it can't expand—it's now in a closed system—fluid filled."

"Pressure and volume," Ranae said. "If the gas can't expand, then its pressure won't change."

"But it's still rising, taking its fixed pressure up the hole as it percolates through the mud."

"That means if a bubble at a thousand psi moves up the hole a hundred feet, then the pressure of the fluid at the shallower depth has to increase, by some amount, to a thousand psi. How much change is that?"

Jessica used her calculator. "A hundred feet, divided by 14.2-pound mud, divided by our friendly constant, equals just less than 70 psi."

"So no big deal."

Jessica sorted memories. Barry and Tanker had fought lost circulation and kicks for days. They'd found nirvana with 14.2-pound mud. Less, and the well kicked. More, it took mud. She wondered if Barry would be willing to close the BOPs and pressure up to 70 psi.

"Seventy could be a hell of a problem," she told Ranae, "and it's certainly not the upper limit."

"So what does the pressure increase hurt?"

"Our weak formations, deep, down by the pay zone. The ones that open up and swallow mud."

"I thought the well was cemented."

Movement in the hallway. Jessica swung her chair. A guy wearing white—housekeeping—carrying towels. She called to him. "See if you can find the company man, Barry Eggerton. Have him page me."

Back to the phone: "Cemented, yes, but we closed the annulus on top of the trapped gas immediately after the cement job. So the cement was green. No pressure integrity. Wide-open annulus all the way down—seafloor to 18,300 feet.

"What about the mud?"

"It's static, but it transmits pressures. So the 70-psi increase goes all the way to bottom."

"And if it's more than seventy?"

"Ranae, whether 70, 100, or 500 psi, any pressure increase may have caused us to lose mud to weak rocks, which would then cause a pressure drop in the annulus."

"Wouldn't the annulus gas then really expand and rise?"

"You bet. And it'd push more mud into the loss zone."

"Does it hurt to lose the mud?"

Something clicked. Jessica's heart pounded as if wired to a spark plug. The rising gas wasn't the problem. The loss of mud wasn't the problem. Their combined effect, though, was destined to cause the annulus pressure to drop.

"With that much pressure drop," she said, afraid to hear her own words, "any sand, or even the pay zone, could be flowing. Kicking."

"Jess, that doesn't sound right. You've got casing in the hole, surrounded by cement, and it's a closed system. The pay's not producing because it's got nowhere to go."

Jessica sat back in her chair as something cold filled the room, clung to her body. "You may be right, but we've got lost-circulation zones above the pay, in the pay, and at the bottom of the pay. There're a number of places for flowing fluid to go—all underground—and we'd never see it. It'd be an underground blowout. Oil and gas producing up through green cement and into the lost-circulation zones. And the only things keeping it underground—" She recalled her show-and-tell on the pipe rack. "—are the wellhead and casing hanger." Like a donut, she recalled. "And the lockdown seal ring, which is designed to ensure—"

The cold solidified. Stopped her words. Stopped her breathing. And the full onerous potential of abandoned gas cleared like a late-morning fog.

"Oh, Jess, I so hope you're wrong," Ranae said. "Talk to Barry. Tell him what you think. Get it fixed."

Jessica cranked her body upright, held the edge of the desk for support. Three words ricocheted through her skull—*lockdown seal ring*.[132] Which had not yet been installed. She booted her chair down the length of the office and slammed the door on her way out.

Racing down the hall, prowling for Barry, she vaguely remembered thanking Ranae and hanging up the phone.

The galley shined. Radiantly clean. More patrons than normal. Lots of clean coveralls. Laughter. Barry's table, empty.

She paged him—gave the galley extension. If she was right, nobody on the rig had a clue.

The sweep second hand on the wall clock ticked away ten seconds.

Nobody except herself. She hoped she was wrong. A fire burned her gut.

Twenty seconds.

She paced. Surrounded by food. Tempted by nothing.

"You okay, Miss Jessica?" the steward asked.

"Yes," she said, staring at the clock—thirty seconds. "I'm looking for Barry—the company man."

"He was here earlier, ma'am, but he left."

She checked the hallway. Checked the clock. Forty-five seconds. Couldn't wait.

"If he comes in," she said to the steward, tossing the words over her shoulder as she raced from the galley, "have him page me."

----

131 BOYLE'S LAW: P x V = C   No calculating necessary, but the simple equation shows there is a relationship between the volume (V) of a fixed mass of gas, and its pressure (P), in that their product is constant (C). In English, this means as the pressure on a gas bubble increases, its volume decreases, and vice versa. It also means that if a bubble in a closed fluid-filled system rises but can't expand (no room to expand), the pressure of the bubble won't change. Note: For simplicity sake, we're ignoring temperature (reference Avogadro's law, if interested).

132 CASING-HANGER LOCKDOWN SEAL RING (LDSR) has two primary purposes: (1) it accepts the sealing mechanism for a future subsea production tree, and (2) with the LDSR in place on top of a casing hanger, the casing hanger cannot be lifted from its metal-to-metal seals. Until the LDSR locks the wellhead in place, the annular seal (casing hanger inside wellhead) is dependent entirely on the weight of the casing. Without the LDSR, the casing hanger can be lifted by a strong upward pull, or pushed upward by pressure from below. The LDSR is sometimes referred to as an LDS—lockdown sleeve.

CHAPTER 46—Grinding Numbers

20 April 2010—5:55 P.M.
3h:54m:00 to Zero Hour

The rig floor sported a gaggle of men crowded together like a cold-day tailgate party. Jessica—the heels of her boots not quite tall enough—bobbed and weaved at the edge of the crowd to spot faces. Tanker, Daylight, and a number of others in a mix of colored coveralls. Barry stood on the far side, facing Tanker. Members of the audience—arms folded or hoisting paper coffee cups—stood silent, while prominent voices cussed and discussed the nuances of ballooning, the bladder effect, and whether the goddamn NPT should be run using drillpipe, like they'd done, or—Barry's voice—using the kill line as mandated by the MMS. Then some kind of shoulder-shaking, head-nodding mumblings that demagnetized the huddle.

Jessica set her sights on one man. The man in charge of the well. She made her way to his elbow and yanked his sleeve.

Barry looked down at her. "What?"

"Five minutes. It's critical."

"I'm busy."

"The well may be flowing."

His eyes got big, then settled on hers. "And you know it *may* be flowing, how?" He cupped a hand to his ear. "No alarms."

She tapped a nail on her tally book. "I'll show you."

"Are we in danger right now?"

She opened her notes. "If not, we're lucky, but if I'm right, it'll get worse."

He lifted his hardhat and scratched his hair into place. Hiked a thumb over his shoulder. "I'll meet you in the unit as soon as we get rigged up on the kill line." He walked away, left her standing.

Preempted by the kill line, Jessica thought. The MMS—not even present—had apparently won the kill-line debate, whatever it was. Had to be critical, she reckoned, since Barry deemed it more important than the well flowing.

Pissed, she did an about-face, crossed the rig floor, and entered the back door of the mudlogging unit.

She used the time to organize her notes. To get ready.

Pages of notes boiled down to four lines. Four lines that defined the problem. She wrote block letters on a clean sheet of computer paper.

Expanding gas (trip, swabbed, nitrogen) in the closed annulus rises and increases annular pressure.

The lost-circulation zone breaks down and takes mud, causing the system pressure to drop.

The pay zone produces into the lost-circulation zone and feeds more gas into the annulus.

The gas pressure under the casing hanger rises and further displaces mud from the annulus.

Until Barry arrived, she could do nothing about whatever might be happening outside the casing. And he hadn't arrived because he was screwing with a test—the NPT—inside the same casing. A test he had defined as critical. Which now involved the kill line.

She checked the time. Wondered where the hell he was. And where the mudloggers had gone after the drill-floor confab. She questioned how she could've fully trusted Barry when they were together, talking numbers, yet not trust what he was doing when out of her sight.

Beyond the liner-versus-long-string debate, she'd had only one bad example—the short CBU and its abandoned trip gas—where she had respected his tell, but had found a flaw in his show.

Maybe she was being too hard on him. Had misjudged his actions. As Ranae had questioned.

Maybe. But her list of four items about the abandoned gas told her otherwise.

She got up from her chair and forced herself to digest the NPT pressure charts. Looked back in time, to the beginning of the test. Then to current time. Drillpipe—420 psi. But so what? And pressure on the kill line. And the choke line. Each pressure on a time line. But they were meaningless numbers without knowing what was hooked up where. Was the drillpipe still full of seawater? Was the kill line open or closed? Were the BOPs open or closed? Barry had talked about using a spacer of 16-ppg mud, made from two kinds of leftover LCM. Where was it? In the drillpipe? In the annulus between the riser and the drillpipe?

A visual learner, she needed schematic drawings showing general plumbing plus all valves, fluids, pressures, and depths. Such drawings made sense to her and were easy to check for logic. For errors. For changes. For actions and reactions.

The drillpipe pressure line looked strange—590 psi and rising. Rising? Why the hell would a static pressure test of *anything* have a rising pressure?

She noted the time—6:13 P.M.—so she could ask Barry.

The door opened.

She swung around in her chair, ready to pounce.

Two mudloggers.

"I'm expecting Barry," she said. "Have you seen him?"

"He's with Tanker and the driller," the senior mudlogger said. "They're getting ready to do a test."

"Where have you guys been?" Jessica asked. "Show me the numbers. Barrels into the well, barrels out."

The supervisor put his hands into his pockets. "Everything balances, Miss Jessica," he said. "We called the boat, got their numbers. They match ours."

"There's nothing much else we can do, ma'am," the younger mudlogger said. "The only mud pit we have for pit-level readings is the trip tank, and it's not nearly big enough for all the mud we'll be getting back from the riser. All that mud's going straight to the work boat."

She wanted to tell the mudloggers the goddamn well was probably flowing underground, but held off. Barry first. She owed him that.

Determined, teeth grinding, she made a small schematic drawing on a clean sheet of paper. A side view of the well. She showed the riser on top of the BOP. The BOP on top of a long string of casing that went to the bottom of the page. Diameters didn't matter—she was interested only in pressures. From the surface she showed a small line—the kill line—following the riser to the underside of the BOP. Down through the riser and through the BOP she drew a tube—drillpipe—that ended an inch below the BOP. She drew a dark horizontal line under the BOP, labeled it *Wellhead*.

She added numbers. The bottom of the drillpipe at 8,367 feet. Seawater—8.6 ppg—in the drillpipe. Mud—14.2 ppg—in the well. The BOP at 5,025 feet RKB. Wellhead at 5,057 feet.

She copied the page. Five copies.

The still-rising drillpipe pressure crossed the 1,000-psi grid line at 6:25 P.M. She knew for sure no pump was causing the rise. One pump stroke, by either Halliburton's pump or the rig's mud pumps, would send the pressure straight up. Quite different than the ongoing steady rise shown on the chart.

She gave the copies of the drawing to the senior mudlogger. "Get these to Daylight. Have him show you exactly how they're currently hooked up.

Is the drillpipe still full of sea water? What's in the kill line? Are the BOPs open or closed? Find out why the drillpipe pressure's increasing. Put the time on the drawing, so when we look at it later we'll know what it means."

The supervisor took the stack of copies.

"Bring me the first drawing as soon as you get it," Jessica said, "then go back to Daylight and wait for the next one. If anything changes, I need a new drawing, with the time of the change noted. You two can trade off as need be, but I want one of you in here and the other on the drill floor to record what's going on, with all changes immediately back to me. Understood?"

Two nods.

Jessica stood from her chair, leaned against the counter. "Guys, if Barry and Tanker and the rig crews aren't counting barrels, other than calling the boat, then we're the only folks out here monitoring the well. I need your help. You in or out?"

"Got it, Miss Jessica," the supervisor said. "We understand." He left.

The second operator scanned charts. "I'm looking back at data, but with the cleaning crew moving mud from one pit to another, it's almost impossible to figure out the volumes. And that screws up any chance we have for measuring barrel changes. Like a couple hundred barrels went to the boat while they used pump strokes to measure how much seawater went into the drillpipe. Then our flow-meter readings for the mud that went from the well into the pits, they're going up and down like—"

He took a breath.

"Do the best you can," Jessica said, having heard no new information, her mind on the NPT. "All I need is *in* versus *out*, and for you to tell me if you see a problem."

She again checked the drillpipe pressure. It had risen to 1,400 psi at 6:35 P.M. and leveled off. Meaning what? Barry the mentor had hammered her with the importance of pressures—wellbore, formation, underbalance, overbalance, kicks, lost circulation. The 1,400 psi—leveled off—was like having the answer to a question, but not knowing the question.

Whatever the cause, the strange pressure had to be a mix of good news and bad. The bad news: the pressure, which should have been static, at zero psi, had *risen* to 1,400 psi over a 35-minute period.

The good news was . . .

Something broke in Jessica's face. Like a dam.

There was no good news.

CHAPTER 47—Teamwork

20 April 2010—6:40 P.M.
3h:09m:00 to Zero Hour

Jessica ignored charts and graphs that covered the walls of the mudlogging unit and studied a single pressure chart. A single black line. Drillpipe pressure.

She had examined the diagram Daylight had marked up. It was simple. With the entire well full of 14.2-pound mud, the bottom hole pressure inside the wellbore would be close to 13,500 psi. Now, drillpipe hung at 8,367 feet. The BOPs were closed around the drillpipe, which was full of seawater. Heavy mud now filled the wellbore *only* from the bottom of the drillpipe to the bottom of the well. The drillpipe pressure gauge—the single black line—showed 1,400 psi.

She used Barry's mud-weight equation to add the numbers. The seawater generated 3,741 psi of hydrostatic head. The mud below the drillpipe contributed 7,334 psi. Add the 1,400 psi. The answer—12,475 psi.

So what? What did it mean?

She backed up. The pressure at the bottom of the wellbore, inside the casing, should have been seawater plus mud. Just seawater. Just mud. So add 3,741 psi for the seawater, plus 7,334 psi for the mud. Total—11,075 psi. That's what the static pressure should have been.

But that wasn't the case. The pressure on bottom, during a 35-minute period, had *risen* from 11,075 psi to 12,475 psi.

Pressures don't increase without a cause. She looked closely at the chart. Strange—something she hadn't noticed. She changed the scale on the screen. Saw spikes. The drillpipe pressure not only had risen 1,400 psi in 35 minutes, it had increased in tiny spiky jumps like a barbwire fence. Small pressure spikes, at first evenly spaced, then increasingly farther apart, along the rising-pressure line.

While her gut churned and her eyes examined the line, she tugged at a deep memory. A memory about her grandmother—paternal. Cooking. With a pressure cooker. A funny pressure cooker that made fantastic pot roast according to her family, though Jessica had eaten only the vegetables. So what had been funny? Why the memory? No—not the food. And not the danger, her grandmother had warned. It was the noise. The little puffs of steam. *Spitting a little tap dance*, her grandmother had called it.

Like the tiny ticks on the drillpipe pressure line. Repetitive pressure spikes on the line that increased to 1,400 psi and then leveled off. The line that represented a bottom hole pressure of 12,475 psi. She crunched the numbers.

At 18,300 feet—the bottom of the well—12,475 psi was equivalent to 13.1 ppg. The number had to be significant.

Where the hell was Barry? She checked the time—7:05 P.M.—and opened her tally book. Her written memory. Searched back to mid-April, found—

The vendor phone beeped. E-mail. Her only thought—not now. Punched her password. Quick look. Text message. *Mom told me, cried. Please come home. ILU Sissy.*

Something inside Jessica wanted to cry too.

But her dad would have said *stay focused.*

She scanned notes in her tally book. Had been looking for what? Reviewed a number of SLDs.[133] Cogitated Barry's axioms, his diatribes, his warnings. Then looked back at hours-old mudlogging charts. Jotted numbers. Collected thoughts. Connected dots. And discovered three ohmagods, two holy shits, and a no frigging way.

Rig phone in hand, she paged Barry. No response for 30 seconds. Paged again just as the pressure-tight back door to the rig floor popped open. Barry walked in.

"We need to talk," they both said.

"Pull up a chair," Jessica said. "You need to hear this."

"I don't have all night."

"The well's kicking outside the casing—want to hear about it?"

No reaction. Then, "I'm listening."

"Your abandoned trip gas, trapped under the casing hanger, broke down the lost-circulation zone and reduced the head enough that the pay zone's flowing. It's an underground blowout."

A long beat. Staring eyes. The tweak of a grin. "And all this is based on me not circulating bottoms-up."

"That for sure, plus new data."

"I hope something credible."

She swung her chair around to the bank of monitors and pointed to the drillpipe pressure chart. "You should have talked to me on the rig floor. Then, it was supposition—now, it's fact."

Book 1

"Ah, from a wild-ass guess to a shred of evidence. Enlighten me."

"Count on it," Jessica said. "What's significant about the NPT drillpipe pressure rising to and leveling out at 1,400 psi?"

"You tell me," Barry said. "I've been looking at drillpipe pressures since noon."

"You once told me an LOT, your casing leak-off test, if run right, could never have a wrong answer. A high-number result meant one thing, a lower number meant something else, and neither was wrong."

"LOT? Where the hell's this going?"

"This isn't about the LOT, but we're going to apply your lesson to the NPT."

Barry gave his watch a hard gaze. "Better hustle or I'm out of here."

"Gladly. Question for you. Does the 1,400 psi mean the NPT failed, or that the result of the test is 1,400 psi? And if so, what exactly does it mean?"

"Ballooning—nothing more, nothing less. An anomaly."

"Anomaly? You mean like something unexpected? How about the up-and-down spikes on the 1,400-psi rise in the observed pressure." She pointed to the chart. "There, from 6:00 to 6:35, like a barbwire fence—up 40 psi, then a two-pointed spike, then up 45, another spike, right up to 1,400 psi. Unexpected? No explanation? Have you ever heard of a pressure cooker with a weighted stopper, sometimes called a rocker?"

"A pressure cooker? Like for food?" He scrunched his eyes. "Are you okay?"

"No. I'm scared—so humor me. Ever seen a weighted stopper on a pressure cooker jump as it releases a puff of steam? Here's a hint—as the pressure in the cooker rises, it lifts the stopper, which releases the pressure, which allows the stopper to drop, until the internal pressure increases again, over and over."

Barry checked his watch, jaw-locked a yawn, shook his head. "You lost me. Where's this going?"

Jessica again pointed to the spikes on the pressure line. "Since there's no lockdown sleeve, I think what you're seeing is the casing hanger lifting off the wellhead and relieving annular pressure."

"What? No way. The casing weighs hundreds of tons."

"It's got lots of help. When you closed the BOPs around the drillpipe and bled the pressure, the pressure throughout the wellbore, including on top of the hanger, dropped by 2,400 psi. And you've got gas rising in the annulus, which causes the pressure to increase under the hanger. One pulling, the other pushing—both up."

"Still hundreds of tons."

"Doesn't have to lift it all," Jessica said. "Just enough to break the seal, rock the hanger, and fart a high-pressure bubble. Repetitious little pressure spikes. About ninety seconds apart, though spike frequency decreases proportional to the pressure rise, which a good engineer who's studied fluid flow might expect." [134]

Barry puffed his cheeks and exhaled through tight lips. "Get logical. A few hundred units of trip gas is *not* going to lift hundreds of tons of steel off the wellhead. Period."

"You left out the possibility of lost circulation and underground flow, which would have means, motive, and opportunity to supply unlimited volumes of annular gas. But keep going—you're zero for one."

"Jessica, the annulus isn't flowing. Nothing's lifting the casing hanger."

"I won't concede, because I don't know. But your NPT confirms there *is* a major problem. If it's not leaking up the backside, through the wellhead, then there's a casing leak, with one logical path. And that has to be from the pay, down the annulus, through the shoe and float equipment, and up the casing."

Barry's face bloomed red. "Jessica . . . what . . . where the hell did *that* come from?"

She jabbed her finger onto the chart. "The 1,400 psi proves it. *Shut-in drillpipe pressure. SIDPP.* Your words. You're seeing formation pressure in the form of underbalance, because something's leaking—from higher pressure outside, to lower pressure inside. It's a kick. A perfect example. And it's been leaking a couple of hours, contrary to everything you've ever said about minimizing kick volume by acting quick."

"Spoken like a petroleum engineer, which you are not."

"You're right. I'm not. But I'm way the hell up the drilling-101 learning curve. Damn sure high enough up the ladder to ask pertinent questions and challenge dumb actions."

Barry held back an almost-verbal *bullshit.* Jerked a look at the chart. Studied it. "Three things," he said, waving a pitchfork of fingers. "First, the 1,400-psi is a bladder effect. Confirmed by Tanker and Daylight."

"Barry, don't go there. This is BP's well, and the well's your responsibility, not theirs—your words. More important, when you bumped the top cement plug with 1,200 psi, you ballooned back five barrels. I was on the cement unit. I saw it." She checked her drawing and notes, then pointed to the drillpipe-pressure chart. "An hour and a half ago, at five fifty-five, you tried to bleed the last of 2,400 psi off the drillpipe, but you got back fifteen barrels of seawater before the pressure got to zero. Fifteen barrels is more than five, and you could have bled a lot more. That's not ballooning; it's a manmade 15-barrel kick, through a leak."

"No way," he said. "Classical ballooning."

She pointed. "Which means we're back to the 1,400-psi build up." She turned to face him. "You ever hear of a self-inflating balloon? No? That's because it's not ballooning—it's a kick. *Drillpipe pressure is king—it doesn't lie.* Your words. Now how about a few engineering facts, instead of bar talk."

"If you want me to leave, I've got plenty on my plate."

"Stay," she said. "You've got two more tries to get it right."

"Second," he said, determination in his voice, his right hand raised with two fingers that suddenly became snake fangs jammed into his left palm. "You say the well might be kicking, through the shoe, flowing up the casing." He stopped, and his face hardened. "Impossible—two reasons. Patio-hard shoe cement, your words, plus you saw the float equipment. Two one-way check valves. Designed for cement to go *into* the well, not the other way around. There's no way a kick can come up the casing."

"Oh? Let's visit the subject, and maybe you can answer a couple of questions, further educate me."

He checked his watch.

She kept going. "First, even though every liner in your well was set on bottom, I noticed on the IADC report the production casing is set at 18,304 feet, above bottom at 18,360. That's 56 feet of rat hole."

"Your point?"

"You've got 14.2-pound mud underneath heavy shoe cement. What's the slurry weigh—sixteen pounds?"

"Heavier—16.7 pounds."

"Worse than I thought. Do you think it's possible that with the help of gravity the 14.2-pound mud from the rat hole is now up in the shoe, and the 16.7-pound cement slurry from the shoe is now down in the rat hole?[135] My guess is it's tough to make a patio out of mud."

Barry's face twisted. "The rat hole's there for a reason—based on experience. And your next point?"

"I'll take that to mean you want to change the subject. So an easy question about the float collar, which is located a hundred-plus feet above the shoe that's filled with mud-contaminated cement."

No reaction.

Okay by her. "You said the float equipment went into the well with the flappers, the check valves, held open by the tube we looked at."

"The auto-fill tube," Barry said.

"Right. And the auto-fill tube was designed to allow flow in both directions, until it was time to close the check valves."

"Those are my words—tell me something new."

"And it's *your* number two, dammit—stay with me." She flipped through her tally book, her face pulsing with heat, found the page she wanted. "You said the auto-fill tube would stay in place, wouldn't *convert*, until you increased the pump rate to—" She glanced at the page. "—five or more barrels per minute."

"As designed by Weatherford," Barry said. "Industry proven. Reliable."

"Then look at this." She pointed to a monitor. "It's a time recording from the Halliburton unit, followed by the rig pumps. It shows two tracks—one for pressure, one for pump rate. You've seen it before."

"Many times. Your point is?"

"You were worried about swab and surge. So you held back on the pump rate. From the time Halliburton installed the cementing head at two-thirty yesterday afternoon, through your pitifully-short CBU with the rig pumps, and until you bumped the top cement wiper plug ten hours later, at thirty-five minutes after midnight, the maximum pump rate was 4.1 barrels per minute. Which is twenty percent less than the minimum required to convert the floats."

Barry stared at the chart.

"And for that period of time when you were displacing cement with the rig pumps, Daylight confirms he never went above the limit you'd given him—fours barrels per minute."

With Barry thinking, she drilled deeper. "Might that mean the auto-fill tube is still holding open the check valves? Surrounded by nothing but mud-contaminated, still-green shoe cement and a hundred million barrels of oil?"

"No. Not right. When we pressured-up to break circulation, to more than 3,000 psi, the surge rate when we finally broke through had to be enormous. Certainly enough to dislodge—convert—the auto-fill tube." He paused. "The flappers are working, Jessica. The well's not kicking up the casing any more than it's kicking up the annulus."

"Barry," Jessica said, pointing to a page in her tally book, "you once told me the NPT is designed to find leaks. No leak, good news. Casing leak, bad news. Because leaks have to be fixed. Not negotiable. Period. And now you're saying the casing's leaking on both ends?"

"That's not what I meant."

"And that's where we differ—I mean every word I say. Zero for two. I can't wait for number three."

Barry waggled his head back and forth. "Then here it is. Since you're hell-bent focused on the drillpipe pressure, I'm sure our next NPT, using only the kill line, will make you feel a lot better."

Adrenaline blew her out of the chair. "*Feel better*? Christ, I'm not some snot-nose kid with a fever who needs to feel better." She'd screamed so hard she thought she'd pass out, had to grab the counter for support.

Barry stepped back.

She caught her breath. Settled. Weighed her adversary. Picked her words. "Okay, dammit, you win. I'm afraid, and I want to feel better, and I'm not as tough as I want to be."

He cocked his head, as if he'd heard something important.

"But none of that matters," she continued, "because this isn't about me—it's about the well. Why? Because the casing's got a leak and you're getting ready to fill the wellbore with water."

He stared, said nothing.

"It's your well, Barry. Your responsibility. I'm not the expert. All I can do is point to data, to things that don't look right, based on what you've taught me. Which means you need to think about what I've said. If I'm right, then you need to get control of this thing, whatever it is, before it eats us alive."

Barry checked his watch. "Recheck your data, Jessica. I'll be back as soon as I rerun the NPT."

He shut the door behind him. A quiet click.

---

[133] SLD—Science Lesson of the Day. Fun learning concepts Jessica got from her dad. On the *Deepwater Horizon*, she added more . . . compliments of her mentor, Barry.

[134] Reference (3)—BP's *Deepwater Horizon* Accident Investigation Report. Internal investigation released to the public on 8 September 2010. Section 2.6 Interpreted the Negative-pressure test. Figure 4—April 20, 2010, Negative-Pressure test (Real-time Data). P. 88 . . . (See page 421 herein).

[135] RAT HOLE—Rat hole is open wellbore (previously drilled hole) left below casing. The O&G industry long ago realized the implications of leaving rat hole under (below) newly installed casing. The industry collectively adopted an RP— a Recommended Practice. As per API RP 65 Section 7.5: ". . . If casing is not run to bottom, the "rat hole" should be filled with a higher weight mud to prevent cement from falling into the rat hole and displacing rat hole mud into the cement column . . ."

CHAPTER 48—Success

20 April 2010—7:25 P.M.
2h:24m:00 to Zero Hour

Barry and Tanker hadn't liked the drillpipe negative-pressure test that eventually produced the 1,400-psi anomaly. Goddamn pressure wouldn't bleed off. Tanker had described it as a bladder effect. Like ballooning. Maybe something to do with the drillpipe. Not worth arguing over. Didn't matter—Barry had an alternative, but only because he recalled that BP's drilling permit to the MMS had stated the NPT would be done through the kill line—not the drillpipe. Barry liked the change—the kill line had control valves on both ends that could be opened and closed, at the rig floor and just under the BOPs.

Barry watched while Tanker's crews rigged up to run the test using the kill line, which all parties confirmed was filled with seawater. Five thousand feet of sea water, on top of 13,300 feet of heavy mud. Daylight opened the relief valve. Less than half a barrel of seawater bled from the kill line. Not much ballooning. Barry reckoned it should have been more, but the kill-line pressure dropped to zero, as it should. Not like with the drillpipe, where the pressure continued to hold at 1,400 psi.

Barry calculated that with zero psi on the kill line, the bottom-hole pressure had been reduced from its normal 13,500 psi to about 12,050 psi. Such low pressure meant the inside of the casing was severely underbalanced compared to annular reservoir pressures; hence, the goal of the negative-pressure test. Daylight closed the kill-line valve, per MMS requirements. The threesome settled in to monitor the kill line for pressure buildup.

Barry watched the gauge. Watched his watch. The pressure remained at zero for 30 minutes—as required. Thirty expensive damn minutes, he figured. No flow from the wellhead. No flow into the bottom of the casing. The Macondo well was secure. Confirmed by the successful test. By the major suck on the well, Jessica would have said.

Barry and Tanker gave each other a thumbs up.

The negative-pressure test—NPT—was concluded and documented by Tanker and Barry as a good test at 7:55 P.M.[136]

The anomalous drillpipe pressure—proved to have been unimportant—remained stable at 1,400 psi during the kill-line test.

Based on the successful kill-line NPT, the riser and wellbore were ready to be filled with seawater down to 8,367 feet. Five minutes later, on Barry's word, Tanker had Daylight open the BOPs, then kick in the rig's mud pumps. That action sent raw seawater down the drillpipe and into the casing at 8,367 feet, where it began the long process of displacing all the fluids from above that depth. Including 3,300 feet of mud from the casing. And the entire content of 5,000 feet of riser—heavy mud, seawater, fresh water, and 16-ppg spacer. It all had to come out. Replaced by seawater.

The M-I Swaco mud engineers would have kept a running count of the barrels received into the mud pits, compared to the amount of seawater pumped, if they'd had access to all the pits, which they didn't. Alternatively, Barry told them to keep an eye on the return-line flow meter.

At 8:50 P.M., expecting the 16-pound spacer to reach the top of the riser, Barry and Tanker slowed the rig's mud pumps. They switched the mud returns to a trip tank, cleaned for the purpose, where they could watch for the heavy spacer. The spacer's arrival meant most of the drilling mud had been displaced from the riser.

Dumping the spacer overboard would be a milestone. Everything that followed—the residual sloppy mix of seawater and mud—could go directly overboard. But only if a sample of the fluid from the riser passed the sheen test—didn't cause an oil slick.

The sample passed the test.

Barry felt good. Tanker and Daylight were back to pumping seawater, cleaning up the casing, BOPs, LMRP, and the riser. All returns going overboard. A good time.

That is, until 9:17 P.M.—the start of pump problems. A common maintenance event. Which got Tanker involved. And Daylight. And the rig electrician.

Which gave Barry the opportunity to visit with Jessica, but only because he had good news. And to let her vent. He headed to the mudlogging unit and walked in at 9:20 P.M. He entered quietly.

A mudlogger looked his way, then turned back to one of the monitors on the wall.

Jessica faced Barry, as if she'd been waiting.

He raised his hands to a defensive position.

"What?" she said. Half question, half demand.

He told her what they'd done, using one of her schematic drawings to make his points. Showed her which valves were open, which ones were closed. Same with the BOP. He punched numbers into a calculator. Showed her how the kill-line pressure had been bled to zero psi. Less than a barrel of ballooning. And how the test had reduced the bottom-hole pressure by 1,450 psi. No flow. No flow back. No pressure buildup—for more than the required 30 minutes.

He folded his arms. "Tanker and Daylight and I are in full agreement. We signed off on the negative-pressure test." He checked his watch. "We've been pumping seawater into the riser, displacing mud, for forty-five minutes."

Nothing. No reaction.

He turned a chair and sat to face her. "I need to go."

Jessica. Disheveled. Stoic. Nobody home.

He was tempted to take her hands. Decided otherwise. "There's no leak, Jessica, and no flow," he said. "And no harm was done by you raising the alarm when you thought we had a problem. I would've been disappointed if you'd ignored your fears."

She nodded a lot. Blinked a lot. Didn't breathe a lot.

"Are you okay?' he asked.

"No," she said. Barely audible. Rubbing her scar. "I see data that scares me to death—you see only what you want to see." She turned her chair, looked toward the pressure charts, and opened her tally book. "Stay if you want, but I'm busy."

Barry kicked the door shut when he left.

---

[136] Reference (3) BP's *Deepwater Horizon* Accident Investigation Report—P. 25. See also Reference (18D)—Appendix G: Negative Test Protocols. Cites BP's Temporary Abandonment Procedure as approved by the MMS 16 April 2010 (MMS-124 (APM)). Steps 2–4 of 8 include: "(2) TIH with 3-1/2 stinger to 8367.' (3) Displace to seawater. Monitor well for 30 minutes. (3) Set a 300' cement plug . . ."

CHAPTER 49—Prepping for Home

20 April 2010—8:45 P.M.
1h:04m:00 to Zero Hour

Transocean's David "Daylight" Stalwart sat in the driller's chair, in control of the rig floor. His derrick hand and three roughnecks huddled near the doghouse, sipping just-perfect coffee. They had joked about not being invited to the big party inside the galley, though they wondered how it could be called a party without dancing girls and live music. And besides, they had work to do—keeping an eye on their boss. Daylight's eyebrow was king—when he raised it their direction it was time to get off their collective asses and get to work.

Daylight counted minutes in his head before he'd have to help the crane operator secure the mud hose after that job was complete. Then, according to Mr. Barry's schedule, they would work with Halliburton to set the cement plug, reverse circulate at the top of the plug, and pull all but the last 3,000 feet of drillpipe. On top of that drillpipe, which would weigh just less than 100,000 pounds, they would install the casing-hanger lockdown sleeve and its inline running tool. That'd be the last trip in the hole. Lock the sleeve in place. Pull the drillpipe and leave the lockdown sleeve behind to do its simple job, forever. After which they'd finally get to pull the riser and LMRP. Then the BOP. Daylight spit and sipped, ready for the long night ahead, followed by a long day.

* * *

The M-I Swaco mud engineers shared a thankless job. Not just while they monitored and counted the barrels of mud being delivered to the workboat, but throughout their months on the Macondo well. Which was just like all the other wells—no better, no worse, always a challenge. They might mix dry powders and liquid chemicals with water on one well to make water-base mud. On another well, the additives—clay, barite, fluid-loss chemicals—were mixed with oil. For Macondo, BP used one of the synthetic oils that had taken over the environmentally-sensitive world market. All products and additives had to meet exacting environmental standards. And the resulting mud, on any given rig, on any given day, on a 24/7 basis, had to be perfect. And not just the density, in pounds per gallon. Perfection also included viscosity, fluid loss, gel strength, to name a few of a dozen required physicochemical properties. Perfection wasn't subjective. The well, the geology, dictated requirements. Something out of kilter with the well? Must be the mud. Blame the mud engineer.

And even the heavy spacer in the riser—the company man's scheme to get rid of the lost-circulation gunk that otherwise had to go to the beach—was expected to show up at the surface any minute. The mud engineers were surprised it'd worked, that it hadn't settled and plugged off the riser. And if it hadn't worked? Finger pointing and blame were common greetings, compliments rare. But at least the rig's mud pits were clean. Transportation on an early boat run the next morning would get them to the beach in time for lunch—boiled crawfish. Maybe crawfish several days in a row, what with Louisiana's rapidly approaching mid-May end of the season.

* * *

The Sperry Sun mudloggers—driven by BP's Miss Jessica, on a rampage about something, they'd confided—pestered Daylight so he'd keep them up to speed on which valves were open, which BOP was closed. But there had been no change for a while. Seawater was being pumped down the drillpipe to displace oil-based mud from the casing down near 8,300 feet and from 5,000 feet of riser, after which the mud went through the flow meter and into the pits. Water in, mud out. When they got to the heavy spacer, they'd stop pumping and test to make sure all the oil-based mud was out of the well—the sheen test. From there, home free. Divert the returns to the sea, starting with the spacer. Fill the rest of the riser with seawater. Then rig crews would pull the riser and BOP—mudloggers not needed.

Which meant the Sperry Sun mudloggers' night was almost done. They planned to have the mudlogging unit cleaned up and ready for another well, whenever and wherever that might be, by noon the next day. Then home.

* * *

The Halliburton operators were at the cement unit, ready to go. Simple job—a couple hundred feet of cement in the 9-7/8-inch casing. As agreed with Barry Eggerton, they would pump the slurry down the drillpipe, at 8,367 feet, and up the casing, with the top of the cement at

8,100 feet, both inside and outside the drillpipe—rightfully called a balanced cement plug. The operators often joked that the procedure—a common necessity—wasn't a good thing to do with drillpipe you ever wanted to see again. So, the tricky part was to pull the drillpipe up the wellbore, out of the green cement plug, to 8,100 feet, and leave the plug behind. Then they had to make sure no cement lingered inside the drillpipe. To clear the drillpipe, they'd reverse circulate—pump mud down the annulus and up the drillpipe. That not only would clear the drillpipe, it would keep them from having to pump a wad of cement up the annulus, through the BOPs. Cement and BOPs—not a good mix.

After the plug job—the last cement work for the well—the crew would stay up as long as necessary to make sure the cement unit was perfectly clean. Cement, even small volumes in dark corners, would not go away easily if ever given the chance to harden. Neither Halliburton operator had checked the logistics schedule for the next day, but they would be ready for whatever transportation was available, and an early morning boat ride wasn't too soon.

* * *

The chief mechanic was down in the engine room—running just two of his six engines. Each 9,775-horsepower diesel engine pushed a 7,000-kilowatt, 11,000-volt generator. Just barely louder than his wife's Bosch dishwasher he'd told her—though he hadn't mentioned he wore ear plugs and padded ear protectors while on the job. He had two engines running because it didn't take a lot of power just to keep the rig on station, with the facility lighted up, and minimal activity on the rig floor. Finishing up the well meant nothing to the mechanic. His job would end only when it was his scheduled time to go home, whether still on the Macondo well, or making way to some geologist's dream location, or after having spudded the next well. The deep rumble of his two babies echoed in his chest. Right. Like twin Bosch dishwashers.

* * *

Tanker Forster, always a busy man, was a very busy man. Multitasking, his CPA son-in-law called it. Tanker called it his three-headed job. *Master*, sometimes called captain, to make sure the rig stayed on station. *Offshore Installation Manager—OIM*, because it was a complex operating facility at sea with 126 personnel aboard, for whom he was responsible. *Toolpusher*, in charge of the drilling facilities, actively working with BP's company man, Barry Eggerton, to ensure the riser displacement process and the remaining temporary abandonment procedures were executed in a safe and timely manner. And on this night, *host*—not a title he aspired to—of the Transocean-BP VIP group that had boarded his rig, their intentions honorable. Seven years for the *Deepwater Horizon* without a lost-time accident. No fingers cut off. No heads bashed. No lives lost. For seven years. He'd hoped a late evening chopper would take the VIPs home. Instead, they had elected to hold a safety meeting to start after dinner, at seven, and to stay the night and chopper home in the morning. Tanker, the VIPs, and others gathered at meeting time and discussed general operations safety issues until just after nine.[137] It was a good meeting, but it had been a long day. Tanker headed to the rig floor. Because he had work to do. A multitasking, very busy man.

* * *

Barry was comfortable displacing the riser. Get it over with. Go to the next step. Finish the abandonment process. But Jessica's diatribe bothered him. Especially the part about the first NPT with its ballooning and the annoying 1,400-psi on the drillpipe. Just before nine, with the safety meeting still in process, he called one of his bosses in Houston. Explained the first NPT and its anomalous 1,400-psi result, followed by the second NPT and its good results—zero psi for 30 minutes. He got the well-experienced engineering feedback he'd expected, which confirmed such disparate results were not possible if the plumbing for the two NPTs had been rigged up right. That was all Barry needed—he confirmed that they were indeed rigged up right on the kill line and that the results for the second NPT were valid.[138]

* * *

Although it'd been two and a half hours since any mud had been pumped down to the boat, the Transocean crane operator maintained his station in case there were problems with the flow line hanging over the side of the rig. From his elevated crane cab, his view of the world was glorious. A clear-sky sunset, the sun down at 7:30 P.M. The seas, though not glassy calm, were as flat as he'd seen them for a while. A breath of breeze—two, maybe three knots. Lights in the distance—from platforms, rigs, workboats, tankers, all the way to the horizon. Each light was pinpoint bright, but outnumbered a million to one by stars and a couple of planets. Including Venus—the evening star—two hours, this time of year, behind the already hidden sun. The waxing moon, to the southwest, would set four hours after Venus. These things he knew, because he was interested, and because he cared. And he enjoyed watching them happen. Over the years, the crane operator had worked other jobs, good jobs, but all it took was one night like what he was seeing to remind him everything had a purpose. He was where he was for a reason. If for nothing else than to enjoy the view before him, as he often did late in the evening, even as Venus quietly dipped closer to the horizon, forty minutes from its inevitable touchdown.[139]

---

[137] References 13 and 14—USCG depositions 28 May 2010, and 7 April 2011

[138] Reference (14)—USCG—7 April 2011, addresses 8:52 P.M. phone call from *Deepwater Horizon* to BP office in Houston.

[139] USNO—Naval Oceanography Portal—A website about the positions of the sun/moon/planets—9:24 P.M. on 20 April 2010—from 50 meters above sea level—New Orleans

CHAPTER 50—Connecting Dots

20 April 2010—9:29 P.M.
00:20m:00 to Zero Hour

One-liners played with Jessica's thoughts— Humpty-Dumpty, a one-eyed man, the village idiot, a naked king.

Data can be misinterpreted, but it doesn't lie.

Barry's marked-up schematic—the one Jessica had used to show 1,400 psi on the drillpipe with the BOP closed—showed the kill line open on both ends and full of seawater. He'd labeled the bottom of the kill line at 5,000 feet—just below the BOP—and had drawn a little circle at the top end of the kill line, like a little gauge, which he labeled *0 psi*. He put a check mark by the closed annular preventer. He'd noted 13,300 feet of 14.2-pound mud below the bottom end of the kill line. He'd calculated the pressure at the bottom of the hole as 12,050 psi, which he and Tanker held—by keeping the kill-line pressure at exactly zero psi—for the required 30 minutes.

What Barry had added to the drawing made sense. A good NPT. BP would be happy. The MMS would be happy.

But something wasn't right. She looked back at her pressure charts and checked the numbers. The pressure on the kill line had indeed been zero from 7:16 P.M. to 7:55 P.M., at which time the kill-line NPT had been declared successful.

She studied the drawing, studied her charts.

And there it was.

In addition to what Barry had *added* to the drawing, he'd *ignored* the notation on the same drawing that showed 1,400 psi on the drillpipe. And from the time the drillpipe pressure had risen to 1,400 psi, at 6:35 P.M., right up to 7:55 P.M.—including the period throughout the kill-line NPT—the drillpipe pressure had *remained* steady at 1,400 psi.

Was it possible that 1,400 psi on the drillpipe, which met all criteria for a casing leak and kick, could share the same wellbore with zero psi on the kill line, indicative of a no-casing-leak NPT?

No. Impossible results. Something had to be wrong.

Which begged a question—the significance of 1,400 psi. Why not zero, like it should have been? And if not zero, why not 400? Or 1,800? Why 1,400 psi? Since the drilling crew didn't cause the 1,400 psi, like with a pump, it had to be the reaction to something. It had to be the *effect* part of cause and effect.

Barry's mentoring and her general understanding of the physical laws of nature told her where to look.

With 1,400 psi on the drillpipe full of seawater, the calculated bottom-hole pressure was 12,475 psi. That was the key. The effect. But why 12,475 psi? Which was equivalent to 13.1 pounds per gallon at 18,300 feet. She'd calculated numbers close to 13.1 ppg before. She flipped through pages of notes.

Found it.

A notation about Ranae Morgan's discovery, with open-hole formation pressures ranging from 12.5 to 13.9 ppg. Jessica's 13.1-ppg calculation fit in the middle of the range.

Jessica studied the numbers. Thought bad words.

No doubt about it. The cause—the source of the energy. The data indicated one of the discovery zones had found a path, a leak, into the wellbore—either up through the annulus, or up through the casing, or both—and had *pumped* the drillpipe pressure up to 1,400 psi. Nice and stable. Nothing flowing. But with no way in hell to bleed off the 1,400 psi. Which Barry had tried to do without success.

Which begged another question. The *kill-line* NPT. It certainly had no view of the high-pressure discovery zone. Quite the opposite. Even with the kill-line pressure bled to zero—which equated to 12,050 psi of bottom-hole pressure inside the casing—the NPT showed no leak. No flow.

She needed a cup of coffee, a jolt of caffeine, a sugar high, but wasn't about to interrupt the VIP party in the palace. She opened and guzzled eight ounces of water from a plastic bottle. Crushed the container into crinkled art. Put the lid back on. Turned inside for strength. Inside, where her dad had never let her down.

He had retired from the navy to begin his second career as a carpenter. He called her rock solid and nicknamed her *Terra—Terra Pherma*—the only person who ever called her that. A stickler for measurements, a pencil behind his ear, he cut her no slack. Live by rules, he taught her. Science lesson of the day—SLD. Doesn't matter from which end you measure—the height can be doors plus drawers, or drawers plus doors, but there's only one answer. One *right* answer, he'd emphasized.

And 12,050 psi at the bottom of the well under the kill line was not the same as 12,475 psi at the bottom of the well under the drillpipe. Because the well had only one bottom.

Which meant something was wrong. Even according to Barry's mentoring—his SLDs in her tally book—those two numbers had to match. Because there could be only one pressure at the bottom of the well, inside the casing. Not negotiable. One answer. The right answer.

Two steel pipes, side by side, both filled with seawater. Which meant, *because* the drillpipe pressure stayed at 1,400 psi during the test, Barry should have found it impossible to bleed the kill-line pressure to zero, for one simple reason.

The two NPT tests results were mutually exclusive.[140]

Which meant something about the kill-line pressure test wasn't right.

Maybe a closed valve had been noted as open.

Maybe the LCM gunk mixed as the heavy spacer had plugged the kill line.

Maybe . . .

A rush of new thought crushed her shoulders, compacted her lungs. The *why* wasn't the immediate problem. Pumping seawater *was*.

She tried to get up. Her legs failed. She fell back. An aura of rushing noises behind her eyes. The taste of bile.

Her dad's words. Trust your life to nobody but yourself. Driving a car. Packing a parachute. And especially offshore.

His words yanked her off the chair, and she jerked the rig phone from its cradle.

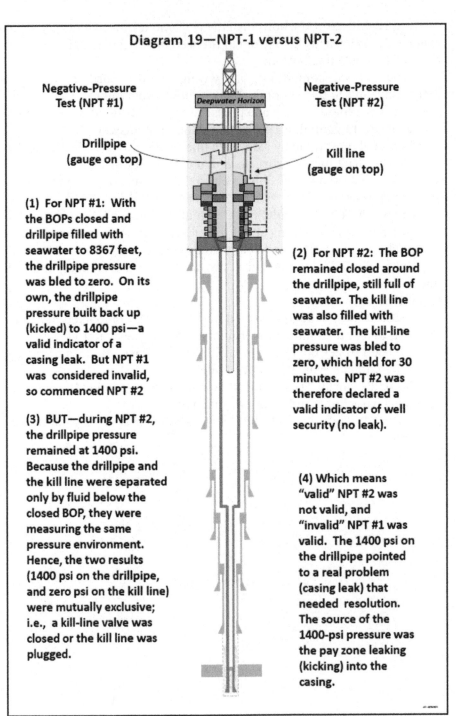

Diagram 19—NPT-1 versus NPT-2

Negative-Pressure Test (NPT #1)

Negative-Pressure Test (NPT #2)

Drillpipe (gauge on top)

Deepwater Horizon

Kill line (gauge on top)

(1) For NPT #1: With the BOPs closed and drillpipe filled with seawater to 8367 feet, the drillpipe pressure was bled to zero. On its own, the drillpipe pressure built back up (kicked) to 1400 psi—a valid indicator of a casing leak. But NPT #1 was considered invalid, so commenced NPT #2

(2) For NPT #2: The BOP remained closed around the drillpipe, still full of seawater. The kill line was also filled with seawater. The kill-line pressure was bled to zero, which held for 30 minutes. NPT #2 was therefore declared a valid indicator of well security (no leak).

(3) BUT—during NPT #2, the drillpipe pressure remained at 1400 psi. Because the drillpipe and the kill line were separated only by fluid below the closed BOP, they were measuring the same pressure environment. Hence, the two results (1400 psi on the drillpipe, and zero psi on the kill line) were mutually exclusive; i.e., a kill-line valve was closed or the kill line was plugged.

(4) Which means "valid" NPT #2 was not valid, and "invalid" NPT #1 was valid. The 1400 psi on the drillpipe pointed to a real problem (casing leak) that needed resolution. The source of the 1400-psi pressure was the pay zone leaking (kicking) into the casing.

Diagram 19

---

[140] See **Diagram 19**—Two Negative-pressure Tests—see page 274 herein. The first test (NPT-1) used drillpipe. The second test (NPT-2) used the kill line. The bottom of the drillpipe and the bottom of the kill line were connected by nothing but fluid, so were measuring the same "fluid-pressure" environment. The two tests yielded different answers. One answer (NPT-1) said the casing had a leak; the other, no leak. Like an accountant adding a checkbook and getting two different answers. Only one answer can be right and the other wrong, or they can both be wrong, but they cannot both be right. The numbers speak for themselves and always point to the right answer—the account never gets to choose. NPT-1 was the right answer—the casing had a leak.

CHAPTER 51—Puzzle Complete

20 April 2010—9:35 P.M.
00:14m:00 to Zero Hour[141]

Having paged four times without a callback, Jessica slammed down the phone.

She yanked open the door and made her way to the rig floor.

Deathly silent.

No Barry. No Tanker. No Daylight.

"Help you, Miss Jessica?" the assistant driller asked.

"I need Barry or Tanker," she said. "Now."

"Ma'am, they either down below working on the mud pumps, or they in the galley. I'd guess below."

She headed down the rig-floor stairs, kept going down, her vision blurry, air hard to come by. Found two guys working on a mud pump, while the adjacent pump cranked away, pumping seawater.

"Have you seen Tanker or the company man?" she asked, gasping.

"Reckon they be on the rig floor, Miss Jessica. Or maybe in the galley."

She jogged to the galley, a mumble per step: bull—shit—bull—shit—bull—

Found the steward at the serving counter.

She tried to talk. Couldn't breathe. Then, "Barry . . . company man . . . where?"

He pointed. "Back in the game room, Miss Jessica. Bunch of guys. Would you like me to fetch him?"

She waved her thanks and marched through the galley, into the room. Heads turned. A visitor or two, maybe, plus a few familiar faces. One was Barry's. She stopped short of the crowd and motioned him over.

Yeah, right crossed his face. His feet moved not an inch.

She took a desperate breath and marched through the gawking crowd. Grabbed a chair, dragged it along. Closed the gap. Leaned on the back of the chair for support. Looked up at the company man. Her mentor.

He stood firm. Looked down at his student.

She sucked a meager breath, found her voice. "The well's flowing, Barry. Stop work! Shut it in now!"

An aura of silence engulfed the room.

"The kill-line test is a failure. Closed valve, plugged line—whatever."

Barry held up his hands, palms to her. "Let's go to the—"

"The pay zone, dammit, is open to the wellbore—that's the 1,400 psi you couldn't bleed off. You're pumping ten barrels a minute into the well, and it's kicking fifty barrels overboard."

Mumbles from the audience.

She focused on Barry. "That's your clue—oil and gas are on the way up. Up the annulus, up the casing, up the riser, doesn't matter. Close-in the well, right now, or you're going to have the biggest goddamn kick you've ever seen, right in your face."

"Jessica—"

"Don't Jessica me, dammit. Go. Run. Call Tanker, Daylight, anybody. Close the frigging BOP now!"

Barry looked to the crowd. "Sorry, guys, I need to—"

Jessica kicked the chair aside and pointed toward the galley door. "Screw theatrics, Barry. Just go!"

Barry shrugged, stepped around Jessica, headed to the exit—

The public address speaker squawked, Daylight's voice, "Rig floor. Well's kicking."

As the announcement repeated, Barry looked back at Jessica, disbelief ripping his face, and mouthed, "Oh, God." He turned hard. Ran. Didn't look back.

Half the crowd followed. Others clustered.

Jessica fought for air, her heart racing, boots heavy. Hyperventilating, she slogged her way toward the BP office and climbed the less-crowded stairwell by the helideck. A door. She pushed it open. Stepped onto the pipe deck.

A geyser of mud and water bloomed and grew around and over the top of the derrick like an old-time photograph. Black and white. Like Spindletop. A ten, she reckoned.

Mind numb, she retrieved her tally book from her pocket and took small steps toward the rig floor.[142]

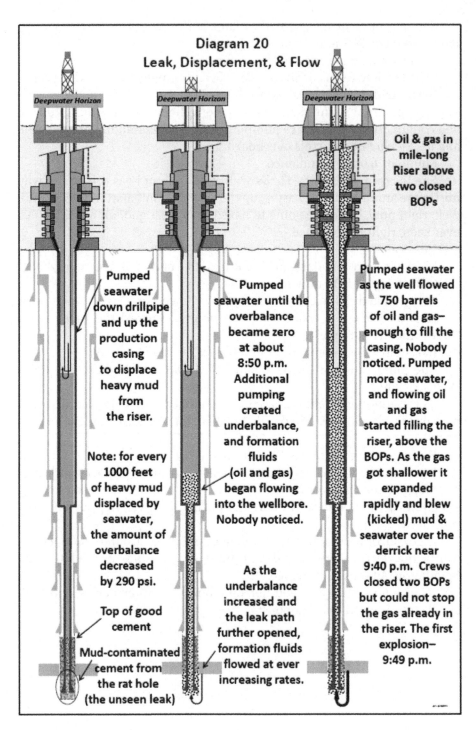

Diagram 20
Leak, Displacement, & Flow

Oil & gas in mile-long Riser above two closed BOPs

Pumped seawater down drillpipe and up the production casing to displace heavy mud from the riser.

Note: for every 1000 feet of heavy mud displaced by seawater, the amount of overbalance decreased by 290 psi.

Top of good cement

Mud-contaminated cement from the rat hole (the unseen leak)

Pumped seawater until the overbalance became zero at about 8:50 p.m. Additional pumping created underbalance, and formation fluids (oil and gas) began flowing into the wellbore. Nobody noticed.

As the underbalance increased and the leak path further opened, formation fluids flowed at ever increasing rates.

Pumped seawater as the well flowed 750 barrels of oil and gas—enough to fill the casing. Nobody noticed. Pumped more seawater, and flowing oil and gas started filling the riser, above the BOPs. As the gas got shallower it expanded rapidly and blew (kicked) mud & seawater over the derrick near 9:40 p.m. Crews closed two BOPs but could not stop the gas already in the riser. The first explosion— 9:49 p.m.

Diagram 20

---

[141] Reference (3)—BP's Internal Investigation Timeline—begins before the 19 April cement job, plus a minute-by-minute breakdown for the last hours of the well.

[142] See **Diagram 20**—(Leak, Displacement, & Flow), see page 278 herein. This series of drawings shows the sequence of events as wellbore pressures dropped during displacement of mud with seawater, including the onset of flow from the oil-and-gas reservoir, and the consequences of such flow going unobserved until oil and gas were well above the BOPs.

CHAPTER 52—Count Down

20 April 2010—9:42 P.M.
00:07m:00 to Zero Hour

Barry and Tanker and a number of others got to the rig floor at the same time. Barry, disoriented by the falling deluge of mud and seawater, couldn't find Tanker. He scrambled man to man. Found assistant drillers. Derrick men. Roustabouts. He found Tanker at the BOP control panel with Daylight. They were huddled, ear to mouth, mouth to ear. Barry moved in, deafened by the roar, ready to help.

* * *

00:05m:00 to Zero Hour

Through yells and sign language, Barry got the message from Tanker. Situation critical. Annular BOP and VBR—closed, but oil and gas already in the mile-long riser, above the BOPs. Diverter—closed. Mud-gas separator—open. Kicking fluids—increasing. Drillpipe pressure—rising. Barry—astounded both by the freight-train of mud blasting through the rig floor, and by the ugly words that slashed through his brain: *Oil and gas on the way up. Up the annulus, up the casing, up the riser, doesn't matter.* Jessica's words.

The captain of the MV *Damon B Bankston*, his workboat covered by heavy mud, radioed the rig. He was told: trouble with the well, move away, five hundred meters.

* * *

00:03m:00 to Zero Hour

Something changed. The noise level increased. Gas. Hissing. Screaming. Beyond hearing. Erupting from the derrick-mounted mud-gas-separator vent. A gas-sensitive alarm blasting into the night, all but silenced by the jet-engine-loud maelstrom. Gas enveloping the derrick. The rig floor. The pipe rack. Cranes. Buildings. People.

* * *

00:02m:00 to Zero Hour

Multiple gas alarms—all directions. The roaring louder, deafening. Vibrations like an earthquake. The facility and derrick creaking and groaning. As if alive. As if trying to expel an unwelcome visitor.

* * *

00:01m:00 to Zero Hour

Gas too thick to breathe. Rig floor packed with hands. Doing. Helping. Barking orders. "Clear the stairs." "Open the goddamn diverter!" "Close the blind shears." A big man, broad shoulders, stood alone, his head bowed in prayer. Gas volume increasing, pressure escalating. The mud-gas separator gagging from the overload, trying to vent gas turned black with oil. Crude oil.

Natural gas and atomized crude oil being sucked into the fresh-air inlets to the rig. Down into the engine room. Feeding the diesel engines like nitromethane in a top-fuel dragster. The engines winding up. Turbocharged. Beyond red line.

Barry spotted a figure on the pipe rack, looking up. Lime green coveralls. He screamed, "Go back!"—couldn't hear his own words. He hurdled steps three at a time down to the pipe deck, slick with oil mud and crude oil, fell hard, and—

* * *

20 April 2010—9:49 P.M.
00:00:00 . . . Zero Hour

The explosion and fireball consumed all in its path. In the first tick of the second hand, the hungry fire fed on plastic, rubber, flesh. The next second, aluminum. Then steel.

Those who could, ran, walked, crawled. Mustered strength. Ignored pain. Survival the goal. Clock ticking.

The second explosion followed the first by ten seconds. The driller's station collapsed. The drill floor evaporated. BOP control lines disappeared. The traveling blocks fell. The mudlogging unit vaporized. Cranes melted. Doors blew open. Walls became doors. The Halliburton unit glowed red—then black. The engine room ceiling fell to the floor, and the floor fell even deeper.

And all went black. Every light on the *Deepwater Horizon*. The supercharged engines—down and quiet. Rig power—off. The galley—dark. Quarters—dark. Walls collapsed into hallways. Ceilings down. Air filled with volatile gas fumes. Personnel scrambling to get dressed. Every living person on the rig, in adrenaline overload, swimming in blackness and tears, defined life by the fight for survival, the love of parents and spouses and kids, the incessant roar, and a litany of where the hells, sonsofbitches, oh Gods.

Barry struggled to his feet—Jessica gone.

Then, emergency lighting, flashlights, in rooms, down smoke-filled hallways—a glow of hope.

* * *

Through the public-address system, the words *Abandon ship* met a willing audience. Barry made his way between muster stations and spotted Jessica among the mass confusion as a number of helping hands got her and others into one of the first lifeboats to be lowered from the rig. On his way back to his assigned station, the guys he met on the main deck—most wearing life vests—were either stunned mute or barking orders, being brave or looking lost, helping the injured or being helped. This, while the deafening, roaring, oil-and-gas-fueled fire melted steel less than a hundred feet away.

For some, the fire and the waiting for a seat on a lifeboat were too much. Twice, Barry could only watch as individuals crawled over handrails, hesitated, and dropped to the sea. He remembered the numbers—a 60-foot drop, two seconds, 35 miles per hour. Fearing for the jumpers' welfare, he imagined the same scene at the aft end of the rig, at other muster stations, though the fire had already claimed major portions of the deck area as no-man's land.

Barry stayed and helped lift injured parties into the second lifeboat and a life raft, all the while twisting and turning to cool his bare head and smoldering body parts.

As the conflagration continued to consume the massive drilling rig and derrick, Barry took one of the last escape-vessel seats. Crowded. Hot. Panicked voices. His knotted guts on fire, he thought of Jessica. The same Jessica who feared deep water, heights, fire.

She would have rightfully called the blowout a ten.

Like Spindletop. Except at sea. And on fire.

He wondered, too, what she might call him.

* * *

For a long hour, then an even-longer second hour, Jessica Pherma shivered on board the MV *Damon B Bankston*, still wearing her torn and blackened life jacket. After midnight, she walked among the survivors, touched hands, and shared tears, like a nurse on a battlefield.

Crew on the *Bankston*'s fast-rescue runabout had collected survivors and delivered them back to the mother ship, then had teamed with Coast Guard helicopters as they hovered impossibly close to the burning rig, their spotlights working the water, checking shadows, looking for survivors. Looking for bodies.

Tanker was missing. As was Daylight. The mud engineers. Floor hands. The crane operator. And others. Jessica kept vigil. Watching boats, watching choppers, afraid to think the worst, to rehash her dad's warning, to speak the words—*nowhere to run.* Yet, rumors and stories about injuries and deaths ricocheted around the cabin and onto the decks, filling the night with manmade white noise, albeit a whisper in the roaring dirge a quarter mile away.

She'd watched as Barry found a seat on the forward port corner of the aft deck. Alone in his private little world, the only sound the roar of the fire in the distance, he sat quietly, his arms wrapped around his knees. Hair fried. Coveralls blackened. Boot on one foot, a soggy, blackened sock on the other. He hadn't moved since. Just stared into space. Yet he nodded when they were face to face.

She closed in, stood at his feet. He looked like he needed a few kind words, perhaps a friend, maybe a hug. Not unlike every demoralized soul who shared the boat, who waited for a miracle, a miracle that would turn back the clock.

A miracle that wouldn't happen.

She knelt in front of her mentor, eyes on eyes. He was a good man. A smart man. Highly experienced. Dedicated to doing his job. Driven by budget and schedule.

His mouth twisted as if looking for a word.

She shifted her body weight, favoring her bloody knees, then pointed over his shoulder toward the rig, burning bright in the night sky. "Your well, Barry. Your well."

He looked through her, didn't answer.

She had more to say, words she'd practiced since leaving the helideck and abandoning the rig. A simple summary of exactly what happened. He could have prevented the leak, but hadn't. He could have found and repaired the leak, but hadn't. He could have made sure the well was secure, but hadn't. He could have noted the well was flowing, but hadn't, until it evolved so extensively as to mimic the irony of Piper—oil and gas on the wrong side of the closed BOP. Which was way too late to stop.

But he already knew such things. Would know them forever. Would know forever that he was the man, the man in charge of managing the well, who could have, should have, saved the day.

Unable to comprehend the horror and complexity of the days and weeks to come—putting out the fire, recovering bodies, mourning the dead, securing the well, unraveling the cause of the disaster—her personal and family issues seemed mere puffs of smoke.

She had no doubt that Barry would get his day in court.

And that she would be on the other side of the aisle.

Jessica Terra Pherma got off her knees. She looked toward the burning remains of the *Deepwater Horizon* and said goodbye to Tanker, and to Daylight, and to both mud engineers, and to her favorite crane operator, and to other missing friends and colleagues, then walked away.

In the wheelhouse, she waited her turn to use the phone.

Wiped her eyes. Dialed.

A raspy voice answered.

"Mom, it's Jessica."

End

EPILOGUE

EPILOGUE—PART 1—

LOOKING BACK

People

Eleven personnel died the night of 20 April 2010, each working to control the well, to save the rig, to save colleagues, to save themselves. None was able to escape the holocaust. Their bodies were not recovered.

Transocean Toolpusher—Jason Anderson—35
Transocean Driller—Dewey Revette—48
Transocean Assistant Driller—Donald Clark—48
Transocean Assistant Driller—Stephen Curtis—40
Transocean Crane Operator—Dale Burkeen—37
Transocean Derrick Man—Roy Kemp—27
Transocean Floor Hand—Karl Kleppinger—38
Transocean Floor Hand—Shane Roshto—22
Transocean Floor Hand—Adam Weise—24
MI-Swaco—Mud Engineer—Gordon Jones—28
MI-Swaco—Mud Engineer—Blair Manuel—56

Transocean's chief mechanic, working in the power-generation room, was severely injured by both explosions. Yet, he climbed from the wreckage and survived his jump to the sea. He wasn't the only jumper.

The captain of MV *Damon B Bankston* and his crew, supported by the *Bankston*'s fast-rescue power boat and a flotilla of other boats that sped to the scene, heroically stayed by the rig and took on survivors pulled from the sea and from lifeboats.

At 23:22, the first US Coast Guard helicopter arrived on location—others followed. Lighting the scene, they lowered rescue swimmers into the water, supplied medical assistance, and flew the most severely injured to the closest hospitals.

One hundred and fifteen personnel survived the tragedy, of whom seventeen were injured, a number seriously. Others were deemed physically okay, their long-term condition to be determined.

The four Transocean and BP visiting VIPs survived, though one was seriously injured.

All BP staff who had been working on the rig survived.

Fictional Characters

The fictional character Tanker Forster carried a big load—three rig-management jobs in one. In real life, he would not have been the lone Transocean leader on the sophisticated *Deepwater Horizon*. The jobs he collectively represented included the Offshore Installation Manager (OIM), the Master (Captain), and at least three toolpushers (senior, night, and day). All the positions he represented changed out every fourteen days. A number of these senior managers were on duty the afternoon and evening of the disaster. One of the toolpushers died.

Fictional driller David "Daylight" Stalwart represented a group of hard workers—day and night drillers, plus day and night assistant drillers, all changing out on staggered three-week schedules. Three of their group died that night, as did a crane operator, a derrick man, and three floor hands, a silent testament to their collective dedication to the job at hand.

The fictional character Barry Eggerton, representative of BP's entire offshore management and engineering staff throughout the well, shouldered the blame for mismanaging and misinterpreting a number of operating and engineering steps during those last few hours on the rig. In the story, Barry took the blame as if he did it all by himself. In real life, BP had a number of representatives on the rig.

Along with a hundred others from Transocean, Halliburton, Sperry Sun, M-I Swaco, Schlumberger, Weatherford, etc., the entire BP drilling and operations crew, offshore and onshore, including the visiting VIPs and others, were called into federal hearings for under-oath depositions that lasted deep into 2011. Depositions are published, and the complete results and findings of the investigations were published and made available to the public before and on 14 September 2011.

It's too bad fictional Jessica Pherma wasn't represented by a real-life character on the rig. Yes, in real life BP had geologists on the rig, but apparently none was key to activities during those last two critical days. Had fictional character Jessica been on BP's payroll, she would have raised a flag and declared "STOP WORK!" (a mandated option available to every person working on the facility) after observing the excessive de-ballooning and "anomalous" pressure data during the first negative-pressure test. BP and Transocean and Halliburton would have agreed with her by acclamation and halted the abandonment procedure. Then they would have worked together to fix the casing leak with a simple remedial cement job. Not high-tech science, not a function of well depth, not related to water depth, not related to the culture of a company, nor due to regulatory oversight—the simple solution would have been arrived at entirely through the fundamentals of Drilling 101, combined with proven, standard, industry operating procedures for offshore drilling rigs around the world.

With the casing leak fixed, there would have been no catastrophe. Yet, the casing leak did not *cause* the blowout. People caused the blowout. That topic is yet to come.

The Aftermath

The half-billion-dollar *Deepwater Horizon* burned throughout the next full day and sank at 10:22 A.M. on 22 April 2010.[143]

By then the media were fully cranked up and smelling blood. Investigations began immediately.

Few adult Americans and other viewers from around the world will ever forget the plumes of oil blasting for almost three months from mangled drilling equipment on the seafloor. Nor will they forget oil-covered wildlife, damaged beaches, out-of-work fishermen, foreclosed homes, or drilling crews being sent home—the industry slapped with a federal ban on deep-water drilling, which was lifted in late 2010, albeit with a hold on new drilling permits.

The Question

That said, the families of the men who died deserve answers. They want to know why.

And fictional character Jessica Pherma was correct when she answered the fictional man's question on the helideck: "Do *you* know what happened?"

She had answered with a simple yes.

Yes, she knew what happened. She knew exactly what happened.

What she knew is delineated below.

---

[143] See **Diagram 21**—Photo—See page 291 herein. Same as cover. Photo Credits to: Richard Braham, U.S. Coast Guard, 21 April 2010.

Diagram 21
Deepwater Horizon--21 April 2010

Photo Credits: Richard Brahm, U.S. Coast Guard, 21 April 2010

Diagram 21

EPILOGUE—PART 2—

INVESTIGATIONS

The cause of the blowout, even as the *Deepwater Horizon* sank on Thursday April 22, was unknown.

Investigations began immediately.

Each of the survivors, as well as those who died, had a vested personal interest in and corporate responsibility for the successful execution of mandated tests that ensured both the safety of the well and the safety of the rig. Toolpushers and company men depend extensively on each other to look out not only for the rig and the well, but for each other and all personnel on the facility. Each survivor was called in to testify as to his or her knowledge of and personal role in the events before and on the evening of 20 April 2010.

Eleven testimonies, unfortunately, came in the form of hearsay only.

Key investigations follow:

House Subcommittee on Oversight and Investigations

On 14 June, congressional representatives Henry A. Waxman and Bart Stupak, joint chairmen of the *House Subcommittee on Oversight and Investigations*, pointed to five procedural and omission problem areas, each triggering additional investigation, justification, and second-guessing. Those areas include: Casing versus liner, Casing centralizers, Mud circulation (CBU), Cement bond log, and the Casing-hanger lockdown sleeve. This Investigation is listed herein as Reference (1).

Joint Investigation (USCG/BOEM)

Additional investigations by the U.S. Coast Guard and Bureau of Ocean Energy Management, Regulation, and Enforcement (BOEM) initiated the *Deepwater Horizon* Joint Investigation (USCG/BOEM). Through depositions with more than a hundred on-the-rig and associated onshore witnesses, investigators asked in-depth questions about BP's "culture," which seems to have driven BP to lead the industry in the number of tragic safety incidents, with additional attention to Transocean's similar culture-driven history. Further, the investigators focused on Macondo-specific questions about well-design considerations, leadership responsibilities, cementing activities, the negative-pressure test, the riser displacement process, and the testing and functioning of blowout preventers. On 22 April 2011, the USCG issued their formal findings: Redacted Report of Investigation into the Circumstances Surrounding the Explosion, Fire, Sinking, and Loss of Eleven Crew Members Aboard the Mobile Offshore Drilling Unit *Deepwater Horizon* In the Gulf of Mexico, April 20–22, 2010, Volume I. The report addresses only at a high level, and with no mention of BP, those rig activities that contributed to the Macondo blowout, yet focuses with detailed scrutiny on the post-blowout fire, control, evacuation, safety systems, and leadership criteria within Transocean and aboard the *Deepwater Horizon*. Note, the BOEM Final Report is listed below. The USCG/BOEM Investigations are noted herein as Reference (2A)

BP Internal Investigation

On 8 September 2010, BP issued the results of their own candid internal investigation. They made available new data not previously seen by the public. The data answered a number of technical questions. Had BP personnel on the rig been as conscientious as the BP investigative team, the blowout would not have happened. This Investigation is listed herein as Reference (3)

The National Academy of Engineering (NAE/NRC)

The National Academy of Engineering/National Research Council (NAE/NRC) committee issued their Interim Report 16 November 2010. The committee issued an "Update" on 1 July 2011. The "Update" referenced the NAE/NRC 16 November 2010 report, with an added statement: "At the request of the Department of the Interior (DOI), a National Academy of Engineering/National Research Council (NAE/NRC) committee is examining the probable causes of the *Deepwater Horizon* explosion, fire, and oil spill in order to identify measures for preventing similar harm in the future. The committee will address the performance of technologies and practices involved in the probable causes of the Macondo well blowout and explosion on the *Deepwater Horizon*. It will also identify and recommend available technology, industry best practices, best available standards, and other measures in use around the world in deep-water exploratory drilling and well completion to avoid future occurrence of such events." The NAE/NRC Investigations are listed herein as Reference (5).

National Commission

In mid-January 2011, results and recommendations were issued by (the president's) National Commission on the BP Deepwater Horizon Oil Spill and Offshore Drilling (see References for access). Two key quotes from the report: (1) from page vii: "The immediate causes of the Macondo well blowout can be traced to a series of identifiable mistakes made by BP, Halliburton, and Transocean that reveal such systematic failures in risk management that they place in doubt the safety culture of the entire industry," and (2) from page viii: "Though it is tempting to single out one crucial misstep or point the finger at one bad actor as the cause of the *Deepwater Horizon* explosion, any such explanation provides a dangerously incomplete picture of what happened—encouraging the very kind of complacency that led to the accident in the first place." This Investigation is listed herein as Reference (4).

Comment by author on (1) and (2): For decades, educated and experienced company men—primarily petroleum engineers, but also mechanical, structural, and geological engineers, as well as re-schooled toolpushers—have accepted responsibility for managing and understanding the mechanics and technology of drilling wells—both the manmade parts and the Mother Nature parts. Most every well ever drilled has the physical capability of blowing out. When a blowout happens, it's not because the industry is complacent. Far from it. Rig owners and operators and rig personnel (the industry) are all too aware of the consequences of failure, the horror side of cause and effect. Hence, a litany of clichés. Zero tolerance for failure. No accident's an accident. Further, though blowouts may be caused by force majeure (in the spirit of lightning, hurricanes, earthquakes, tsunamis, terrorism), they are not caused by Mother Nature on a deep-geology rampage. They are caused by neither company culture nor the safety culture of the industry. Nor are they caused by complacency in regulatory-oversight. No—a blowout (loss of well control) is caused most often by the failed actions and inactions of one or more key persons on a rig—the individual(s) responsible for and in charge of managing all aspects of the well being drilled. With few exceptions, especially on challenging, difficult, expensive (i.e., offshore) wells, that person is the well-site leader, also known as the company man.

BOEM—Final Report

The final BOEM report (see USCG/BOEM above) was issued 14 September 2011. Its title: The Bureau of Ocean Energy Management, Regulation, and Enforcement (BOEM): Regarding the Causes of the April 20, 2010, Macondo Well Blowout. This final report is listed later as Reference (18A). An excellent and comprehensive report. All relevant topics to be discussed in two sections that follow: (1) Possible Contributors to the blowout, and (2) Direct Contributors to the blowout.

Nevertheless, key to the report is a statement in Article C— Temporary Abandonment, Kick Detection, & Emergency Response, Section XVIII, Conclusion: Pp 200: "BP personnel and Transocean personnel failed to conduct an accurate negative test (NPT) to assess the integrity of the production casing cement job."

Comment by author: The BOEM statement quoted above about a critical topic is grossly misleading—as follows. The NPT is designed to show which of two cases exists: (1) The casing evidences no leak—all is well, continue with abandonment, or (2) Something is leaking—stop the abandonment procedure—investigate and repair the leak as necessary— then repeat the NPT. Contrary to the quoted BOEM statement (". . . failed to conduct an accurate . . . (NPT) . . ."), the documented NPT obtained on 20 April between 16:54 and 18:35 (including the "anomalous" 1,400 psi) was NOT a failed test. The test (often referred to as NPT #1) was accurate. It behaved exactly as designed. The unambiguous NPT #1 results indicate the casing was leaking, with direct pressure communication to high-pressure formations in the annulus. Unfortunately, senior rig personnel in charge of the well considered the "accurate NPT" and its valid test results to be "anomalous," with disastrous consequences. Namely, the casing leak remained undetected during the subsequent abandonment process, even as the well began to flow, undetected, until too late—and became BP's Macondo blowout.

Det Norske Veritas (DNV)

A Joint Investigation Team of the Departments of the Interior and Homeland Security was charged with investigating the explosion, loss of life, and blowout associated with the *Deepwater Horizon* drilling rig failure. As a part of this overall investigation, Det Norske Veritas (DNV) was retained to undertake a forensic examination, investigation, testing, and scientific evaluation of the blowout preventer stack (BOP), including its components and associated equipment used by the *Deepwater Horizon* drilling operation.

The objectives of the investigations and tests were to determine the performance of the BOP system during the well control event, any failures that may have occurred, the sequence of events leading to failure(s) of the BOP, and the effects, if any, of a series of modifications to the BOP Stack that BP and Transocean officials implemented. The DNV Investigation is listed herein as Reference (6)

National Commission—Chief Counsel's Report

National Commission on the BP *Deepwater Horizon* Oil Spill and Offshore Drilling—Chief Counsel's Report—2011. Released 17 Feb 2011. Included as Reference (20), below.

Transocean

Transocean—MACONDO WELL INCIDENT—Transocean Investigation Report—Volume I and Volume 2—June 2011. Included as Reference (21), below.

EPILOGUE—PART 3—

POSSIBLE CONTRIBUTORS

The effects of BP's Macondo blowout were and continue to be tragic and staggering for the entire population of the Gulf Coast community. No one wants such a catastrophe to ever happen again. With that goal in mind, we need to address the *causes* of the Macondo blowout. As we do, we'll watch for those technical and operating procedures the industry needs to fix, as well as those proven technical and operating procedures that will forever depend on proper execution by responsible rig leaders industry wide.

The good people on the *Deepwater Horizon* did not have the advantage of such in-depth analysis noted in the numerous investigations cited above and as used throughout *THE SIMPLE TRUTH*. They did not have the advantage of hours of 20/20 hindsight by teams of engineers and attorneys. What they did have was education, experience, responsibility, and a ticking clock.

In that spirit, the following items are possible contributors to the blowout.

Production Casing

A number of design and operating issues, had they been executed otherwise, likely would have led to the normal temporary abandonment of the well; i.e., without the blowout. High on that list is the choice to run the full string of casing, rather than the more-expensive short liner with a tie-back.[144] Of note: with a mechanically competent liner in place and pressure tested, the tieback could have been put off for a year or more; i.e., until the well was reentered for completion.

Barry Eggerton and Jessica Pherma covered all the right casing-versus-liner reasons in the story, the most important being the ease of testing the liner, and the tie-back, and the pressure security associated with the confirmed isolation of the pay zone from the rest of the wellbore. Having elected to run the long string of production casing,

BP was left with only one barrier between the pay zone and the seafloor—the footage of cement above the pay zone. Though the full-casing option is less common in industry than the liner and tie-back, it can be made to work by ensuring: (1) the casing is mechanically secure throughout (top, middle, and shoe), and (2) the cement job (and float collar, as appropriate) provides all necessary annular isolation and pressure integrity. Production-casing integrity can be assured only if leak-related problems with the casing and cement are fully resolved.

And one might wonder—isn't the wellhead also a form of annular isolation? Normally, yes, both for the long string and for the liner/tie-back option, but only if the casing-hanger lockdown sleeve is installed, which it wasn't.

Abandoned Gas

Industry Recommended Practices—based on problem avoidance—call for circulating bottoms-up (100 to 150 percent of annular volume) prior to cementing the casing. BP pumped approximately 10 percent of the Macondo annular volume.

Note: when trip gas is left behind prior to running casing, the auto-fill float collar allows mud *and trip gas* to be pulled into the casing as it's being run. Which means the trapped trip gas must be pumped *down* to the shoe before it can begin its voyage up the annulus to the surface.

Fictional Jessica Pherma may have sounded paranoid, but the repetitious (barbwire) ticks on the drillpipe pressure chart—Jessica compared them to pressure-cooker hisses—are real. Events of Macondo's casing-hanger liftoff are visible (on a mudlogging chart) as multiple, short-duration, pressure-relief bursts on top of the *anomalous* pressure-buildup curve to 1,400 psi (during the first negative-pressure test).[145]

The events are significant not only because the casing-hanger lockdown sleeve had not been installed, but also because they show there was pressure in the annulus, under the casing hanger. How much pressure would have been required for lift off? BP reports the necessary pressure at 260 psi to 620 psi.[146] The abandoned trip gas in the annulus, rising through 14.2-ppg mud, can easily generate such low pressures. Further, in Reference (18a), which mentions the BOEM's "Buoyancy Casing Analysis," P. 14-15 describe "Condition 5," wherein wellbore conditions are mimicked by NPT-1, concluding that "... *the casing can lift.*" And that's with no uplift from abandoned gas.

Post-blowout investigations of the Macondo wellbore in September-October 2010 used direct intervention to target an understanding of the upper wellbore and what might have leaked, failed, flowed, etc. This work was done with the BOPs in place. The casing hanger appeared correctly seated in the wellhead. The production casing passed a positive pressure test. The annulus below the wellhead was wireline logged, indicating no "free gas." A contractor perforated the production casing near 9,100 feet with 14.3 mud in the casing. Based on the lack of U-tubing (casing to annulus), investigators concluded there was no light-weight crude oil in the annulus. All this data indicated Macondo oil and gas did not blowout up the production-casing annulus. Furthermore, the data support the previous conclusion that the path of the oil and gas had been from the pay zone, down and around the shoe, and up the production casing.[147]

Such conclusions warrant comment. First, early evidence during NPT-1 (the pressure spikes on the 1,400-psi pressure buildup line[148]) indicates the casing hanger did lift (burp) when the pressure on top of the hanger was radically reduced during the test. However, the finite volume of abandoned trip gas trapped under the hanger had to have diminished with each "burp." Any remaining trip gas had an opportunity to "belch away" during the months the well blew out; hence, no "free gas" was found under the wellhead.

Second, though the "spike" argument may sound trivial, a greater implication follows. As noted above, no new gas or crude oil found its way up the annulus during the blowout. Yet, such migration would not have been a surprise. To the contrary, BP had predicted the chance of such "production-related" migration long before the discovery, as evidence by installation of rupture/burst disks in the 16-inch casing.[149] Further, Halliburton cautioned BP that without a good cement job (enhanced, for example, by an adequate number of centralizers) the well had the potential for severe gas migration.[150] Such gas migration would have been up the annulus.

Third, the intervention tests outlined above confirm the good news: The 16.7-ppg lead slurry—with or without the nitrified cement located across the pay zones—did its job. It set up. It created isolation. It prevented fluid flow. It prevented gas migration. Nothing moved up the annulus. And even during the maximum annular pressures caused by the rising trip gas—evidenced by the pressure-cooker effect—no lost-circulation zone broke down, further confirming cement isolation.

The bad news is the well did flow—from the discovery zone, down the annulus, around the shoe, and up through the shoe track and float collar. Which means we need to look further into the cement job.[151]

Cement

Note: *concrete* contains cement plus sand and gravel; oil-field cement contains neither sand nor gravel. Laying a concrete patio on a hot summer day requires fast action and a number of skilled workers. Now picture an example job that's 3-1/2 miles from the cement truck, inside a pressurized bunker at 13,000 psi, heated to 260°F, with no skilled workers on the receiving end. Not quite as simple as a patio.

Operators like BP and cement service companies like Halliburton, working together, do the best they can to design the perfect cement job for the unique conditions of each well. Casing design is important. A single size of casing is the most efficient relative to cement displacement and wiper plugs. Conversely, BP used three sizes—a 6-5/8-inch-inch landing string, on top of 9-7/8-inch casing, on top of 7-inch casing. The downside of the three sizes of pipe is the increased likelihood the cement just under the top wiper plug gets contaminated by previously pumped fluids. That cement would have been the high-expectation, shoe-track cement between the float collar and the shoe.

But a more onerous phenomena awaits the shoe-track cement, as discussed below—Shoe Cement.

Further, for best results, the production casing or liner is sized to fit inside the previous casing, while leaving plenty of annulus in the open-hole wellbore for the cement yet to come. BP ran 7-inch production casing inside the previous 9-7/8-inch liner—8.63-inch inside diameter—which had been set at 17,168'. Below that depth, the well had been drilled and under-reamed to 9-7/8-inch diameter down to 18,126', then drilled to 18,360' with an 8-1/2-inch drill bit. This means the bottom 177 feet of the production casing annulus had room for a cement sheath only three-quarter-inch-thick. Above that "tight" zone, the annular cement was about 1-1/2-inch thick. A string of 5-1/2-inch production casing would have yielded a larger annulus, with more room for cement, and better odds of success.

Moreover, for best cement results, operators centralize the pipe (BP was short on centralizers), use plenty of cement (BP minimized the volume), pump at turbulent rates (BP restricted pump rate), ensure the casing rat hole is filled with mud at least as heavy as the cement (BP left

14.2-ppg in the rat hole), and monitor for losses (BP missed an 80-bbl loss during the mixing and displacement of the cement).

Then cementer and operator pump the cement down the casing and up the annulus and give it time to set up. Together, operators and cementers expect success—good hard cement where it's supposed to be. But together, they also expect there will be failures—defined as anything other than good hard cement where it's supposed to be.

Whether Macondo's cement job was expected to be perfect (as designed), or there was a statistical chance of the cement being bad (based on the noted deficiencies), BP was mandated to test the post-cement-job final product to ensure cement integrity and competent casing. BP had access to two key tests—the CBL (cement bond log) and the NPT (negative-pressure test), as well as a number of ways to remediate a discovered deficiency (cement squeeze, cement plugs, bridge plug, etc.).

The CBL is designed to look specifically at cement integrity—yet says nothing about pressure integrity of the annulus. Further, if the same production-zone cement job (and results) had been applied to a liner, the pressure barriers at the top of the liner (liner-hanger seals and cement plug on top of liner lap) would have precluded, even without a tie-back, the need to repair the production-zone cement until the well was ready for production—years later. As it was, with the long string of wide-open production casing, the associated cement-related tests necessary for integrity confirmation had to be pressure perfect from day one. BP failed to spot the 80-barrel gross loss of returns during the cement job, which should have been justification for running the CBL. The CBL would have led to remediation (not as related to annular cement quality, but because the casing leak would have been found). That said, post-well investigations indicate the 80-barrel loss occurred during the entire displacement process, with less than 3 barrels lost while the cement was actually being pumped up the annulus.[152] Nevertheless, with BP missing the *significant* 80-barrel loss during a 50-barrel cement job and canceling the CBL, BP missed the opportunity for remediating the casing-leak problem.

Moreover, the NPT is not designed to look at the quality of a cement job—it targets the pressure integrity of the post-cement-job production casing. Further, the NPT results (the first test) pointed definitively to a casing leak.

All BP had to do was discover the leak and fix it.

That said, what leak? How did the oil and gas get into the casing shoe?

Float Equipment

Though important, such equipment can be problematic, especially in a deep well; hence, a good general rule is to keep it simple. The dual-float, auto-fill float collar, on the bottom end of a tapered casing string, was anything but simple. The operator wants the floats to hold, as the floats (check valves) isolate the annulus from the otherwise wide-open wellbore until the cement sets up. Two Macondo problems got minimal attention. First, the extreme pressure necessary to initiate circulation (before the cement job) through the otherwise-open float equipment, and the subsequent lower-than-expected circulating pressure. Debris on top of the float collar likely caused a plugging problem, which was cleared only with high pressure. Given that high pressure, BP management feared the casing had been damaged, but the cementers later bumped the top plug with 1,200 psi, which showed any such casing damage (still possible) had to be below the top of the float collar (where the plug bumped). More likely, the floats were damaged by forcing the debris out of the floats with high pressure, resulting in the lower-than-expected circulating pressure.

Second, minimal attention was given to ensuring the circulating pumps reached at least the minimum rate necessary to convert the floats. BP's maximum documented rate of 4 bpm was at least 20 percent less than the 5–7 bpm required to convert the floats. If the cement was green, contaminated, or otherwise bad outside the casing, and the floats were not converted (holding backpressure), the wellbore was wide open to the annulus. In this case, the cement would backflow into the wellbore after the cement job.

But when the 1,200 psi was bled off the casing (after the top plug bumped), the greater-hydrostatic-head annular cement did not u-tube from the annulus back into the casing. Had such backflow occurred, BP would have known the floats were not converted and the float-collar flappers were leaking. Nevertheless, subsequent investigations by BP and others indicate the calculated backpressure from the 51-barrel nitrogen-infused cement job was only 38 psi (and would have been less if any of the 80-bbl loss had been heavy cement). This small pressure, in an 18,300-foot-deep well, is equivalent to only 0.04 ppg (too small to measure, and otherwise known as *balanced* pressures). Which means the *lack* of backflow was not a good indicator of float-collar pressure integrity.[153]

Shoe Cement

To ensure the casing hanger reached the wellhead before the casing shoe reached the bottom of the well, the production casing was set off bottom by almost 60 feet. The resulting rat hole (under the bottom of the casing) contained 4 barrels of 14.2-ppg mud.[154] The shoe-track cement inside the 7-inch casing shoe, up to the bottom of the float collar (188 feet), totaled about 7 barrels. The shoe-track cement slurry weighed 16.7 ppg.

Outside the casing, bottom-up, was some of the same 16.7 ppg cement slurry, plus 14.5-ppg nitrified cement, which together—from the shoe up to the bottom of the deepest pay zone (about 100 feet)—totaled only 2 barrels.

Either of two Recommended-Practice industry standards would have changed the picture: (1) set the casing closer to bottom, (2) fill the rat hole with heavier mud (i.e., 17 ppg) before running the casing.[155] BP did neither, with the resulting problem: two fluids, heavy water-based cement on top of less-heavy, oil-based mud, do not stay that way for long. The fluids exchange places.[156] The heavy cement from inside and outside the casing starts falling into the rat hole, and the light-weight mud in the rat hole starts climbing up the casing into the shoe-track, and up the annulus toward the pay zone.

And the resulting mix, inside and outside the casing, is no longer good cement, regardless of the number of centralizers.

The industry calls mud-contaminated cement inside the shoe a *wet shoe*. In the annulus it's called *contaminated cement*. On a CBL (cement bond log) it would appear as *no cement*.

Remediation of a wet shoe and mud-contaminated cement in the annulus is called a cement squeeze job.

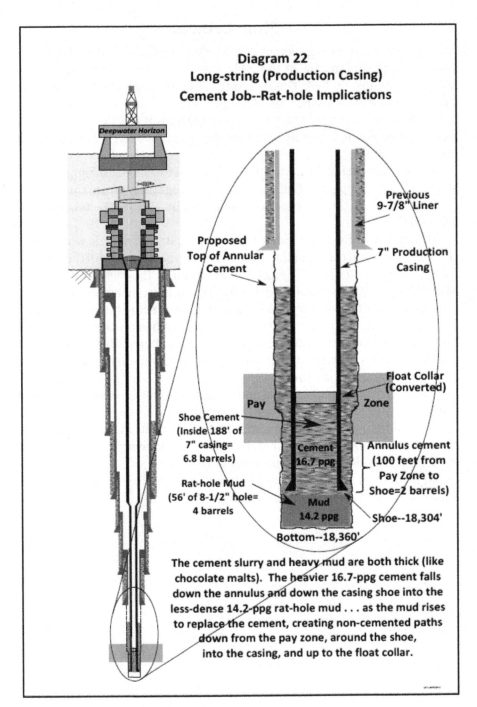

Diagram 22
Long-string (Production Casing)
Cement Job--Rat-hole Implications

Previous
9-7/8" Liner

Proposed
Top of Annular
Cement

7" Production
Casing

Float Collar
(Converted)

Pay   Zone

Shoe Cement
(Inside 188' of
7" casing=
6.8 barrels)

Cement
16.7 ppg

Annulus cement
(100 feet from
Pay Zone to
Shoe=2 barrels)

Rat-hole Mud
(56' of 8-1/2" hole=
4 barrels

Mud
14.2 ppg

Shoe--18,304'

Bottom--18,360'

The cement slurry and heavy mud are both thick (like
chocolate malts). The heavier 16.7-ppg cement falls
down the annulus and down the casing shoe into the
less-dense 14.2-ppg rat-hole mud . . . as the mud rises
to replace the cement, creating non-cemented paths
down from the pay zone, around the shoe,
into the casing, and up to the float collar.

Diagram 22

The Leak

The above criteria set the stage. The non-converted and/or damaged float collar, plus a mud-cement mix in both the casing shoe-track and the lower annulus, created the un-cemented open pathway (the leak) for formation fluids from the pay zone, down 100 feet of annulus, up the casing shoe track, through the open float collar, into the casing, and to the surface.

Fortunately, at the end of the cement job (as rat-hole mud was quietly exchanging places with the shoe and annular cement) the well was still full of 14.2-ppg mud, which provided plenty of hydrostatic head to keep the formation fluids at bay. With the well static and stable, BP had two shots at discovering the leak: the cement bond log and the negative pressure test.

All they had to do was find the leak and fix it.

Blowout Preventers

BOPs provide the valves and pressure lines necessary for well control. Chronologically, if a formation kicks, and the BOPs fail to work, and rig crews are unable to control the well with failed BOPs (i.e., by pumping heavy mud or cement into the wellbore), then kicking formation fluid will eventually displace the wellbore and flow freely to the surface—by definition, a blowout. After the Macondo disaster, there were a number of questions about the *Deepwater Horizon* BOPs. Had they been inspected in a timely manner? Had the batteries been maintained? Had proper notification been made to the MMS when a control pod was found to be leaking hydraulic fluid? Were the BOPs actuated by rig crews after mud blew through the rig floor? Did the BOPs work? The seemingly obvious answer to this last question is no, based on 86 days of oil and gas venting from mangled equipment on the seafloor.

Nevertheless, the US Department of the Interior contracted Det Norske Veritas (DNV), a world-class maritime certifying authority, to forensically test the BOPs after they had been carefully salvaged from the ultimately secured Macondo well and taken to the NASA Michoud facility in New Orleans. Four significant findings answer some of the key questions noted above: (1) the upper annular preventer was found to be closed around the 5-1/2-inch drillpipe. *Note: the annular was closed throughout the NPTs, but was opened for the displacement process, which means crews closed it when the well visibly kicked,* (2) the upper variable-bore rams (VBRs) were found to be closed around the drillpipe, (3) the drillpipe trapped between the closed annular preventer and the closed VBRs was found to be elastically buckled by the upward force of the uncontrolled flow of formation fluids, and (4) the blind shear rams (BSRs), which are also located between the annular preventer and the VBRs, had been actuated for closure, but could not complete the close-shear-seal cycle because of the buckled, off-center drillpipe the shear rams encountered. The BSR had closed either automatically when power to the BOPs was lost (after either the first or second explosion), or later during manual intervention using an ROV—remote-operated vehicle, a small submarine. Because the BSR was unable to fully close and cut the drillpipe and seal the well, the well continued to flow through the drillpipe until finally "top killed" on 15 July 2010.[157]

BOPs are designed as redundant, brute-strength, multi-option, pressure controllers. When wells kick and rig crews take action, BOPs respond with integrity, and the control of even the most intractable kick is likely just a matter of time. BOPs are somewhat like brakes on a car. We get brakes inspected so we can trust they will work in an emergency. But too often, drivers fail to see trouble, fail to apply their brakes, until too late. Think—seventy mph on the interstate with a stopped 18-wheeler over a rise, or Thelma & Louise in mid-air over the Grand Canyon. The *Deepwater Horizon* BOP, had it had a brain, might have wondered about the freight-train load of mud, oil, and gas blowing through its guts at an accelerating rate—for more than a half hour—with no driver jamming the brakes. That wasn't supposed to happen. Industry-wide, every leader of rigs mandatorily trains to detect kicks as early as possible—like checking drilling breaks for flow—and shutting-in the well without hesitation. Nobody trains, or expects, or should ever allow a kick to progress from a 12,000-psi pay zone near 18,000 feet to blow hydrocarbons over the crown before closing the BOPs. When two of the Macondo BOPs (one annular and one VBR) were finally actuated, they closed around the drillpipe as they were supposed to. But the runaway oil and gas, already above the BOPs, had a head start. The explosively expanding gas in the riser met no obstacle, somewhat akin to the popular Mentos/Diet Coke experiment—except the offshore Coke bottle was the 21-inch-diameter, mile-long riser.[158] The gas kept coming, carrying crude oil with it, through the rig floor and over the derrick.

Further, even as the annular BOPs closed, the rapidly moving kick fluid sliced and eroded its way into the body of the steel drillpipe, precluding the sealing mechanism.

Moreover, with the annular BOP closed above a (rapidly eroding) drillpipe tool joint, the still-high-velocity kicking fluid lifted the next deeper joint of drillpipe and crammed and buckled it between the closed annular and variable-bore rams. Such deformation occurred within the cavity of the blind shear rams, which were disabled in the process. Like the speeding car, brakes locked, tires smoking, trying to stop, skidding into the truck—a non-brake-related terminal tragedy.

Kudos to the rig hands who fought the kick, who closed the BOPs, who jammed on the brakes, who fought to the end—they did the best they could under the circumstances.

And kudos to the BOP—it, too, did the best it could under the circumstances.

What If?

Given the casing had a leak, none of the above items is a positive contributor to the sequence of events that *caused* the blowout. Yet, few could argue that: (1) *IF* the well had been completed with a liner (including tests to ensure liner-cement-pressure integrity), with or without a tie-back; or (2) *IF* the 80 barrels of lost circulation during cement displacement had led to a CBL, which would have led to remediation; or (3) *IF* the 7-inch casing had been set deeper or the rat hole had been filled with heavy mud, allowing the high-integrity cement in the shoe track to set up and do its intended job; or (4) *IF*, when the first step of NPT-1 showed the casing did not have pressure integrity, standard leak-remediation measures had been implemented; or (5) *IF* LCM had not been pumped into the riser (and fallen into the BOP) so as to give NPT-2 a chance to reveal the leak so it could be remediated, *THEN* the blowout would not have happened.

Items (1)-(5) are moot points.

The leak was still there. The well flowed. The blowout did happen.

---

[144] Reference (1)— House Subcommittee Letter to BP's CEO—dated 14 June 2010. " ... The liner-tieback option would have cost $7 to $10 million more and taken longer ... but it would have been safer because it provided more barriers to the flow of gas up the annular space ..." Also References (7) and (8): BP's management and engineering decisions about the production casing changed—first a long string, then it was the liner and tie-back, then back to the long string.

145 Reference (3)—BP's *Deepwater Horizon* Accident Investigation Report. Internal investigation released to the public on 8 September 2010. Section 2.6 Interpreted the Negative-pressure test. Figure 4—April 20, 2010, Negative-Pressure test (Real-time Data). P. 88 . . . (**Author's note: see page 421 herein**)

146 Reference (3)— BP's *Deepwater Horizon* Accident Investigation Report. Internal investigation released to the public on 8 September 2010—P. 73: (1) An analysis using worst-case conditions (non-cemented casing, no friction, and fully displaced seawater above the drillpipe) indicated that approximately 260 psi of additional pressure in the annulus could initiate lift off. (2) A more realistic analysis . . . required 620 psi.

147 Reference (18A)—BOEM—Article IV, Possible flow Paths, P. 61–74. " . . . concluded that hydrocarbon flow during the blowout occurred through the . . . production casing from the shoe track (the casing between the float collar and the shoe) as a result of float collar and shoe track (cement) failure."

148 PRESSURE BUILDUP and PRESSURE SPIKES— Reference (3). P. 88, Item 2.6, Figure 4, Annotated Item 8: "Drillpipe pressure slowly increases to 1400 psi." Shown in graphical form, this is mudlogging data from the rig . . . (full scale copy on page 421 herein). Petroleum engineers assess O&G formations by analyzing the changing rate of pressure buildup when the target formation is allowed to flow into a low-pressure test chamber, like drillpipe. The "1,400 psi" pressure recording on Macondo is a typical pressure-buildup curve showing communication with the flowing formation. Further, the curved line exhibits atypical spikes, like short-duration power surges, related to casing lift off. As the pressure flattened off toward 1,400 psi, the frequency of the spikes also decreased, as one might expect.

149 Reference (3)—BP's Wellbore schematic—P. 19. Note installation of rupture/burst disks at 6046, 8304, and 9560 feet.

150 Reference (18A)—Article III, Section C: Gas Flow Potential, P. 46–49.

151 Reference (3)—BP's Internal Investigation—Item 5.4, P. 78. Additionally, Reference (18A) BOEM—Article III: Cementing, P. 40–60.

152 Reference (16)—USCG Hearing 24 August—Articles 424–426—Halliburton cementer on *Deepwater Horizon*—testifying about 80-barrel loss noted on Sperry Sun mud log during the 51-barrel cement job, which had not been noticed on the rig. Nevertheless, additional investigation, Reference (18A), P. 71, indicates that fluid losses during the period cement rounded the shoe and was being pumped up the annulus totaled less than 3 barrels. Note—if the rate of

loss of mud/cement after the cement job matched the rate of loss during cement displacement, the cement could have "fallen" down the hole into the loss zone; hence, the need for a CBL.

153 Reference (3), P. 71. The 38 psi was estimated by Halliburton's pre-cement-job model.

154 See **Diagram 22** (see page 304 herein) in this chapter for pictorial of rat hole at bottom of Production Casing.

155 Reference (18A)—BOEM—Article III Cementing, Section G Industry Standards, P. 57—As per "API RP 65 Section 7.5 . . . If casing is not run to bottom, the "rat hole" should be filled with a higher weight mud to prevent cement from falling into the rat hole and displacing rat hole mud into the cement column . . . "Refer to **Diagram 22** for implications of rat hole (page 304 herein).

156 DENSITY SEGREGATION OF FLUIDS—A good visual example is a Lava Lamp, where two colored fluids with minutely different densities, triggered by a heated lamp, exchange places at a hypnotizing pace. Oil and vinegar also separate, though much faster. Now picture two fluids—water-based cement slurry at 16.7 ppg, on top of oil-based mud at 14.2 ppg. Unless the cement flash sets, the density segregation is inevitable. The Macondo cement was heavily retarded to ensure it stayed a slurry for a number of hours. Note: visuals of Lava Lamps are available at numerous Internet sites.

157 Reference (6): Det Norske Veritas (DNV) Forensic Examination of *Deepwater Horizon* Blowout Preventer. 20 March 2011.

158 MENTOS© / DIET COKE© experiment. An explosive discharge of carbon-dioxide foam from the candy-Coke mixture is referenced as a visual for the much-more-significant volumes of expanding gas erupting from the massive Macondo riser above the closed BOPs and over the 242-foot-tall derrick. Visuals for the experiment are available from numerous Internet references.

EPILOGUE—PART 4—

DIRECT CONTRIBUTORS

In addition to the above potential contributors, a number of other design and operating issues contributed directly to the blowout. For the purpose of clarity, we'll call them *high-alert* issues. *High-alert*, in that each event, silent but visible, was subject to being quashed by the company man (his paid job responsibility), or by any other engineer, toolpusher, driller, mudlogger—on *high-alert*. Every senior technical and operations person working on such a facility should be on *high-alert* at all times. It's in their education and training. It's in their job scopes. It's in their respect for the job. It's in their instinct for survival. Unfortunately, no *high-alert* person stepped forward on the evening of 20 April 2010 to challenge a number of facts. Facts in the form of hard evidence. Evidence of an escalating problem. Evidence on gauges and graphs. Highly visible. Readily seen. None complex. All mechanical. Simple engineering. Easily explained. All pointing to common problems begging for repair, even as the clock ticked down. Yet, rig management—contrary to industry-wide proven procedures—ignored the simplest of pre-emergency red-flag evidence, as well as subsequent confirming evidence, even as the growing crisis escalated toward its explosive unveiling. That's unacceptable for the oil-and-gas industry, and it's unacceptable for skilled offshore experts, especially for company men and drilling engineers.

Accordingly, without a *high-alert* person to step forward, the direct *causes* of the blowout are as noted below.

First Negative-Pressure Test—(NPT-1)

At about 5:55 P.M., as part of the first negative-pressure test (NPT-1), the BP-Transocean-Sperry Sun-Halliburton NPT test team, with BP fully responsible and in command, opened a valve to bleed about 2,400 psi of trapped U-tube pressure from 8,367 feet of drillpipe that had been filled with seawater, now under a closed BOP.

Here, an example: Household garden hose. Spray nozzle closed. Open the faucet. The hose swells up (balloons), but water goes nowhere. Close the faucet. Now pressure is trapped in the hose (closed on both ends). Open the nozzle—water sprays out as the pressure in the hose declines (the hose de-balloons). How much water sprays out? Let's say a pint.

The drillpipe, full of seawater, protruded down into the wellbore, below closed BOPs. The entire 18,300-foot-long steel tubular system (drillpipe plus casing) was fluid-filled, pressured-up to 2,400 psi, and closed on both ends. *Like the garden hose.* The pressured system, when de-pressured to zero at the surface (a 2,400-psi drop throughout the wellbore), might have been expected to bleed back (de-balloon) about five barrels before reaching and holding at zero psi (recall the casing bled back 5 barrels after the top plug had been bumped with 1,200 psi). Fifteen barrels later, they got the pressure near zero, but could not keep it near zero.

Fifteen is the key word.

If you open the nozzle on the hose and get back more than the expected pint—like a few gallons—and even then it keeps coming, it suggests the faucet is still open, or the faucet is leaking. What to do? Close the faucet, or fix the leak.

Bleeding *fifteen* barrels shouts: Something's leaking! Something *beyond ballooning* is bumping the pressure and supplying fluid to replace the void created by the dropping drillpipe pressure. This means that before 6:00 P.M. the wellbore had yielded 15 barrels out the top. More onerous, the wellbore remained full, which means it had *filled itself* with 15 barrels of unknown fluid (mud, green cement, and/or formation fluid), through a leak, there being no other possible source.

Further, in whatever form the leak existed, the 15 barrels effectively scrubbed open the pathway, precluding any chance of green cement eventually setting up and blocking the leak.

The company man—the man in charge of the well—had one viable response: stop the NPT, find the leak, fix the problem. Period. No blowout. That's the way it's supposed to happen. But no *high-alert* person on the rig noticed the anomaly, recognized the problem as a leak, raised a flag. Nobody said: The casing has a leak—we need to fix it.

And the answer was as simple as watering flowers with a dysfunctional garden hose—a leaking faucet on one end, a nozzle on the other.

The 15-barrel event occurred four hours and four minutes before the first explosion.

Rising Drillpipe Pressure

Instead, the NPT team gave up on problematic NPT-1 and its "anomalous" volumes and pressures. They turned their attention to rerunning the negative-pressure test (NPT-2) using the kill-line, instead of the drillpipe, as stipulated and submitted in BP's well plan to the MMS. Yet, BP's goal remained unchanged: prove the production casing had pressure integrity so rig crews could pull the BOPs.

Early during the kill-line test (NPT-2), from 6:00 P.M. to 6:35 P.M., the pressure recorded on the then-ignored NPT-1 drillpipe (still filled with seawater, still under the closed BOP), steadily rose to 1,400 psi, where it stabilized. The climb in drillpipe pressure, linked into the otherwise-static fluid-filled wellbore system, meant fluids from the *annulus* had continued to bleed (likely another five barrels) into the casing through the leak. The 2400-psi *underbalance* created by NPT 1 allowed formation fluids to flow through the leak and into the fluid-filled system, until *balance* was reached. Balance occurred with 1,400 psi on top of 8,367 ft of seawater, under the closed BOP.[159] The mud-weight equation says 1,400 psi, on top of 8,367 ft of seawater in the drillpipe, on top of about 10,000 ft of 14.2-ppg mud below the drillpipe, matches the formation pressure (about 12,500 psi) of the deep oil-and-gas reservoir (pressure equivalent to about 13 ppg).

BP did not meet its goal. The casing did not have integrity. It had a leak.

Of note, the 1,400 psi has a drilling-related equivalent. It's the same as shut-in drillpipe pressure (SIDPP), which is normally taken immediately after a drilling kick is detected and the BOPs are closed. For a drilling kick, SIDPP on top of the known footage of clean drilling mud (in the drillpipe) adds up to the pressure of the kicking formation. The only difference with the "NPT-1 kick" is the wellbore had two fluids—a known footage of seawater (in drillpipe) on top of a known footage of heavy mud (in the casing). Add those two calculated heads to the 1,400 psi and get the pressure of the kicking formation.

No *high-alert* person further considered the *anomalous* NPT-1 drillpipe pressure. No *high-alert* person made the simple calculation. No *high-alert* person saw the leak.

The drillpipe pressure peaked and held steady at 1,400 psi beginning three hours and fourteen minutes before the first explosion. At that time, the well was still full of 14.2-ppg mud and was under control.

Second Negative-Pressure Test—(NPT-2)

Neither the BP company man (men) nor any other member of the NPT team noticed the rising NPT-1 drillpipe pressure with enough understanding (and concern) to raise the alarm. Perhaps because they were busy with NPT-2, using the kill line (filled with seawater) in their second attempt at a successful negative-pressure test. But this is good, because the second test, NPT-2, would no doubt lead them to the leaking casing. And they did get good results. The kill-line pressure bled to zero, as it should when the casing system has total pressure integrity.[160]

But there were problems (see Footnote 159). First, had the kill line been filled with seawater while the BOPs were open, the U-tube pressure would have been about 1,450 psi (available pressure reduction for NPT-2). This step was not taken during NPT-2. Further, when the well kicked 1,400 psi up the drillpipe, it should have kicked 450 psi on the kill line. Such a kick did not manifest on the kill line. Third, during NPT-2, crews pumped into the kill line (6:40-6:43 P.M.) to ensure it was full of seawater, only to inject into the wellbore (formation) at 400-500 psi (slightly above reservoir pressure, equivalent to the missing kick pressure). Fourth, crews bled about 100 psi of residual pressure from the kill line at 7:15 P.M. and held it at zero psi for 30 minutes, indicating wellbore security, albeit while the drillpipe still showed the in-progress 1,400-psi kick.

A number of the NPT-2 problems are a result of more than 400 bbl of LCM spacer pumped into the lower riser and BOP. The LCM: (1) masked the kill-line kick, (2) allowed injection down the kill line and into the formation, and (3) allowed the kill line to be bled to zero psi, otherwise impossible given that the well was kicking with 1,400-psi on the drillpipe.

Unfortunately, BP's rig leaders and the rest of the NPT team failed to recognize the mutual exclusivity of NPT-1 and NPT-2 test results and declared the kill-line test *valid* at 7:55 P.M.[161] And the casing still had an *undetected* leak. Fortunately, although the drillpipe and the kill line were full of seawater, the entire wellbore was full of 14.2-ppg mud (had they opened the BOP, the well would have been dead). The well was still stable. Nothing was flowing. But the leak was there.

BP did not meet its goal. The casing did not have integrity. The second NPT was wrongly declared valid—indicative of pressure-competent casing—one hour and fifty-four minutes before the first explosion.

Well Kicking

The following activities are all based on the INCORRECT official declaration that the NPT-2 test results were valid, and that BP had met its goal to prove the casing was secure. It was apparently comforting to all parties to know that the casing would not leak when the upper wellbore was eventually filled with seawater, and the BOPs and riser were pulled. The declaration of NPT-2 validity (ALBEIT INCORRECT) opened the door to displacing the wellbore with seawater to 5,000 ft (though ultimately displaced to 8,367 ft). Such displacement would reduce the amount of heavy mud inside the casing, and was expected to happen without incident, as proved by the NPT-2 simulation.

Though no cement plug had been set in the well, displacement of mud with seawater at 8,367 ft commenced immediately. *High-alert* expectation—the volume of seawater pumped into the well was expected to match the volume of heavy mud leaving the well. Simple stuff—a critical drilling concept. More seawater into the wellbore than the amount of mud out of the wellbore—lost circulation. More out than in—the well's kicking. BP expected neither to happen—the casing was sound. Yet, BP kept the mudloggers on alert for mud-pit alarms, and it was good to know that Transocean's drillers were always watching. Any red flag, any visual or verbal alert, by anybody, was to be investigated.

But BP had a problem—they wanted the oil-based mud cleaned out of the mud pits before the rig was imminently released, the next day, to another BP project, the Nile. So the mud pits got a good cleaning—while the Macondo wellbore was being displaced with seawater. The cleaning crew constantly transferred mud from one pit to another. Incremental gains and losses of mud in the pits were meaningless. Additionally, seawater volumes could be estimated based only on pump strokes. Which meant the barrel-per-barrel balance between seawater in and mud out was ineffectively monitored, at best. And when the mud being discharged was determined at about 9:10 P.M. to have passed a non-polluting "sheen test," additional mud returns went *unmonitored* into the sea.[162]

During the displacement process, the pressure environment in the wellbore changed. No longer was the casing full of 14.2-ppg mud (less footage of mud and more footage of seawater, with every pump stroke). This means the fluid pressure at the bottom of the wellbore was decreasing, while the formation pressures outside the casing remained high. Hence, the amount of *overbalance* was dwindling by the minute. And the casing leak was still there.

Calculations (reported in Reference 3, BP's Internal Investigation) indicate that when about 4,800 feet of the annular casing/riser had been displaced with seawater above 8,367 ft, the diminished head of mud and seawater in the wellbore no longer overbalanced the pressure of the discovery zone.[163] At that single moment in time, the pressures were balanced—with a crucial reality.

As soon as more seawater was pumped, the well became *underbalanced* and began to flow oil and gas through the casing leak. That was near 8:52 P.M. The flow was helped, exacerbated, by every additional barrel of seawater pumped during the displacement process.[164]

The science of such flow should remind us of why drillers check drilling breaks for flow and shut-in kicking wells as soon as possible. The oil-and-gas *flow rate* the night of 20 April was approximately proportional to the underbalance—the amount of pressure *difference* between the higher-pressure formation outside the casing and the lower pressure inside the casing. This means the flow started as a trickle. But with each barrel of seawater pumped, helped by flowing barrels of oil and gas (all displacing heavy mud from the riser), the amount of underbalance increased, which means the *flow rate increased* at accelerating rates.

Just before 9:00 P.M., charts on the rig (Reference 3, Fig 8), though complicated by pit-cleaning activities, show riser returns exceeded the seawater pump rate—positive evidence the well had begun to flow. Further, as formation fluids pushed heavy mud up the casing and into the lower drillpipe annulus (creating more backpressure), the drillpipe pressure began to increase—more evidence that the well was flowing.

About 9:10 P.M., the seawater pumps were stopped to allow execution of the sheen test for the heavy LCM spacer (no sheen). Charts show the well continued to flow with the pumps shut off. Nevertheless, beginning at 9:14 P.M., displacement was resumed and the heavy LCM spacer and all other fluid returns from the well were discharged, unmonitored, directly to the sea. At that time, with or without the pumps injecting seawater, the well had been underbalanced and flowing for about 22 minutes.

By 9:30 P.M., the well had gained almost 300 barrels (Reference 3, Fig 18), which, in the next ten minutes, increased exponentially to about 1,000 barrels. This means reservoir fluids—oil and gas—had filled the casing up to the seafloor and were flowing into the riser above the BOPs. The flowing fluids, eventually driven by rapidly expanding gas in the lower-pressure riser, would reach the rig floor in minutes.

No BP, Transocean, Halliburton, or other *high-alert* person noticed the flow and/or stepped forward to shut down the ongoing operation. [165]

The well had begun to flow fifty-seven minutes before the first explosion. The *unmonitored* overboard discharge of heavy mud—even as the well flowed up the wellbore—commenced thirty-five minutes before the terminal event.

A Major Clue

All BP, Transocean, and contract management and technical staff went on *high-alert* status when heavy mud erupted through the rig floor. The well was flowing. Kicking. Crude oil and rapidly expanding natural gas were flowing up the wellbore, and had flowed for long enough time, unnoticed, to already be in the mile-long riser, *above* the BOPs. The chain-reaction eruption of deep expanding gas in the riser blew mud and seawater through the rig floor and over the derrick. Drilling crews rightfully actuated the BOPs, but the flow was so intense as to cut through the steel drillpipe and buckle the drillpipe in the BOPs. *High-alert* criteria called for opening the high-volume overboard diverter system, sending volatile fluids away from the rig, away from people. Instead, the flow was sent through the low-volume mud-gas separator (MGS). And when the expanding gas—onerous gas—reached the surface, it blew from the MGS vent, located up the side and on top of the derrick, and enveloped the rig floor and surrounding areas.

The electrical power system inhaled gas and destroyed the diesel generators. The facility lost power. Within seconds, the gas exploded, then exploded again even more violently, and the Macondo well evolved into a full-fledged blowout. Actuation of the EDS (emergency disconnect system) after the second explosion came too late.[166]

Not Enough Time To Act ?

The first explosion was at 9:49 P.M., which was:

(1) Four hours and four minutes after the first, irrefutable, casing-leak clue—the unnoticed or ignored bleed-back of fifteen barrels during the first drillpipe NPT.

(2) Three hours and fourteen minutes after the second, overwhelming, casing-leak clue—the non-manmade build-up of drillpipe pressure during the NPT to 1,400 psi.

(3) One hour and fifty-four minutes after the third casing-leak clue—the mutual exclusivity of the two test results, leading to the invalid kill-line NPT being declared valid.

(4) Fifty-seven minutes after the fourth casing-leak clue—that the well was creating volume, flowing through a leak, even as the growing discharge was later sent overboard, unmonitored.

Pick a clue. Call to action. Solve the problem. Fix the well. It's what the entire offshore oil-and-gas industry does. It's what the industry is about. No exceptions. Zero tolerance for failure.

Unfortunately, nobody on the *Deepwater Horizon* stepped forward. None from BP. None from Transocean. None from a service company. Until too late.

Eleven men died trying to control the out-of-control well.

Control lines to the closed blowout preventers were severed by back-to-back explosions.

The violence of the blowout buckled the drillpipe and kept the BOP shear rams from closing.

Actuation of the EDS, the emergency disconnect system, came too late.

The burning rig, still attached to the riser and BOP, sank a day and a half later.

The riser fell to the seafloor.

The well blew oil and gas into the Gulf for 86 days.

And it was all preventable. Had just one person stepped forward. Like Barry—the company man—the man in charge of the well. Or one *high-alert* person, like Jessica Terra Pherma, to slam her tally book on the table at six o'clock, or seven o'clock, or eight o'clock, or nine o'clock and say *Hey, guys, shut this sonofabitch down, the goddamn well's kicking*!

But nobody did.

---

[159] Reference (3). P. 88, Item 2.6, Figure 4, Annotated Item 8: "Drillpipe pressure slowly increases to 1400 psi." (Full scale version of chart is on page 421 herein.) Timeline indicates that on 20 April 2010, pressure started rising at 6:00 P.M. and peaked at 1400 psi at 6:35 P.M.

[160] Reference (13): USCG deposition 28 May 2010—Validation of NPT #2 by Transocean toolpusher. See also Reference (18A)—The Bureau of Ocean Energy Management, Regulation, and Enforcement (BOEM): Regarding the Causes of the April 20, 2010, Macondo Well Blowout. Issued 14 September 2011. Article VII, Item D, Well Integrity Testing—P. 88-97

<u>161</u> Reference (14): USCG deposition 7 April 2010—During 8:52 P.M. call from rig, BP drilling management in Houston declared the two NPT results as mutually exclusive (either pressure trapped or valve not lined up). Nevertheless, BP's rig managers, responsible for the well, continued to declare NPT-2 valid, as they had since 7:55 P.M. For additional commentary see Reference (18A)—Item VII, D, Well Integrity Testing, P. 88–97.

<u>162</u> Reference (18A)—MMS regulations required operators to use the best available and safest technology to monitor and evaluate well conditions and to minimize the potential for a well to flow or kick. The regulations also required the operator (emphasis by author) to ensure that the toolpusher, operator's representative, or a member of the drilling crew maintains continuous surveillance on the rig floor from the beginning of drilling operations until the well is completed or abandoned, unless they have secured the well with a BOP, bridge plug, cement plug or a packer. P. 99.

<u>163</u> Reference (3)—BP timeline—P. 25—model indicates the well went from overbalance to balance about 8:52 P.M. Which means the well was flowing during the sheen test, and subsequently when returns were being dumped overboard.

<u>164</u> Reference (18A)—The Bureau of Ocean Energy Management, Regulation, and Enforcement (BOEM): Regarding the Causes of the April 20, 2010, Macondo Well Blowout. Issued 14 September 2011. Article VIII, Section A—Kick Detection Methods and Responsibilities.

<u>165</u> Reference (18A)—The Bureau of Ocean Energy Management, Regulation, and Enforcement (BOEM): Regarding the Causes of the April 20, 2010, Macondo Well Blowout. Issued 14 September 2011. This is BOEM's FINAL report on the BP Blowout. Section XV, D (P. 190): Stop Work Authority. ". . . no stop work authority (by any party) was implemented on the day of the blowout despite the fact that . . . numerous anomalies . . . might have caused such authority to be invoked."

<u>166</u> Reference (15)—USCG deposition 27 May—Offshore Installation Manager (OIM) about attempt to actuate the EDS.

EPILOGUE—PART 5—

THE CAUSE

BP'S Macondo Blowout—Conception To Birth

BP ran and cemented the production casing but unknowingly caused a leak (a non-cemented pathway) from the pay zone into the bottom of the casing by: (1) allowing rat-hole mud to gravity segregate (rise) and contaminate the heavier cement slurry inside the lower casing and the lower annulus, and (2) damaging and/or not converting the float-collar check valves in the lower casing.

BP tested the casing (negative pressure test, NPT-1) to confirm its pressure integrity, but exacerbated the leak in the process by flowing formation fluids from the annulus, through the leak, and into the wellbore. Further, BP failed to recognize that such backflow and a subsequent pressure buildup (a kick) on the drillpipe were valid, in that they evidenced and confirmed the existence of the leak. Conversely, BP declared the test results to be anomalous and NPT-1 to be invalid.

Using the kill line, BP retested the casing (NPT-2) to confirm its integrity and got good results. BP accepted the "good results" of the second test as proof the casing was secure (no leak). Nevertheless, BP failed to recognize that injection into the kill line during NPT-2 proved annular communication, and that lost circulation materials in the BOP plugged the kill line and invalidated NPT-2 results.

Nonetheless, based on the "good results" of NPT-2, albeit invalid, BP (prior to setting a cement plug in the casing), directed crews to displace heavy mud from the upper half of the well with seawater.

Simultaneously to the displacement process, and in preparation for BP's next well, crews were directed to clean out the mud pits, which eroded the ability of rig and contract personnel to measure and compare volumes of fluids into and out of the wellbore (necessary for lost-circulation and kick-detection purposes).

No person on the rig noticed the well was flowing—evidenced by dynamic drillpipe pressures before, during, and after the sheen test—

even as oil and gas filled the production casing. Oil and gas then flowed through the open blowout preventers and into the riser. In minutes, high-velocity oil and rapidly expanding gas kicked through the rig floor and over the crown of the derrick.

Rig crews closed two BOPs, located a mile below the rig. Oil and gas from inside the riser—*above* the closed BOPs—continued to erupt over the derrick, enveloped the rig, and exploded and burned, killing eleven. The high rate of fluid throttling through the closed BOPs eroded its way into the steel drillpipe. The blind shear rams were jammed open by deformed drillpipe. The flow, through the drillpipe, did not stop. The emergency-disconnect system failed to release the rig.

The burning *Deepwater Horizon* sank a day and a half later.

BP's Macondo blowout continued for a total of 86 days and spilled an estimated five million barrels of oil into the Gulf of Mexico.[167]

BP'S Macondo Blowout—The Simple Truth

BP leadership caused a leak, failed to see the leak, declared the well secure, caused the well to flow, and failed to see the flow until too late.

---

[167] See **Diagram 23** (page 321 herein)—Graphic of The Cause of BP's Macondo Blowout, from Conception to Birth. Also see on following pages— Epilogue, Part 6, Recommendations.

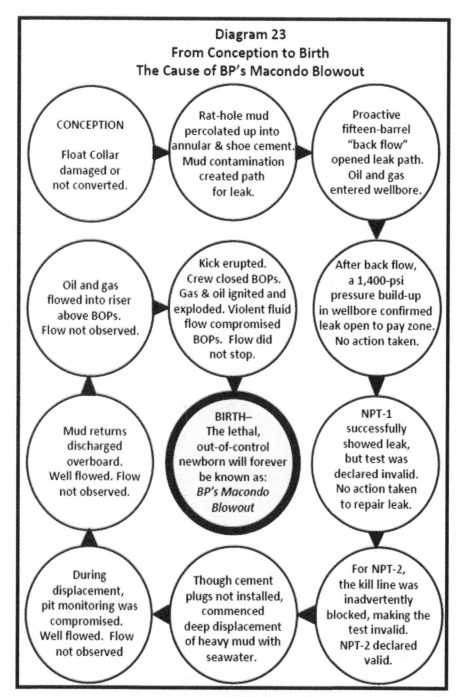

Diagram 23

EPILOGUE—PART 6—

RECOMMENDATIONS

Without the benefit of do-it-again scripts like for Bill Murray in *Groundhog Day*, or Adam Sandler in *50 First Dates*, we are forced to use clear-vision hindsight to look back at events on the rig and investigate the data, the decisions, and the industry recommended practices that would have saved the day.

Notes refer to **Diagram 23** (page 321 herein).

Float Collar Damaged or Not Converted

The float collar is designed with check valves (flappers) to prevent annular fluids from flowing into the otherwise-wide-open casing (See **Diagram 14**, page 202 herein). If annular fluids (cement or formation fluids) flow back into the casing, it means the cement hasn't yet set *and* the float collar has failed. Allowing *any* annular fluids to backflow into the casing will likely destroy the cement job.

Instead, WOC—wait on cement—until the shoe cement (below the float collar) sets and passes pressure tests. If so, proceed with the negative-pressure test (NPT).

If WOC fails, repair annular and shoe cement as necessary, then proceed with NPT.

Rat-Hole Mud Percolation

Rat-hole mud percolated into annular and shoe cement. Mud contamination created path for leak.

WOC would not have worked for Macondo's mud-contaminated cement. Such contamination began even during the cement job (see **Diagram 22**, page 304 herein).

Mud percolation is a common problem, with an easy solution, *before the casing job*. The solution is an industry recommended practice (RP), used around the world. As noted and quoted in API RP 65 Section 7.5 ". . . If casing is not run to bottom, the "rat hole" should be filled with a higher weight mud to prevent cement from falling into the rat hole and displacing rat hole mud into the cement column . . ." Such percolation likely destroyed the cement job in the bottom of the well, regardless of the condition of the float collar. Recommendation: As per the above RP, spot a pill of heavy mud in the bottom of the well prior to running the casing.

Excessive Proactive Backflow

Proactive fifteen-barrel "backflow" opened the leak path. Oil and gas entered the wellbore.

Any backflow greater than actual ballooning (i.e., 5 barrels of casing expansion) can only be formation fluids in combination with the failed float collar (See **Diagram 14** page 202 herein). For Macondo, 15 barrels was greater than the capacity of the shoe and annulus between the bottom of the pay zone and the float collar (**Diagram 22** page 304 herein).

Such backflow—any volume in excess of ballooned casing—must not be allowed. Instead, recognize float-collar failure and then WOC, as above. If such backflow takes place, it must be treated as an induced kick (through an open path), which can be repaired or isolated only with a drill bit, bridge plug, and/or additional cement.

Pressure Build-Up

After the backflow incident, the 1,400-psi pressure build-up confirmed the leak was open to pay zone. During this period, the pay zone flowed more fluid into the casing (raising the total to *more* than 15 barrels). Though this was a kick, no action was taken.

The classical response to an underbalanced kick (albeit man-made) would be the immediate initiation of well-control activities. Fortunately, for the NPT *simulation*, all they had to do was *open* the BOP to kill the well, prior to cement remediation and/or installing a mechanical plug.

NPT-1 Declaration

NPT-1 successfully showed the leak, but the test was declared invalid. No action taken to repair the leak. (See **Diagram 19** page 274 herein).

Casing tests are designed to show *existing* conditions; i.e., the entire casing system is good, or something is leaking. Either result is valid.

If something's leaking, find the leak and repair as necessary.

NPT-2 Declaration

For NPT-2, the kill line was inadvertently blocked with LCM, making the test invalid. Yet, NPT-2 was declared valid. (See **Diagram 19** page 274 herein)

When faced with counterintuitive, mutually-exclusive results (NPT-1 and NPT-2), neither choice is right, until all data is examined and the single reality is revealed and confirmed.

Such examination would have revealed NPT-1 as the viable test, and/or would have led to a repair (unblocking) of the kill line for NPT-2. Either way, the result would have again shown the casing leak.

Find the leak and repair as necessary.

Displacement of Heavy Mud With Seawater

Though cement plugs were not installed, BP commenced deep displacement of heavy mud with seawater. (All the following Recommendations refer to **Diagram 20**, page 278 herein)

As soon as the NPT (the *simulation* of what was to come) correctly confirmed that the casing and cement system was structurally sound at reduced (negative) pressures, there was no engineering reason not to set the federally-mandated cement plug(s) and bang on it (them) with a bit.

Even based on invalid NPT-2, such cement plug(s) would have pressure-isolated the deep wellbore (with its bad float collar and failed cement) from the subsequent displacement process.

Recommendation: Only with valid NPT that shows no leak, set and test cement plug(s), prior to displacing wellbore mud with seawater.

Pit Monitoring

During displacement, pit monitors and flow alarms were compromised. Rig personnel are paid to monitor pit levels and flow alarms 24/7 (looking for differences between fluids in and fluids out), but they cannot do their jobs if simultaneous activities (cleaning mud pits) take away their observation tools. Yet, even knowing there were minimal standard pit/flow monitoring capabilities, personnel failed to note that dynamic drillpipe pressures constantly showed the well to be flowing. The flow was not observed. (see **Diagram 20**, page 278 herein.)

Recommendation: Clean mud pits only after completion of temporary abandonment activities. Ensure all responsible personnel understand the physics and mechanics of flow and are prepared to respond accordingly.

Overboard Discharge

Mud returns discharged overboard. Well flowed. Flow not observed. (**Diagram 20**, page 278 herein)

The overboard discharge of non-quantified fluid volumes from an active well—other than during a well-control event (i.e., through overboard diverter lines, as designed)—is not an industry recommended practice, for good reason.

Alternatively—measure the volume of the fluid to confirm well security, then discharge the mud.

Kick Fluids Above the BOPs

Oil and gas flowed into riser above BOPs. Flow not observed. (**Diagram 20** on page 278 herein)

This was an extreme case of unobserved *massive* flow into the wellbore, the antithesis of "*catch kicks quickly to minimize kick volumes.*" The internal volume of the production casing was about 750 barrels. The internal volume of the riser was about 1500 barrels. After more than 750 barrels of oil and gas (a massive kick!) had flowed into and filled 2-1/2 miles of casing, the riser started filling with oil and gas, above the BOPs. Once the gas started expanding at shallow depths, *in a volatile bubble-point chain reaction*, all remaining fluids (mud and seawater) were explosively ejected from the riser, up through the diverter and rig floor, and over the derrick. Natural gas and crude oil arrived next.

Recommendation: Preventative solutions include any one or all of the above noted measures—each a standard industry practice. Kicks must be caught early, as evidenced by the extreme consequences at Macondo.

Kick Broaches Rig Floor

Kick erupted. Crew closed BOPs. Gas & oil ignited and exploded. Violent fluid flow compromised BOPs. Flow did not stop. (**Diagram 20**, page 278, herein).

For the hand they were dealt, in the little time they had *after* the kick erupted, crew members did the best they could. Eleven paid the ultimate price.

For the in-progress violence of the flow (high rate, high pressure) at the time the BOPs were finally actuated, the BOPs did the best they could.

As noted in all the above, any one or a number of preventative alternatives would have precluded failure, directed repairs, led to confirmed security for the well, and ensured an industry-standard and socially-expected outcome for all persons on board.

The Birth

Without such proven preventative measures, the tragedy was born. The lethal, out-of-control newborn was aptly named *BP's Macondo Blowout.*

REFERENCES

(1) Congress of the United States, House of Representatives, Subcommittee on Oversight and Investigations. Letter to BP CEO Tony Hayward dated 20 June 2010, signed by Henry A. Waxman Chairman, and Bart Stupak Chairman, Subcommittee on Oversight and Investigations— an early formal declaration that BP made errors and omissions prior to the blowout. Categories of concern included BP's schedule-induced shortcuts, cost-driven design choices (casing) that sacrificed safety, cement-design considerations (centralizers) that sacrificed cement integrity, cancellation of an evaluation tool (CBL) for cement integrity, incomplete clearance of gas from wellbore (CBU) prior to cement job, installation sequence of casing-hanger lockdown sleeve as potential contributor to blowout. The complete letter can be found at: http://energycommerce.house.gov/documents/20100614/Hayward.BP.2010.6.14.pdf

(2A) United States Coast Guard and the Department of the Interior's Bureau of Ocean Energy Management (BOEM—previously called the MMS), Joint Investigation. The committee deposed more than a hundred personnel from the rig (BP, Transocean, Halliburton, Schlumberger, and a number of other service companies) and their office-based line managers. Individual depositions can be accessed within the primary file: http://www.deepwaterinvestigation.com/go/site/3043/

(2B) Further, on 22 April 2011 the USCG issued: Redacted Report of Investigation into the Circumstances Surrounding the Explosion, Fire, Sinking, and Loss of Eleven Crew Members Aboard the Mobile Offshore Drilling Unit *DEEPWATER HORIZON* In the Gulf of Mexico, April 20–22, 2010, Volume I. The entire report can be found at:
https://homeport.uscg.mil/cgi-bin/st/portal/uscg_docs/MyCG/Editorial/20110914/2_DH%20Volume%201_redacted_3.pdf?id=39b8430fcadaffb9a72d2db7d9053e78fb2247dc

(3) BP's *Deepwater Horizon* Accident Investigation Report. This is BP's Internal investigation released to the public on 8 September 2010. This internal investigation tabled engineering data not previously made public. Selected data include casing setting depths and sizes, as well as a detailed timeline of key events and hard data for cementing, pressure-test, seawater displacement, and post-blowout consequential events. The report can be found at: http://www.bp.com/liveassets/ bp_internet/globalbp/globalbp_uk_english/incident_response/STAGING /local_assets/downloads_pdfs/Deepwater_Horizon_Accident_Investigati on_Report.pdf

(4) National Commission on the BP *Deepwater Horizon* Oil Spill and Offshore Drilling. President Obama "charged the Commission to determine the causes of the disaster, and to improve the country's ability to respond to spills, and to recommend reforms to make offshore energy production safer." Final report issued 11 Jan 2011. http://www.oilspillcommission.gov/ See also: http://www.gpo.gov/ fdsys/pkg/GPO-OILCOMMISSION/pdf/GPO-OILCOMMISSION.pdf

(5) National Academy of Engineering/National Research Council (NAE/NRC) committee. Charged to examine the probable causes of the *Deepwater Horizon* explosion, fire, and oil spill in order to identify measures for preventing similar harm in the future. Interim report made public 16 November 2010. The NAE/NRC issued an "Update" 1 July 2011. See three references at:
    http://www.nationalacademies.org/includes/DH_Interim_Report_fi nal.pdf
    http://books.nap.edu/openbook.php?record_id=13047&page=1
    http://www.nae.edu/activities/20676/deepwater-horizon-analysis.aspx

(6) Det Norske Veritas (DNV)—Final Report for the United States Department of the Interior Bureau of Ocean Energy Management, Regulation, and Enforcement, Washington, DC 20240: Forensic Examination of *Deepwater Horizon* Blowout Preventer. 20 March 2011. The in-depth assessment includes an operational, mechanical, electrical, and hydraulic biopsy of the salvaged components—BOPs, LMRP, control pods. The complete report can be found at: http://www.deepwater investigation.com/external/content/document/3043/1047291/1/DNV %20Report%20EP030842%20for%20BOEMRE%20Volume%20I.pdf

(7) Macondo—Rig and Well Specs—BP-Production Casing—BP-HZN-CEC022118, which recommends the liner and tie-back. Document also references pay-zone depths and pressures and depths of lost circulation zones above and below the pay. No access, as document states *Confidential treatment requested.*

(8) Macondo—Rig and Well Specs—BP-Production Casing— BP-HZN-CRC022145, which states the full Casing string is again the choice, and the liner is now an option. No access, as document states *Confidential treatment requested.*

(9) Macondo Well Casing—Long String—Cement—Ops—BP-HZN-CEC017621. This is BP's well plan for the 9-7/8 x 7 production casing and cement job. References criteria for swab and surge, circulating bottoms-up (CBU), cement bond log (CBL), negative-pressure test (NPT), drillpipe for the NPT, the casing hanger lockdown sleeve, the top cement plug, etc. Document states *Confidential treatment requested.* For additional information about the Lockdown Sleeve decision, and the unseating of the casing hanger from the wellhead (the spikes on the pressure-build-up curve), see Transocean Appendix F, Reference (21), below.

(10) Schlumberger.MC 252. Timeline. References timing of arrival and departure orders for CBL team on the 19th and 20th April 2010. Full report also available at: http://democrats.energycommerce.house .gov/documents/20100614/Schlumberger.MC.252.Timeline.pdf

(11) BP's *Deepwater Horizon*— Accident Investigation—Static Presentation—Slides. Similar in part to Reference 3 (above), but in slide/pictorial form. The entire presentation can be found at: http://www.bp.com/liveassets/bp_internet/globalbp/globalbp_uk_engli sh/gom_response/STAGING/local_assets/downloads_pdfs/Deepwater_ Horizon_Accident_Investigation_static_presentation.pdf

(12) Macondo casing specifications: Noted in a number of BP documents, including BP's *Deepwater Horizon* Accident Investigation Report (Reference 3, Figure 2). Full report at: http://www.deepwater investigation.com/posted/3043/MacondoWellSchematic.796427.pdf

(13) USCG deposition 28 May—Senior toolpusher—P. 280–281. Testimony by senior toolpusher about commencement of second NPT, and subsequent validation of its results. Individual depositions can be

accessed within the primary file: http://www.deepwater investigation.com/go/site/3043/

(14) USCG deposition 7 April 2011—BP VP—P. 71–72. Testimony by BP VP about impossibility (mutual exclusivity) of the two NPT results, as documented in response (BP-HZN-BLY00125475) from BP office management to BP rig management an hour prior to the blowout. Individual depositions can be accessed within the primary file: http://www.deepwaterinvestigation.com/go/site/3043/

(15) USCG deposition 27 May—Offshore Installation Manager— P. 10-11. Testimony by OIM relative to post-explosion attempt to actuate the Emergency Disconnect System (EDS). Individual depositions can be accessed within the primary file: http://www.deepwater investigation.com/go/site/3043/

(16) USCG Deposition 24 August 2010, P. 250–257, testimony by Halliburton about cement modeling results, nitrified cement, the need for 21 centralizes to minimize annular cement-mud channeling, and the well's potential for severe gas flow if fewer centralizers were used. Individual depositions can be accessed within the primary file: http://www.deepwaterinvestigation.com/go/site/3043/

(17) WSJ—Monday, May 10, 2010. "Rig Owner Had Rising Tally of Accidents." Article by Ben Casselman. References BP/Transocean VIP visit to *Deepwater Horizon* 20 April 2010 to celebrate the rig's completion of seven years without a lost-time accident. See reference (14), above, for additional information about the VIP visit.

(18A) The Bureau of Ocean Energy Management, Regulation, and Enforcement (BOEM): Regarding the Causes of the April 20, 2010, Macondo Well Blowout. Issued 14 September 2011. This is BOEM's FINAL report on the BP Blowout. This report includes "Buoyancy Casing Analysis" dated September 2010, as well as a number of Appendices, below. See: http://www.boemre.gov/pdfs/maps/DWHFINAL.pdf

(18B) See also (for all Appendices): http://www.boemre.gov/ DeepwaterHorizonReportAppendices.htm

(18C) See also Appendix B, Review of Operational Data Preceding Explosion on *Deepwater Horizon* in MC 252, John Rogers Smith,

Petroleum Consulting LLC, submitted to the MMS 1 July 2010. http://www.boemre.gov/pdfs/maps/SmithCover.pdf

(18D) Appendix G—BP's NPT Protocols— http://www.boemre.gov/ pdfs/maps/AppendixG_MacondoProduction.pdf

(19) Turley, John A.: "A Risk Analysis of Transition Zone Drilling," Society of Petroleum Engineers, SPE 6022, 1976. See: http:// www.osti.gov/energycitations/product.biblio.jsp?osti_id=7225619

(20) National Commission on the BP *Deepwater Horizon* Oil Spill and Offshore Drilling—Chief Counsel's Report—2011. Released 17 Feb 2011. The committee was disbanded 11 March 2011. A comprehensive and credible investigative report on the topic. For full report, see: http://www.oilspillcommission.gov/sites/default/files/documents/C21 462-408_CCR_for_web_0.pdf

(21) Transocean—MACONDO WELL INCIDENT—Transocean Investigation Report—Volume I—June 2011. A comprehensive and credible investigative report on the topic. See: http://www.deepwater.com/_filelib/FileCabinet/pdfs/00_TRANSOCEA N_Vol_1.pdf and http://www.deepwater.com/_filelib/FileCabinet/pdfs/ 12_TRANSOCEAN_Vol_2.pdf Further, for additional information about the Lockdown Sleeve decision, and the unseating of the casing hanger from the wellhead (the spikes on the pressure-build-up curve), see Appendix F: http://www.deepwater.com/_filelib/FileCabinet/pdfs/ 18_TRANSOCEAN_App-F.pdf

(22) Halliburton—9.875" X 7" Foamed Production Casing Post Job Report. 20 April 2010. See: http://www.boemre.gov/pdfs/maps/ HalliburtonReport10.pdf

(23) New York Times photo gallery—"Firestorm Aboard the *Deepwater Horizon*." See: http://www.nytimes.com/interactive/ 2010/12/26/us/20101226_spill_slideshow.htm Copyright 2010—The New York Times Company.

BOOKS ABOUT BP'S MACONDO BLOWOUT

A number of nonfiction books about BP's Macondo Blowout and its aftermath entered the market beginning in late 2010. Those noted below have two things in common: seven of eight covers show a picture of Transocean's burning *Deepwater Horizon*, and each has minimal (if any) specific text relating to the engineering and operating cause of the disaster. To their credit, the authors of a number of the more-recent books annotated with clarity (based on USCG depositions) the names of key people pre-blowout, and who did what, who said what, who got injured, who the heroes were, and who perished. None, of course, could define with certainty the tactical actions or involvement of the eleven front-line personnel who died during the disaster.

While *THE SIMPLE TRUTH* is a character-driven story that targets the drilling of the well and the Cause of the event, the majority of the noted authors cover in depth an assortment of Effect-related aspects of the blowout, including the response, the kill, the environment, jobs, politics, Big Oil, press releases, photo galleries, energy independence, government oversight, and BP's history and culture.

(1) Achenbach, Joel. *The Race to Kill the BP Oil Gusher*, New York, Simon & Schuster, 2011. Book available at: http://www.amazon.com/Hole-Bottom-Sea-Race-Gusher/dp/1451625340

(2) Cavnar, Bob. *Disaster on the Horizon*. White River Junction: Chelsea Green Publishing Company, 2010. Book available at: http://www.amazon.com/Disaster-Horizon-Stakes-Deepwater-Blowout/dp/1603583165

(3) Freudenburg, William R., and Gramling, Robert. *Blowout in the Gulf*. Cambridge: MIT Press, 2010. Book available at: http://www.amazon.com/Blowout-Gulf-Disaster-Future-America/dp/0262015838

(4) Holland, Jack. *Blowout: In Plain Sight.* Lexington: Tarpon Productions, 2010. Book available at: http://www.amazon.com/BLOWOUT-Plain-Deepwater-Horizon-Coverup/dp/1453661247

(5) Lehner, Peter, with Deans, Bob. *In Deep Water: The Anatomy of a Disaster, the Fate of the Gulf, and Ending Our Oil Addiction.* New York, The Experiment, 2010. Book available at: http://www.amazon.com/In-Deep-Water-ebook/dp/B004KNWVJ6

(6) Steffy, Loren C. *Drowning in Oil—BP and the Reckless Pursuit of Profit.* New York, McGraw-Hill, 2010. Book available at: http://www.amazon.com/Drowning-Oil-Reckless-Pursuit-Profit/dp/0071760814

(7) Stanley Reed and Alison Fitzgerald. *In Too Deep: BP and the Drilling Race That Took it Down.* Hoboken, Bloomberg, 2011. Book available at: http://www.amazon.com/In-Too-Deep-Drilling-Bloomberg/dp/0470950900

(8) John Konrad and Tom Shroder. *Fire on the Horizon.* New York: Harper, 2011. Book available at: http://www.amazon.com/Fire-Horizon-Untold-Story-Disaster/dp/0062063006

(9) Carl Safina. *A Sea in Flames.* New York: Crown Publishers, 2011. Book available at: http://www.amazon.com/Sea-Flames-Deepwater-Horizon-Blowout/dp/0307887359

(10) Joseph A Tainter and Tad Patzek. *Drilling Down: The Gulf Oil Debacle and Our Energy Dilemma.* Springer, 2011. Book available at: http://www.amazon.com/Drilling-Down-Debacle-Energy-Dilemma/dp/1441976760

Other books will follow.

ABOUT THE AUTHOR

J. A. (John) Turley grew up in San Diego, California. Advanced degrees in petroleum engineering and ocean engineering from the Colorado School of Mines and the University of Miami and a three-year petroleum-engineering professorship at Marietta College preceded his oil-and-gas-industry career. His two decades of offshore drilling- and project-management responsibilities with a major U.S. energy company began with the Gulf of Mexico, evolved to the North Sea, were bolstered by executive education at Harvard Business School, and led him to be named manager of worldwide drilling. After a number of years as the company's senior technical officer, he elected to retire early to focus on writing.

THE SIMPLE TRUTH, narrative nonfiction, is Turley's debut publication. Others will follow.

Author's Website:
http://www.JohnTurleyWriter.com

Book description:
J. A. Turley, Author
THE SIMPLE TRUTH: BP's Macondo Blowout,
Littleton, Colorado, The Brier Patch, LLC, 2012.

Book availability:
Hard-copy (paperback) and E-book available through Amazon

Book 1
The Simple Truth
ends here

Book 2
From the Podium
begins here

BOOK 2

From the Podium:
The Cause of BP's Macondo Blowout

From the Podium:
The Cause of
BP's Macondo Blowout

Nonfiction
by

J. A. Turley

The Brier Patch, LLC
Littleton, Colorado, USA

Book 2

Published by:
The Brier Patch, LLC
P.O. Box 184, Littleton, CO 80160-0184, USA

ISBN-13: 978-0-9858772-3-1 (paperback)
ISBN-13: 978-0-9858772-5-5 (eBook)

This is a work of nonfiction based on a true incident—BP's 2010 Macondo Blowout in the Gulf of Mexico aboard the Transocean *Deepwater Horizon* drilling rig. Information and diagrams used throughout have been extracted from the author's data-driven book *The Simple Truth: BP's Macondo Blowout* (published in 2012). Public data about companies, equipment, the well, and the rig, though modified by the author for ease of reading, form the basis for the work. Opinions are the author's.

Editing by Thomas N. Locke, with my sincere thanks.

Acknowledgements

I researched and wrote for two years before publishing *The Simple Truth: BP's Macondo Blowout* in September 2012. Soon thereafter, an energy company invited me to present my research findings in Houston, Texas. That successful event opened the way for me to make dozens of technical and keynote presentations around the world. Those presentations form the basis for this publication, *From the Podium.*

My sincere thanks go to each entity and host who welcomed my message, including Colorado School of Mines, Marietta College, Murphy ExP, Marathon Oil Company, the Society of Petroleum Engineers (SPE) in Houston, the Evangeline (Lafayette, Louisiana) Sierra Club, University of Louisiana at Lafayette, Louisiana State University, NETL/DOE, Chevron, Offshore Process Safety Conference, SPE Dallas, University of Oklahoma, Tulsa University, LAGCOE, SPE Mid-Continent, NPR Tulsa, Pennsylvania State University, West Virginia University, Southern Ohio Oilman's Association, IADC/SPE Dallas, Lafayette Geological Society, American Association of Drilling Engineers, Montana Tech University, University of Leoben Austria, Decom World, SPE Amsterdam, and Encana Services Company.

After almost fifty such presentations, I was invited to be an SPE Distinguished Lecturer (DL) for the 2015-2016 season and to make my presentation (same topic) to SPE sections on a global basis.

I am especially grateful to the thirty SPE sections and their volunteer leaders who hosted me during my DL travels to St. John's, Newfoundland and Labrador; Boston, Massachusetts; Washington, DC; Canton, Ohio; Lansing and Traverse City, Michigan; Tyler, Texas; Hobbs, New Mexico; Texas A&M University; Denver Colorado; Abu Dhabi and Dubai, UAE; Bahrain; London and Great Yarmouth, England; Aberdeen, Scotland; Madrid, Spain; Lloydminster, Canada; San Ramon, California; Bartlesville and Duncan Oklahoma; Lafayette, Louisiana; and Tuscaloosa, Alabama.

I finished my tour with a trip to Perth, Adelaide, Melbourne, and Sydney, Australia; New Plymouth, New Zealand; and then to Brisbane, Australia, where I made my last SPE-DL presentation on May 19, 2016.

For Jan

Forever my love,
my best friend,
and always my CFOOE

Contents

Introduction

On April 20, 2010, the major London-based energy company BP plc was blasted into the headlines by a disastrous blowout in the Gulf of Mexico aboard a deep-water drilling rig named *Deepwater Horizon*.

At the time, I had retired from my engineering and management career in oil and gas and had started a new venture—writing fiction. My learning curve was steep, but I won a couple of contests and wrote four novels. In the middle of drafting my fifth book, I heard about the hours-old catastrophe in the Gulf, the thought of which threatened to consume me. I immediately quit writing and focused on the disaster.

Why? Because that tragic event led to what I've now long described as *one of the most lethal, costly, manmade environmental disasters in history*, which will forever be known in the industry as *BP's Macondo Blowout*. But because the blowout took place aboard Transocean's *Deepwater Horizon* drilling rig, the media addressed the disaster as BP's *Deepwater Horizon* blowout and oil spill. In fact, the 2016 Hollywood film about the tragedy is named, simply: *Deepwater Horizon*.

A *blowout* is defined as the loss of control of a drilled well, wherein naturally occurring formation fluids—oil, gas, and water from sediments thousands of feet below the seafloor—flow into the drilled well and to the surface without control. Further, *Macondo* is BP's chosen nickname for the deep geologic structure targeted by the exploration well; hence, it was called the "Macondo well," and, without ambiguity, "BP's Macondo blowout."

BP's Macondo blowout killed eleven men. Further, the half-billion-dollar *Deepwater Horizon* burned, capsized, and sank in the mile-deep Gulf. An estimated five million barrels (200 million gallons) of crude oil spilled into the Gulf during a media feeding frenzy that lasted eighty-six days.

The cleanup operation, using a toxic dispersant, hid the floating crude oil but put the Gulf in jeopardy for years to come. Costs skyrocketed, eclipsing $60 billion—pushed by failed businesses, open-ocean cleanup, coastline and estuary rejuvenation, and litigation-related fines and penalties that, as of 2018, do not yet include the results of ongoing civil trials.

So, where do I, John Turley, fit in? Offshore operations, and drilling operations in general, occupied much of my professional career. In retirement, I was the neighborhood "oil guy," which led family and friends—inundated by TV and internet news—to ask, "John, what happened to that rig in the Gulf?"

My inability to answer that single question led me to give up fiction writing and start digging. I returned to the world of research and gathered publicly available data wherever I could find it. I listened to and reviewed thousands of pages of United States Coast Guard depositions, investigative reports, engineering procedures, published studies, and court transcripts. The more I looked for and found hard data, the more hooked I became at filling in the blanks and completing the puzzle as to what caused the blowout.

During the interim, as the media continued to saturate the listening/watching/reading world with the best information they'd gleaned from pontificating experts who often had vested interests, I made a two-part decision.

First, I committed myself to ignoring finger-pointing, politics, opinion, he-said-she-said, oily beaches, media headlines, court findings, company cultures, journalists, attorneys, and—albeit with great difficulty—the people involved, whether victims, survivors, or associated leaders.

Second, I would focus on only one thing—hard data. Specifically, I would gather available engineering and operating data from the rig and from the well that had been captured by the entities involved. The data would tell me the story I needed to know. It's the kind of story engineers can put to good use when working to help minimize the chance of ever losing control of another well.

My mantra, while gathering data and since, has been and remains: Only if we understand and care about the cause of BP's Macondo blowout will we know why it should not have happened and why it should never happen again.

The data I found were compelling, but I knew it would mean little to those friends and family who still wanted to know "what happened to that rig in the Gulf?" To answer their question, I let the hard data define a chronologically accurate skeletal framework, which allowed me to write a book—*The Simple Truth: BP's Macondo Blowout.*

A key element of my writing goal was to include and use all the data I'd gathered and to explain the technical cause of the disaster for readers across the entire oil and gas arena. But I wanted to do it in a way that, for example, my non-engineering father might have appreciated.

For that reason, and because eleven key personnel from the rig perished that dreadful night and could not speak for themselves, I wrote the book as narrative nonfiction. The data are real, but the story is written as a novel with made-up characters—surrogates for those who survived and those who died.

More than a hundred literary agents turned down my query letters and manuscript submissions for *The Simple Truth*. Many didn't respond, perhaps because they didn't like my writing. Others turned me down using reasons like "untold litigious liability." It seems they wanted nothing to do with the risk of an unknown author telling *The Simple Truth* about a major disaster.

So, I self-published *The Simple Truth* in September 2012. I worked with Amazon to publish paperback and ebook versions.

To date, sales are nice, comments are positive, and I'm in the black. But it's also important for me to look back to when an industry leader read my book and called me in late 2012. As an executive for a major energy company, he invited me to attend a corporate HES (health, environment, and safety) conference to present the results of my research findings. And so began my third career as a professional speaker about Macondo.

As referenced in my acknowledgements, I made seventy-five presentations about BP's Macondo blowout across the U.S. and international to energy companies, professional societies, technical conferences, petroleum engineering universities, civic groups and environmentalists. After my last presentation in Brisbane, Australia, I declared an end to my seventy-three-month, passion-driven, self-imposed Macondo mission.

Yet it hasn't been a clean break because at virtually every technical presentation somebody asked for a copy of my slides. I always apologized before I turned them down—because I had an obligation to make my slides available to SPE so that SPE could publish the slides on their website. I have fulfilled that obligation, so SPE members can now go to the SPE-DL site and see the slides.

But—those slides contain none of the text included herein. Accordingly, this document—*From the Podium: The cause of BP's Macondo Blowout*—is the only complete, annotated, footnoted resource about my research-related presentations and recommendations.

Note: Footnotes in *FROM THE PODIUM* are numbered sequentially from the end of *THE SIMPLE TRUTH*.

Presentation Format

This presentation—*From the Podium: BP's Macondo Blowout*—includes my slides, text, and Q&A from four years of presentations, including my 2015-16 SPE-DL road trip.

Audience members who scheduled an hour and who participated in extensive Q&A got very close to hearing the full story, as told here. Audiences scheduled to listen for only forty-five minutes got less. Some Q&A sessions garnered a question or two; others lasted an hour or two. Significant Q&A topics are included herein.

Though this written text is comprehensive, there are a number of topics in *The Simple Truth* that could not be covered in *From the Podium*. In general, if a topic was not *germane* to understanding the physical *cause* of the blowout, I've left it for *The Simple Truth* to tell the more-comprehensive, fully footnoted, story. However, I have included a few of the 160-plus footnotes and references from *The Simple Truth*, each of which may help readers to better understand the *cause* of the blowout.

This document, especially when paired with *The Simple Truth*, is intended for the general reading public, but it is particularly important to petroleum engineering faculty and students and to all others in the O&G industry (including members of SPE) who care about and want to understand the *cause* of BP's 2010 Macondo blowout in the Gulf of Mexico.

Nevertheless, and for the record, the following is not an SPE presentation, nor is it presented here as an SPE document.

Accordingly, whether you are a layman or a technical expert, if you care about the *cause* of the blowout, I hereby invite you into my world.

FROM THE PODIUM:

The Cause of
BP's Macondo Blowout

The Presentation

After being introduced, I shake hands with the host, go to the podium, and pick up the mic.

I use no notes.

A title slide is already on the screen.

My presentation follows.

FROM THE PODIUM—GOAL

- *Assess* petroleum-engineering, operations, and drilling data from
  BP's 2010 Macondo blowout in order to define and understand the *cause* of the disaster, and
- *Apply* lessons learned to future wells

SLIDE 1

Assessing and *applying* are key words in my presentation.

Our goal will be to trace and understand the failure mechanisms that led to and *caused* the Macondo blowout[168], so that we can apply what we learn to future wells. We're going to do this by looking at data, all data, from a petroleum engineering perspective.

Let's begin by looking at the Macondo well prior to the blowout.

---

[168] BLOWOUT—Formation fluids flowing up a well into the environment without control. Blowout, by definition, means lack of well control; hence, the emphasis on kick-control (well-control) training. Industry wide, highly experienced rig crews (toolpushers, drillers, company men) react immediately to resolve every well-control event. When such actions fail, for whatever reason whether human or mechanical, the result can be a blowout.

MACONDO—THE PLAN

- **Drill an Exploration Well in the Gulf of Mexico (Mississippi Canyon Block 252)**
- **Spud October 2009**
- **Depth: Approximately 20,000 ft (6,100 m)**
- **Water: Approximately 5,000 ft (1,525 m)**
- **Target: Geological structure (called Macondo)**
- **Target Depth: Below 17,000 ft (5,200 m)**

SLIDE 2

In late 2009, BP (the operator) began drilling its 20,000-foot exploration well in a mile (about 5,000 feet) of water, utilizing an anchored rig named Marianas. That rig was ultimately damaged by a hurricane and replaced by the *Deepwater Horizon* in 2010.

The target was a deep geologic structure (nicknamed *Macondo)* below about 17,000 feet subsea. The nickname allowed for private discussions in public places, without identifying the well. But it also meant the well became the *Macondo* well, and the ultimate disaster became the *Macondo* blowout.

So—let's jump deep into the well, to about 17,000 feet.

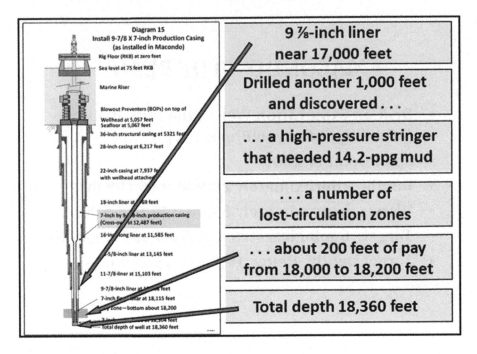

SLIDE 4

Note: The tiny words in the *diagram* (on the left, above, noted as **Diagram 15**) are extracted from *The Simple Truth* and are not specifically important to the presentation. But—the big words on the right side do matter. Readers are invited to review details of all diagrams as extracted from *The Simple Truth* (referenced throughout), including a full-size copy of **Diagram 15**, page 206 herein.

* * *

Slide 4 shows the floating rig (*Deepwater Horizon*) and the Macondo well. The rig, using GPS and thrusters to stay on location, was considered a *vessel underway*; hence, it had a captain and was regulated by the US Coast Guard (USCG). And that's why the USCG (rather than the federal MMS)[169] led the original post-blowout depositions of survivors and personnel related to the Macondo well and the blowout.

A note of interest: Since the *Deepwater Horizon*[170] was a powered vessel, like a ship, its name in print is always in italics. Conversely, the Marianas, an anchored rig, is not considered a powered vessel at sea; therefore, its name is not italicized.

Though not part of this presentation, see **Diagram 1** from *The Simple Truth*, page 22, herein, for a photo of the Transocean Marianas, and **Diagram 7**, page 103, herein, for an early photo of the *Deepwater Horizon*.

Now, back to **slide 4** (and, yes, there is no slide 3, as replaced by two lines of text in slide 4).

The floating rig was connected to the Macondo well by a 21-inch-diameter marine riser, also known as a drilling riser (capacity about 1500 barrels). The riser was connected to the blowout preventers (BOPs), which were firmly connected to the wellhead (casing head), and in turn to the structural casing strings that penetrate the seafloor. All subsequent drilling activities took place through the drilling riser.

The 18 ¾-inch, 15,000-psi BOP stack, solidly fixed to the wellhead, was topped by two 10,000-psi annular BOPs (though one was de-rated to 5,000 psi to accommodate stripping 6⅝-inch drillpipe). Below the annular BOPs were three VBRs (variable-bore rams), a blind-shear ram (BSR), and a casing shear ram. The lower VBR was dedicated to act as a test ram. (See **Diagram 5**, BOP schematic, page 68 herein)

As drilling progressed, using synthetic-oil-based mud, multiple strings of casing were run to just below 17,000 feet (see **slide 4**, the upper big arrow). After drilling out and testing the 9⅞-inch shoe to 16.0 ppg (pounds per gallon), the operator then took a few days to drill the next thousand feet of wellbore, with several interesting findings.

Early on, a ten-foot-thick stringer of sand required 14.2-ppg mud. The operator would have liked the drill-ahead mud weight to be heavier, but persistent lost-circulation zones below the sand stringer proved to be sensitive to increased mud weights and required massive doses of lost-circulation material (LCM)[171] to control mud losses.

And then good news: The operator drilled a 200-foot-thick oil-and-gas discovery "pay zone" (see **slide 4**, the lower arrow). The pressure of fluids in the pay zone proved to be 1,000 psi underbalanced (less than) than the pressure exerted by the 14.2-ppg mud column[172].

With the 10-foot sand stringer under control, the lost circulation zones plugged, and the pay zone over pressured (with mud) by 1,000 psi, the operator drilled an additional 160 feet of wellbore below the pay zone to look for additional hydrocarbons, and to ensure adequate footage of wellbore below the pay zone for well-logging and casing operations.

---

[169] MMS—MINERALS MANAGEMENT SERVICE—The regulatory authority for offshore operations in federal waters. By October 2010, the MMS was reorganized and renamed the BOEMRE (Bureau of Ocean Energy Management, Regulation, and Enforcement), which was then replaced by the BSEE (Bureau of Safety and Environmental Enforcement) and the BOEM (Bureau of Ocean Energy management). Which means the MMS is no longer extant.

[170] The Transocean *Deepwater Horizon* was built in 2000. The dynamically positioned (DP) semisubmersible was also referred to as a MODU—Mobile Offshore drilling Unit. It was rated for 10,000 feet of water and a well-depth of 30,000 feet. The rig measured 396 feet by 256 feet with a 242-foot-tall derrick. Because rig was DP while hovering over a fixed location on the seafloor (the well), the Coast Guard considered it a "vessel underway," which is required to have a Master (captain). Additionally, because the facilities were massive, with accommodations for 130 personnel, federal regulations called for an Offshore Installation Manager (OIM). An on-duty toolpusher was in charge of all drilling facilities and rig personnel associated with drilling the well. The company man worked for BP, and was responsible for managing the well—engineering, logistics, data, decisions, services, costs, and executing procedures.

[171] LCM—Lost circulation materials. Anything designed to plug leaking, fractured, mud-swallowing formations. Shredded cotton and ground-up walnut hulls have been replaced over the years with sophisticated lab-manufactured plugging agents. A slug or "pill" of the material, pumped down to drillpipe and into the annulus, is quite often successful. Or not.

[172] MUD WEIGHT EQUATION: $P = MW \times D \times 0.052$. Here, P is pressure, in pounds per square inch (psi), MW is mud weight, in ppg (pounds per gallon), and D is depth in feet below the top of the mud column (important to use true vertical depth). The constant—0.052—keeps the units straight for feet, ppg, and psi.

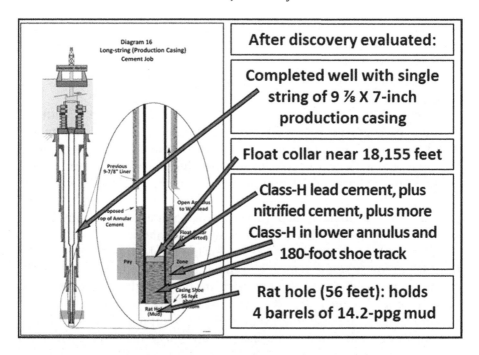

SLIDE 5

After extensive well-logging and formation-testing activities, the operator made a bit trip to bottom and then ran a single string of 9⅞-inch X 7-inch production casing[173] (see top arrow in **slide 5**). The operator used a 5,000-foot-long drillpipe[174] workstring to get the top of casing down to the wellhead, where the casing was hung with a casing hanger. (For a scaled-up version of **Diagram 16**, see page 218 herein.)

The casing was landed 56 feet above the bottom, and for good reason. Specifically, in deep-water operations, the casing hanger, located on top of the casing, must reach and seat inside the casing head, at the seafloor, *before* the bottom of the casing reaches the bottom of the well. The opposite would be disastrous—requiring the too-long casing string to be pulled from the well and sent back to town for the threaded connections to be redressed.

The 56 feet of open wellbore below the casing (see lowermost arrow in **slide 5**) was called the *rat hole*, an area of importance to the rest of this discussion.

A twin-flapper float collar (see the second big arrow) had been installed in the Macondo production casing string, near the middle of the pay zone. The flappers in the float collar were designed to act as one-way check valves. If and when they were closed, flow *up the casing* would not have been possible.

Note: See the schematic of the float collar, **Diagram 14**, page 202 herein.

But there was a catch. The casing was run (as designed) with the flappers blocked open, so that, as the casing was lowered into the wellbore, mud[175] filled the casing from the bottom up. Then, as planned, the open flappers were "converted" into high-pressure check valves. *Conversion* was initiated (as typical) by increasing the fluid flow through the float collar (while circulating bottoms up or while displacing the cement) to a predetermined rate (barrels of mud per minute). The *converted* flappers then became one-way check valves.

Two kinds of cement[176] were used for the production casing (see the third big arrow, **slide 5**, pointing to the gray stuff inside and outside the casing). The lead slurry (uppermost in the annulus) was 16.7-ppg Class H cement, which was followed by lightweight nitrified cement (generally across the pay zone), and finally by additional 16.7-ppg Class H cement as the tail slurry. The Class-H tail slurry filled the lowermost annulus and filled the inside of the bottom 180 feet of casing, called the shoe track (see the third big arrow). The shoe track is the interval of casing between the float collar and the guide shoe (bottom end of the casing). The combined heavy and lightweight slurries in the annulus were designed to ensure the cement column did not exceed 14.2 ppg, because of the lost-circulation zones.

So—with casing and cement in the well to protect and isolate the pay zone, the next operation was temporary abandonment.

* * *

Why temporary abandonment?

Producing hydrocarbons from a deep offshore well is radically different than producing from shallow-water and onshore wells, either of which can produce oil and gas in weeks or months after drilling is complete.

Conversely, in deep water, the general plan would include drilling one or more delineation wells[177] to better define reservoir geometry and the size of its reserves. Then a subsea development team, using the acquired data, would design the seafloor production facilities, pipelines, and receiving stations necessary to accommodate production of crude oil and natural gas. And then all that equipment would become the focus of a major construction and installation project. Hence, from the time a deep-water exploration discovery[178] is made, years go by until the field first produces hydrocarbons, called *first oil*.

Hence, the then current need for *temporary abandonment* of the Macondo well.

------

[173] CASING STRING—the entire length, comprised of many joints of casing (each about 40 feet long) screwed together. As the well gets deeper, additional casing strings will be run and cemented in place. Each successive casing string must be smaller diameter than the already-installed casing string. The final casing for Macondo (9⅞-inch X 7-inch) was run as a single continuous section.

[174] DRILLPIPE—(or drill pipe)—a joint of drillpipe is a single length of steel tubing (normally 5-, 5 ½-, or 6⅝-inch diameter), usually 30 feet long. Drillpipe has larger diameter couplings (tool joints) for strength and wear resistance.

[175] DRILLING FLUID (Mud)— The simplest drilling fluid is water, but it doesn't carry cuttings very well unless its viscosity is increased. Mud density (measured in pounds per gallon—ppg) provides fluid pressure, or hydrostatic head. Water-based muds use water, clay for viscosity, and barite for density, as minimum ingredients. Synthetic-oil-based muds (SOBM) are environmentally superior muds synthesized from vegetable oils, sugar alcohols, or sugar glucose. In complex geology and deep wells, SOBM is often necessary to make the hole "slicker" to help keep the drill bit from getting stuck.

[176] CEMENT—All casing strings placed in drilled holes need cement to keep them in place and to isolate rocks, fluids, and pressures. Cement slurry is pumped down the inside of the casing and up the outside (the annulus). By adding chemical retarders, the cement slurry is kept pumpable (called green cement, with the consistency of a chocolate malt) for a pre-determined number of hours, to allow time for pumping and for resolving problems.

---

[177] DELINEATION WELL—After a successful discovery well, the operator's next well(s) in the area would be drilled to help delineate and quantify the discovery zone (the geometry of the accumulation and amount of oil and gas). These answers are needed before the operator can proceed with the design and economics of a development plan; hence, the normal three years or more from discovery, through delineation, through development wells, and the installation of development facilities, before first oil.

[178] DISCOVERY—A discovery is a well that contains commercial quantities of recoverable oil and gas. If not, it is a dry hole.

**AFTER THE CASING AND CEMENT JOB,
TEMPORARY ABANDONMENT WOULD INCLUDE:**

① **Prepare well for testing and the rig for abandonment**

② **Positive and negative pressure tests—prove wellbore secure or remediate as necessary**

③ **Install Lockdown Seal Ring (LDSR)**

④ **Set and test cement plug**

⑤ **Displace riser with seawater**

⑥ **Pull BOPs and Riser**

⑦ **Release the Rig**

SLIDE 6

That means, as noted in **slide 6**, the cased-and-cemented Macondo well was ready to be tested and secured (temporarily abandoned), so the rig could be released—an event scheduled for *the next day*—and sent to its next contracted location.

① The next location for the *Deepwater Horizon* was to have been a previously drilled well that required water-based mud. But Macondo was using a synthetic-oil-based mud. Therefore, a planned activity for *preparing the rig for temporary abandonment* included displacing all the mud from the drilling riser, and then sending all the recovered mud, plus several thousand additional barrels from the mud pits, to a workboat for onshore disposal. That exercise also included getting rid of several hundred barrels of leftover, 16-ppg, water-base, lost-circulation materials (LCM). Getting rid of the mud and LCM was done simultaneous to other Macondo temporary-abandonment operations.

② Two important pressure tests were designed to ensure, prior to releasing the rig, that there was no leak in the wellbore.

First, a 2,700-psi *positive-pressure test* (from the BOPs down to the top of the float collar, near 18,000 feet) successfully proved there was no leak from inside the casing to outside the casing. Then, a *negative-pressure test* successfully showed that, with reduced pressure inside the wellbore, there was no leak from outside the casing to inside the casing.

③ The lockdown seal ring (LDSR) is designed in conjunction with the casing hanger and casing head. The LDSR[179], when attached to a 5,000-foot-long drillpipe workstring and lowered through the drilling riser to the casing hanger, would lock and seal the casing hanger in place, forever preventing uplift and leaks. For a schematic of the LDSR, see **Diagram 17**, page 222, herein.

④ Even though the well was cased with steel pipe and surrounded by annular cement, with another 180 feet in cement in the shoe track, and given that the float collar contained two one-way high-pressure check valves, *that still was not enough security*. So the well plan called for one last cement plug in the casing, below the seafloor. This plug would be 300 feet long, and would be allowed to set-up and harden, after which it would be tested with a drill bit to ensure its structural and pressure integrity.

⑤ With the well thus declared to be secure, it was then planned to displace the drilling riser with seawater and capture and recover its contents—about 1,500 barrels of oil-based mud—which could not be dumped into the sea.

⑥⑦ Given that the well was pressure tested and declared secure, and the riser's mud was displaced with seawater, the last steps were to be strictly mechanical—pull the riser and BOPs, and then release the rig.

---

[179] CASING-HANGER LOCKDOWN SEAL RING (LDSR) has two primary purposes: (1) it accepts the sealing mechanism for a future subsea production tree, and (2) with the LDSR in place on top of a casing hanger, the casing hanger cannot be lifted from its metal-to-metal seals. Until the LDSR locks the wellhead in place, the annular seal (casing hanger inside wellhead) is dependent entirely on the weight of the casing. Without the LDSR, the casing hanger can be lifted by a strong upward pull or pushed upward by differential pressure (when the pressure above is less than the pressure below).

AFTER THE CASING AND CEMENT JOB, TEMPORARY ABANDONMENT WOULD INCLUDE:

① **Prepare well for testing and the rig for abandonment**

② **Positive and negative pressure tests—declare wellbore secure or remediate as necessary**

③ **Install Lockdown Seal Ring (LDSR)**

④ **Set and test cement plug**

⑤ **(While) Displace riser with seawater**

THE WELL BLEW OUT

SLIDE 7

UNFORTUNATELY—DURING THE PROCESS OF DISPLACING THE RISER WITH SEAWATER, THE WELL BLEW OUT.

What? How can this be? The operator seemingly followed the above steps, ran good casing, got a good cement job, obtained good pressure tests inside and out, and left high-integrity flow barriers in the wellbore.

But the fact is, as shown in **slide 7**, the well did blowout while the drilling riser was being displaced with seawater. The unexpected violent eruption of oil and gas from the riser, through the rig floor, and up inside the derrick, *seemingly* came from nowhere, without warning. Explosions and fire followed. Eleven men died. The rig burned and sank. The months-long oil spill that followed was an environmental disaster.

Bottom line—a fundamentally sound temporary-abandonment procedure turned into a tragically lethal catastrophe. And that should not have happened. Therefore, something had to be wrong with the above list, and every step must be examined.

But first—given that the near-term post-blowout world was fraught with litigious, political, personal, corporate, and unbounded media hype seemingly with no end in sight, a few ground rules were established.

Like—ignore finger pointing, he-said-she-said, politics, litigators, deep pockets, companies, and pontificating experts. Further, establish a no-committee, no-deadline, no-vested-interest solo focus to *assess rig data* to define the *cause* of the Macondo blowout.

Therefore, starting at the top of the list (**slides 6 and 7**) and working down the page, it's time to look at rig data and other hard evidence that may or may not have been critical to the entire operation.

A good starting point (because it was critical to *Prepare the well for testing and the rig for abandonment*) includes data that define the casing and cement job.

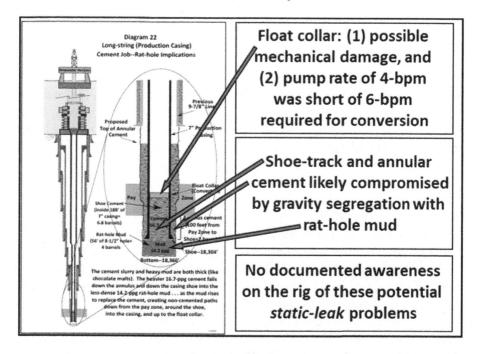

SLIDE 8

Recall that the float collar (the top big arrow in **slide 8**) needed to be "converted" to have its flappers function as dual high-pressure check valves. The manufacturer had performed (pre-blowout) laboratory tests using Macondo's 14.2-ppg oil-based mud and determined that a 6-bpm (barrel-per-minute) minimum pump rate was required to produce the pressure drop inside the float collar necessary for conversion. (Note: For a schematic of float-collar conversion, see **Diagram 22**, page 304, herein.)

Yet records show that after the casing was installed, the pump rate into the casing, while circulating mud and displacing cement, never went above 4.1 bpm. That means the float collar was never converted, and the open flappers never acted as high-pressure check valves. And the path up the wellbore was wide-open.

There was also contention (during court testimony) that residual rock cuttings may have entered the casing (as often happens) while the casing was being lowered into the open-hole section of the deep wellbore.

Such cuttings may have settled in the guide shoe, which then needed more than 3,000 psi to break circulation (dislodge the cuttings).

Further arguments claimed that such high pressure may have even breached (made a hole in) the casing just below the float collar. These points form the basis for a later discussion (**Q&A #2**), but the data confirm this: the well ultimately flowed *upward* through the float collar, so the flappers had to have been open, and therefore had not been converted.

The data pertaining to cement in the shoe track (below the float collar) are also important. Think of oil and vinegar, where separation is common, where the less-dense fluid (oil) floats on top of the heavier fluid (vinegar). The same thing happens in wells with rat holes, especially if the mud in the rat hole is less dense than the cement slurry above it (in other words, the slurry in the shoe track and the annulus). For Macondo, the 14.2-ppg mud in the rat hole was significantly less dense than the 16.7-ppg Class-H cement slurry in the lower annulus and inside the 180-foot-long shoe track.

This is not a surprise phenomenon—American Petroleum Institute (API) RP (Recommended Practice) 67 Section 7.5[180] specifically identifies and addresses the problem and recommends that a heavy "pill" of mud (a few barrels) be placed in the rat hole prior to running the casing, so that gravity segregation will not occur. This was not done on Macondo. (This topic is covered in **Q&A #3**, to follow.) Such predictable gravity segregation in the Macondo well allowed lighter-weight (14.2-ppg) oil-based mud to percolate upward from the rat hole into the 16.7-ppg cement in the bottom of the annulus, where it was only one hundred feet to the bottom of the pay zone. Further, the same percolation phenomena allowed mud from the rat hole to simultaneously migrate upward into the casing and contaminate the 180 feet of 16.7-ppg shoe-track cement.

Bad news: There were two potential sources of a static leak path between the annulus and the wellbore: (1) the unconverted (open) float collar, and (2) the oil-based-mud-contaminated cement in the lower annulus and inside the casing shoe track. A third potential leak path was through breached casing (to be discussed in **Q&A #2**).

Good news: The potential leaks were *static* because the wellbore was still full of 14.2-ppg mud, which yielded a hydrostatic pressure that was 1,000-psi overbalanced compared to the pressure of the reservoir. This was also good news because the next steps in the abandonment procedure called for *testing for leaks* and repairing as necessary.

* * *

So, it was time two pressure tests—positive and negative.

The 2,700-psi positive-pressure test was accomplished by closing the fully functional (important later) blind-shear rams (BSR)[181] and using the cement-unit pumps to raise the pressure below the BSR and down into the wellbore. Such pressure extended from under the BSR down to the top cement wiper plug, which was on top of the float collar. The cement plug had a solid-core, so the increased wellbore pressure above the plug could not get past (below) the float collar. The pressure was held for thirty-minutes. (Note: for a schematic of the Macondo BOPs, see **Diagram 5**, page 68, herein.)

The test called for the low pressure first, then the higher-pressure thirty-minute test. All data indicate the positive-pressure test successfully proved there was no leak from inside the casing to outside the casing (between the BSR and the top cement plug—on top of the float collar).

So, it was time to get ready for the negative-pressure test (NPT).

----

[180] RAT HOLE—Rat hole is any section of open wellbore (previously drilled hole) left below casing. The O&G industry long ago realized the implications of leaving rat hole under (below) newly installed casing. The industry collectively adopted an RP—a Recommended Practice—to handle a common problem. As per API RP 65 Section 7.5—"If casing is not run to bottom, the *rat hole* should be filled with a higher weight mud to prevent cement from falling into the rat hole and displacing rat hole mud into the cement column . . . "

[181] BLIND SHEAR RAMS—The BSR's two steel blades face each other. When actuated, they close, cut, and seal the wellbore. Rated for 15,000 psi. They will cut drillpipe but will not cut through a tool joint or heavy casing. To ensure the blind shear ram will work when needed, the driller keeps track of the depth of critical tool joints (in the area of the BOPs). If necessary, for example during a kick, the driller will raise or lower the drill string to get tool joints away from the blind shear rams.

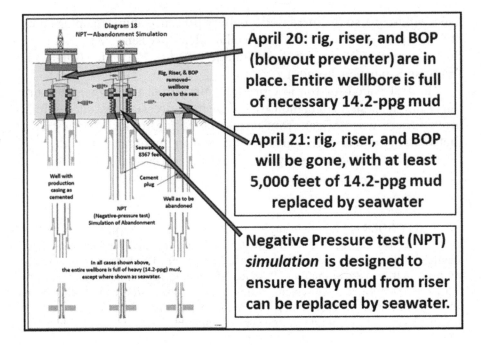

April 20: rig, riser, and BOP (blowout preventer) are in place. Entire wellbore is full of necessary 14.2-ppg mud

April 21: rig, riser, and BOP will be gone, with at least 5,000 feet of 14.2-ppg mud replaced by seawater

Negative Pressure test (NPT) *simulation* is designed to ensure heavy mud from riser can be replaced by seawater.

SLIDE 9

Look at **slide 9**, the far-left schematic shown in **Diagram 18**. The 18,000-foot wellbore was full of 14.2-ppg mud. The well was dead—1000-psi overbalanced. Nothing was leaking (despite the *potential* static leaks).

A full scale of **Diagram 18** is on page 232 herein.

Given the passage of time, the schematic on the far *right* of **Diagram 18** depicts the Macondo well as early as "tomorrow." There was to be no BOP, no riser, and no rig—and the well would have lost at least 5,000 feet of 14.2-ppg mud (from the riser) and replaced it with 5,000 feet of 8.6-ppg seawater. With such radical changes, there was a need to know (before pulling the BOP and riser) if the associated loss of hydrostatic pressure (almost 1,500 psi, from the BOP down to the float collar) would cause a leak and allow external fluids to enter the wellbore.

Alternatively, disconnecting and lifting the BOPs to look for leaks would not have been a good idea.

Better yet, the operator would *simulate* the loss of hydrostatic head with a *negative pressure test* (NPT)[182] by making the wellbore think the top 5,000 feet of mud had been replaced by seawater.

Important—even before taking the first step, there were two possible results from the successfully implemented negative-pressure test (NPT):

(1) **Good news**: If during the NPT simulation (with pressures throughout the wellbore significantly reduced) the wellbore showed no leak for thirty minutes—then the casing (having already passed the high-pressure test) would be deemed pressure secure, and the temporary-abandonment process could proceed; or

(2) **Also good news**: If the NPT simulation revealed there was an active leak anywhere from the annulus (outside the casing) into the reduced-pressure wellbore, then the temporary-abandonment process would be delayed, and the required repair job (likely with cement) would commence immediately. It would have been *good news*, not because there was a leak, but because *identifying and remediating* any such leak was mandatory—and it was exactly the reason for the negative-pressure test.

So, the next step was the negative-pressure test—the NPT.

---

[182] NEGATIVE-PRESSURE TEST (NPT)—For deep-water wells being temporarily abandoned, it is necessary to pull the BOPs when the well is done. But this means all the mud in the riser (above the seafloor), which had been necessary to control the well, will be replaced by seawater. To ensure the casing is mechanically secure before the BOPs are pulled and the heavy mud is lost, an NPT is designed to simulate the replacement of riser mud with seawater. If a casing leak is detected via the NPT, the fix is mandatory and may call for repairing and securing the wellhead or drilling out and perforating and squeezing cement into the area of the leak (at a casing connection or at the casing shoe).

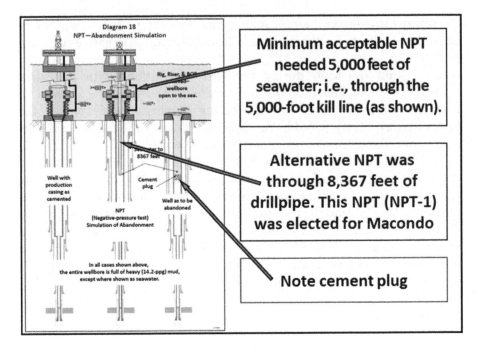

Diagram 18
NPT—Abandonment Simulation

Minimum acceptable NPT needed 5,000 feet of seawater; i.e., through the 5,000-foot kill line (as shown).

Alternative NPT was through 8,367 feet of drillpipe. This NPT (NPT-1) was elected for Macondo

Note cement plug

SLIDE 10

The drilling riser had a number of external, built-in, high-pressure lines, one of which was the kill line. For clarity, the kill line was added (above) to **Diagram 18** (see page 232 herein), as per the upper big arrow.

The $3\frac{1}{16}$-inch-diameter, 15,000-psi kill line reached from the rig floor to the BOP—about 5,000 feet. By filling the kill line with seawater and then closing the BOP (above the kill-line opening), the 18,000-foot-deep wellbore saw 5,000 feet of seawater on top of 13,000 feet of heavy mud. This mixed-fluid column *simulated* what the wellbore (from the sea floor down to the float collar) would see if the BOPs were lifted in the 5,000-foot-deep Gulf. Such a test (an NPT using a 5,000-foot-long kill line) would have met the approved regulatory requirements for testing the Macondo well.

However, data indicate the operator elected to run an even-more-rigorous NPT, using 8,367 feet of drillpipe (second big arrow in **slide 10**) rather than 5,000 feet of kill line. The middle pictorial in **Diagram 18**, above, shows the drillpipe hanging through the BOP, down to 8,367 feet.

The stated purpose of the deeper NPT was to accommodate the installation of the LDSR (lockdown seal ring). The LDSR (which locks and seals on top of the casing hanger inside the casing head) had to be either pushed down, or pulled down, with 100,000 pounds of load. The operator elected to use about 3,000 feet of heavyweight drillpipe below the wellhead to *pull down* on the LDSR, which required 3000 feet of room below the casing head.

The final cement plug (bottom arrow in slide 10 (**Diagram 18**) above) was therefore planned to be installed and tested below that depth (near 8,300 feet) to provide the necessary clearance for the LDSR installation.

All that planning meant the negative pressure test (NPT) would utilize the 8,367-foot-long drillpipe workstring (and not the 5,000-foot kill line).

For convenience, we call the 8,367-foot negative-pressure test *NPT-1*.

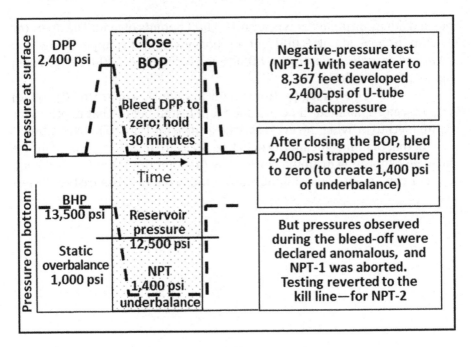

SLIDE 11

Slide 11 is a graphic pictorial of NPT-1 pressures (using 8,367 feet of drillpipe).

The upper half of the graph (dashed line) generalizes the pressure-gauge readings at the top of the drillpipe (the gauge was located at the cement unit), and the lower half of the graph (dashed line) generalizes simultaneous pressures at the top of the float collar (with a *hypothetical* pressure gauge located at the float collar).

In the lower graph, note that the bottom-hole pressure (BHP), with a full wellbore of 14.2-ppg mud, is about 13,500 psi. Also, note that the main reservoir pressure (the pay zone) is about 12,500 psi (the difference was the 1,000 psi of overbalance mentioned earlier).

In the upper graph, the observed drillpipe pressure increased from zero (0 psi) to 2,400 psi as the 8,367-foot workstring was filled with seawater. This trapped pressure was "backpressure"—a measure of the u-tube effect of the 14.2-ppg mud column outside the drillpipe and the 8.6-ppg seawater on the inside.

Note that the bottom-hole pressure (BHP) at the float collar (lower graph) did not change as the drillpipe was filled with seawater.

But then the fully functioning (important later) annular BOP was closed around the drillpipe. The closure isolated 5,000 feet of heavy mud in the drilling riser (outside the drillpipe) from the rest of the wellbore. The closure also created a single, closed, fluid-filled system comprised of 2,400 psi at the rig floor (on the top of the drillpipe), above 8,367 feet of seawater (inside the drillpipe), on top of the remaining column of 14.2-ppg mud (about 10,000 feet from the bottom of the drillpipe to the float collar). This single fluid column (plus the 2,400-psi on the drillpipe) continued to exert a bottom-hole pressure (BHP) of about 13,500 psi at the float collar. Note that the pressure at the float collar did not change when the BOP was closed.

Given that the wellbore was a closed and fluid-filled system, if there was a 100-psi decrease in the 2,400 psi of trapped pressure (on the drillpipe), every pressure in the wellbore would have dropped by 100 psi (decreasing pressure is shown on both graphs). This proactive reduction was accomplished by opening a valve at the cement unit and allowing a small volume of seawater to escape the drillpipe (like releasing air from a balloon). Then the exercise was repeated, and each reduction of trapped pressure on top of the drillpipe had a corresponding reduction of pressure inside the drillpipe and throughout the wellbore below the closed BOP.

When the operator back-flowed additional increments of seawater and dropped the trapped pressure by a total of 1,000 psi, the pressure inside the wellbore at the top of the float collar was then equal to the reservoir pressure (though the wellbore and the pay zone were separated by casing, cement, and the float collar). This was noteworthy, because at that time the pay zone was neither overbalanced nor underbalanced compared to the wellbore pressure inside the casing.

The operator then continued dropping the drillpipe pressure by back-flowing additional small volumes of seawater, step-by-step (barrels out, pressure down), until the goal of zero (0 psi) was reached (a total pressure reduction of 2,400 psi).

At zero (0 psi) on the cement-unit surface gage, the wellbore bottom-hole pressure (BHP) at the float collar was 1,400-psi *underbalanced* compared to the pay zone (located outside the casing).

The regulatory requirement for Macondo's NPT called for dropping the pressure to zero (0 psi) and observing it for thirty minutes. The goal was to determine if the reduced pressure stayed at zero (0 psi), which would indicate no leak, or if the shut-in pressure increased above zero (0psi), which would indicate a leak.

But there was a problem during the test.

Once the trapped backpressure on the drillpipe was dropped from 2,400 psi to about 200 psi, it would not decrease further. Even after opening the cement-unit valve and allowing as much as 15 barrels of seawater to flow from the drillpipe to the cement unit, the drillpipe pressure would not go to and stay at zero (0 psi).

Because the drillpipe pressure would not bleed to zero (0 psi) even when the drillpipe was opened, and because 15 barrels of seawater had flowed back to the cement unit during the five-minute period ending at 6:00 P.M., the valve at the cement unit was closed (at 6:00 p.m.) and not reopened.

These phenomena were declared (on the rig, and in USCG depositions) to be the result of "the bladder effect." For more on this topic (see **Q&A #1**).

NPT-1 was therefore considered *anomalous* (flawed) and the test was *aborted*.

The Macondo negative-pressure-test procedure was therefore switched from the 8,367-foot drillpipe test (*NPT-1*), to the 5,000-foot kill-line test (deemed *NPT-2)*.

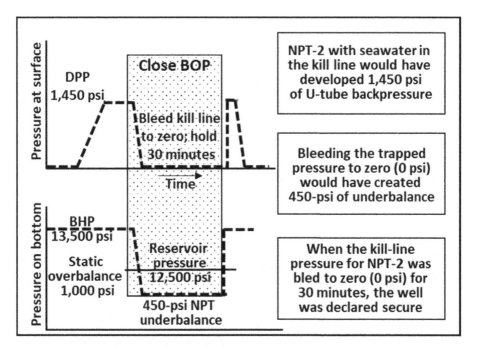

SLIDE 12

If NPT-2 had been run using the NPT-1 procedure, then **slide 12** (top of diagram) would have represented the kill-line pressure at the rig floor. And the bottom half of the diagram would have shown the pressure at the float collar for NPT-2, using the 5,000-foot kill line.

The 5,000-foot kill line, when filled with seawater, would have exhibited 1,450-psi of trapped u-tube pressure (5,000 feet of 14.2-ppg mud in the riser acting on the bottom of the kill line, which contained 5,000 feet of seawater).

Then, with the BOP closed, bleeding the 1,450 psi to zero (0-psi) would have generated 450 psi of underbalance. This test, when completed, would have met regulatory requirements, whether it showed no leak (casing integrity with no flow from the outside the casing), or it showed an increase in pressure, which would have indicated a leak that needed to be repaired.

Good news: The kill-line pressure was successfully dropped to zero (0 psi), where it held steady for the required thirty minutes. That meant the well could be, and was, deemed secure, ready for the rest of the temporary-abandonment procedure, including displacing the riser with seawater to recover the heavy mud for onshore disposal.

Bad news: The well was known to *not* be secure (because it later blew out). Therefore, something had to be wrong with one or both of the negative-pressure-test procedures, results, and/or conclusions.

That translates to a mandatory need to go back and look deeper at all aspects of NPT-1 and NPT-2.

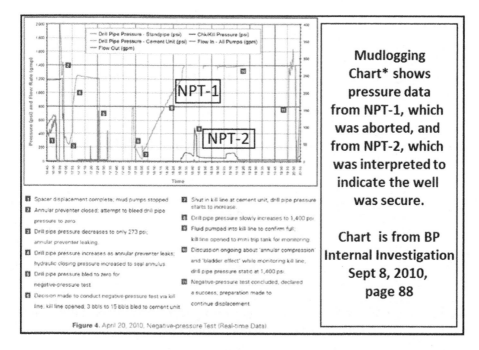

Figure 4. April 20, 2010, Negative-pressure Test (Real-time Data)

Mudlogging Chart* shows pressure data from NPT-1, which was aborted, and from NPT-2, which was interpreted to indicate the well was secure.

Chart is from BP Internal Investigation Sept 8, 2010, page 88

SLIDE 13

To do this, actual mudlogging pressure charts, as recorded offshore, in real time, must be assessed. These charts were, fortunately, generated on the rig and sent wirelessly to onshore computers[183]. After the blowout, the operator made the charts available to the courts and to the public through its own internal investigative reports. A full-scale version of the **Mudlogging Chart**[184] from BP's Internal Report is on page 421, herein.

The time/pressure mudlogging chart (**slide 13**) shows measured surface pressures (vertical axis) for NPT-1 drillpipe pressures and subsequent, simultaneous NPT-2 kill-line pressures. Each horizontal pressure line is 200 psi, and the time labels are five minutes apart.

For NPT-1, a key area of interest is the short diagonal line between 5:55 and 6:00 P.M. This is the recording of that last 200 psi that would not bleed off for NPT-1. More on this later.

But the data to the left (prior to 5:55 P.M.) is also of interest because it appears to have set the stage for an important decision yet to be made. In short, when the annular BOP was first closed around the drillpipe for NPT-1, and the pressure *under* the annular BOP was severely reduced (as per the NPT procedure), heavy fluid from the riser leaked *down* past the rubber sealing element (the bladder) of the annular BOP and pushed seawater back up the drillpipe. And yes, the fluid level in the riser had dropped accordingly.

No big deal—standard operating procedure—the driller increased the closing pressure on the annular BOP. Then it happened again. Same leak, same result.

When the annular-BOP closing-pressure was ultimately increased enough to withstand the 2,400-psi pressure *reduction* (under the BOP) required for NPT-1, no additional fluid leaked past the BOP (and the riser stayed full). More on this later in a discussion about "the bladder effect" (and in **Q&A #1**).

Now, back to the short diagonal line at 5:55 P.M. In the next slide, the same data are shown on an increased scale.

---

183 BP and its operating partners (in offices) had access to the real-time (offshore) data through *Insite Anywhere*, BP's electronic data system owned by Halliburton. The system provides (and records) real-time flow-in and flow-out data, gas analysis data, pressures, and other drilling data.

184 This chart is from BP's *Deepwater Horizon* Accident Investigation Report. The internal investigation was released to the public on September 8, 2010. Section 2.6 interpreted the negative-pressure test. This chart, referenced as Figure 4—April 20, 2010, Negative-Pressure Test (Real-time Data), appeared in the report on P. 88 (see page 420 herein).

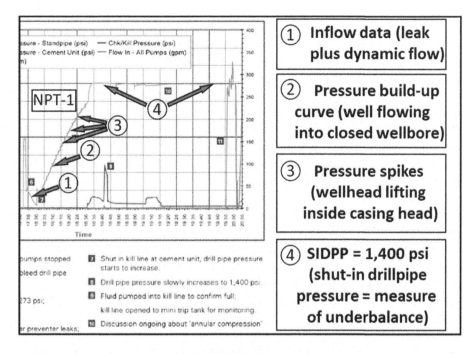

SLIDE 14

The #1 arrow in **slide 14** points to a short diagonal line from 5:55-6:00 P.M. That line is the drillpipe pressure recording when Macondo first began to flow at 5:55 P.M. During that five-minute period, with the drillpipe open, the well *flowed* 15 barrels of seawater from the drillpipe to the cement unit, and the pressure refused to go to zero (0 psi).

That was not supposed to happen.

The 15 barrels of unidentified fluid (the deep source) that displaced 15 barrels of seawater had to have come from somewhere. Two choices—from the riser or from the wellbore. But the fluid level in the full riser was static. Fifteen barrels in five minutes is 4,000 barrels a day. The data say the well—the pay zone—was flowing at 4,000 BPD.

Proof that the reservoir was flowing is based on rig data, as follows.

At 6:00 P.M., the short-duration *vertical* line indicated the valve at the cement unit was closed (no more flow to the cement unit). But the flowing reservoir (not yet affected by the 6:00 P.M. closure of a valve 3-1/2 miles away) continued to flow into the under-pressured *closed* wellbore, which increased the wellbore pressure, as if blowing up a balloon.

The ever-increasing wellbore pressure (recorded at the top of the drillpipe) produced a typical *pressure-buildup curve* (arrow #2).

Such pressure-buildup curves, with Macondo NPT-1 an excellent example, are often produced and analyzed during the testing of a well (for example, during a drill stem test, or DST) to quantify the flow characteristics of a producing reservoir.

As flow continued into the closed Macondo wellbore and the wellbore pressure increased, the drawdown on the reservoir and the flow rate from the reservoir decreased. Finally, when the reservoir had fully pressured-up the wellbore and flow had stopped (at 6:35 P.M.), the built-up pressure at the surface was 1,400 psi. (See arrows coming from #4.) That constant, stable pressure was a measure of the original underbalance between the fluid column in the wellbore and the discovery pay zone.

For clarity, consider a common drilling example. If an operator took a kick (formation fluids entered the wellbore), closed the BOP, and observed 1,400-psi on the drillpipe, the pressure would be noted as the SIDPP (shut-in drillpipe pressure)[185]. The SIDPP would be a measure of how underbalanced the wellbore was at the time of the kick.

Likewise, the 1,400-psi stabilized pressure observed during NPT-1 was the SIDPP for the in-progress Macondo kick. The 8,367 feet of seawater on top of 10,000 feet of heavy mud was 1,400 psi underbalanced compared to the 12,500-psi pay zone, which was in open communication with the wellbore; hence, the well flowed—kicked.

The NPT-1 pressure-buildup data provided irrefutable evidence that the source of flow was the 200-foot-thick discovery pay zone.

* * *

Further, reference is made to several pressure spikes on the pressure-buildup curve (See arrows from #3). The data (spikes) indicate that something in the pressured system was moving—like a valve opening and closing. Here, the "something" was the casing hanger, where the radical reduction of pressure (2,400 psi) *under* the closed annular BOP and *above* the casing hanger unseated (uplifted) the casing hanger in short bursts, as if belching. Each "belch" was recorded as a spike on the pressure-buildup curve.

Importantly, this unseating of the casing hanger was hard data that meant the LDSR (lockdown seal ring) had not yet been installed. This situation did not contribute to the blowout yet to come, but it did play a role in delaying the kill procedure after the blowout.

* * *

The hard data portrayed in **slide 14** showed solid evidence of annular communication, of flow, and of the well kicking.

Yet, as stated earlier, there was (heated) discussion on the rig (shortly after 6:00 P.M.) that the data generated by NPT-1 (in **slide 14**) proved the test to be "*anomalous.*"

The argument, in part, was that "the bladder effect" (leaking annular BOP bladder) caused the flowback (u-tube) of 15 barrels up the drillpipe, which led to NPT-1 being aborted.

To complicate matters, there were serious world-class petroleum engineers who pontificated on an unrelated form of the "bladder effect" associated with rising gas bubbles, gas trapped under the annular preventer, temperature-expansion criteria, etc. But that level of engineering sophistication was likely beyond the scope of the same-name "bladder-effect argument" on Macondo.

Instead, the term was used loosely on the rig referring to the hours prior to 5:55 P.M. (refer to **slide 13**) when the closing pressure on the annular was insufficient, more than once, and mud from the riser leaked past the annular bladder and expelled seawater from the drillpipe.

In these instances, as mud from the riser leaked past the BOP the fluid level in the riser dropped. But when 15 barrels of seawater were expelled from the drillpipe from 5:55 to 6:00 P.M., the fluid level in the riser did not fall.

Three criteria dispel the leaking-BOP argument: (1) the riser remained full of mud, (2) the pressure-buildup curve was a measure of the rate of flow from the reservoir, and (3) the 1,400-psi SIDPP could have been generated *only* be the reservoir, which was 1,400 psi underbalanced to the wellbore.

More on the "bladder effect" in **Q&A #1**.

Nevertheless, based on the argument, a key outcome prevailed—NPT-1 was aborted and NPT-2 was commenced using the kill line.

---

[185] SHUT-IN DRILLPIPE PRESSURE—When a kick is detected and the BOPs are activated (which shuts-in the well), drillpipe and casing pressures are measured and recorded. The shut-in drillpipe pressure (SIDPP) is key to characterizing the kicking formation, wherein the bottom-hole pressure of the kicking formation is the simple sum of the SIDPP and the pressure exerted by the drilling mud (and other fluids as appropriate) in the drillpipe at the true-vertical depth (TVD) of the kick. The shut-in casing pressure (SICP) can be high and erratic, due to the unknown volume and type of formation fluids (oil, water, and especially gas) moving up the wellbore.

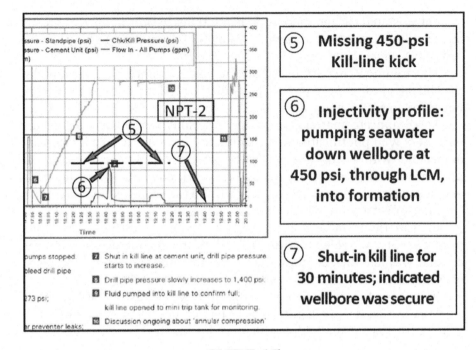

SLIDE 15

NPT-2 utilized the kill line, which was full of seawater. As noted in **slide 15** (with the valid NPT-1 data still recording in real time), the NPT-2 data showed two anomalies, though neither (apparently) got enough attention to shut down the test.

The **first** anomaly—NPT-2 was missing the 450-psi kick (arrows from #5), which was the kill-line equivalent of the 1,400-psi NPT-1 kick up the drillpipe.

So, the question was this: How could a very real kick through the wide-open 8,367 feet of drillpipe not manifest as a kick through the wide-open 5,000 feet of kill line? Maybe a kill-line valve was closed.

Whether or not there was a closed valve, something did block the kill line, preventing the kick from being seen at the surface.

So—back to **slide 6**, which provided the steps to be taken for temporary abandonment. Recall the need to get rid of the oil-based mud as well as several hundred barrels of water-based lost-circulation material (LCM).

Records show the viscous slug of LCM[186] was pumped *down* the drillpipe and *up* into the lower riser (above the BOPs), where it would act as a spacer between the heavy mud—to be eventually displaced from the riser—and the seawater that would do the lifting.

But there was a problem.

Recall that prior to 5:55 P.M., there were a couple of occasions (at the start of NPT-1) when the closing pressure on the annular BOP was inadequate and mud bypassed the annular BOP and displaced seawater from the drillpipe. The closing-pressure problem was ultimately fixed, but the occasions of heavy mud from the riser dropping and bypassing the leaking annular meant the LCM moved *deeper* into the riser and ended up *in the BOPs* rather than *above the BOPs*.

By design, LCM plugs holes in rocks. But it will also plug holes like the small inside diameter of the $3\frac{1}{16}$-inch, 15,000-psi kill line, the bottom end of which was in the BOP, nested in the slug of LCM. With the bottom end of the kill line plugged by LCM, the 450-psi "push" of the kick apparently could not manifest at the surface.

Or, as previously noted, perhaps the bottom kill-line valve was closed.

This leads us to the **second** NPT-2 anomaly, which occurred when seawater was pumped into the top end of the kill line (and into the wellbore) to ensure the kill line was full of seawater. That would have been a small-volume test, as the fluid-filled, closed wellbore had passed the 2,700-psi positive pressure test (down to the top cement wiper plug on top of the float collar).

Nevertheless, during the act of pumping seawater into the kill line (see arrow from #6), the pressure increased rapidly (as expected), but (unexpectedly) broke back—rapidly decreased on its own—at about 500 psi *before* the pump was turned off. The break-back pressure-drop appeared equivalent to the break-back pressure-drop one might see during a cement squeeze job. So, even though the kill line should not have been able to take whole fluid, it did.

And that meant the valve at the bottom of the kill line was indeed open. Further, it meant the fluid from the kill line had found an outlet and was being pumped (lost) "somewhere"—like outside the kill line, down the wellbore, and back into the kicking (albeit shut-in) pay zone.

But that begs two questions:

(1) What happened to the LCM that blocked the kill-line kick?

(2) How can fluid be pumped into the formation given the successful results of the 2,700-psi positive pressure test?

First, the 450-psi driving pressure of the kick likely was not enough to pump LCM *up the kill line*, but 500 psi from the top was able to push LCM *away from the kill line* and transmit pressure into the wellbore.

Second, had the top cement wiper plug remained in place on top of the float collar, it would not have been possible to pump past it, into the formation. But the top plug did not remain in place, as it had been lifted away from (above) the float collar by the 15 barrels of pay-zone flow during NPT-1. This meant that when the kill-line pressure exceeded about 450 psi and transmitted its pressure to the wellbore, the pressurized fluid (now overbalanced) found an exit point—back into the formation—turning the Macondo well into an injection well.

The data confirm that with the missing 450-psi kill-line kick, and with "injection-well" communication into the annulus, and with there still being a 1,400-psi kick in progress on the drillpipe—*the entirety of NPT-2 was invalid.*

Nevertheless, with the (unobserved) help of the LCM plug, the kill line showed zero (0 psi) pressure at the surface, and when opened at 7:16 P.M. *discharged no seawater* (see **slide 15**), for more than the requisite thirty minutes; hence, the test was declared to have successfully shown no leak into the casing.

In other words, NPT-2 was declared **valid** and the well was declared **secure**.

And that meant it was time to get on with the temporary abandonment procedure. Recall in **slide 6** that the remaining steps to be taken included the following: ③ install the LDSR, ④ set a cement plug in the casing, ⑤ displace the riser with seawater, ⑥ pull the BOPs and riser, and ⑦ release the rig.

Data indicate step ⑤ began next, without steps ③ and ④.

This meant the displacement of the riser took place before the LDSR had been set (recall the spikes on the pressure-buildup curve, **slide 14**, arrows from #3). It also meant the 8,300-foot-deep cement plug (see **slide 9**, middle of right-hand schematic) had not been set, which was designed to further isolate the deep wellbore from the seafloor and the rig.

Instead, the next activity (after declaring NPT-2 successful) included two key steps: opening the BOPs and displacing the riser with seawater.

Opening the (annular) BOP immediately killed the well, since all the heavy mud in the riser once again became an integral part of the 18,000-foot column of heavy wellbore fluid. In fact, the kicking pay zone, even though still open to the wellbore, was immediately 1,000-psi overbalanced by the total mud column.

The Macondo well was dead.

The simple act of opening the BOP, which killed the well, was the upside of the negative-pressure test *simulating* the replacement of the upper wellbore with seawater.

But that good news was short lived.

---

[186] LCM AS SPACER—The operator had asked the contracted mud service company to make up at least two different batches, or "pills," of lost circulation material—one commercially known as Form-A-Set and the other known as Form-A-Squeeze. These (leftover) materials were combined for use as a spacer during displacement of the riser. The combined material weighed 16 ppg. A total of 424 barrels of 16-ppg spacer was pumped into the well and displaced with seawater, placing the spacer 12 feet above the BOP. The operator chose to use the (combined) LCM pills as a spacer so as to avoid having to dispose of the unused material (onshore) as hazardous waste—because the ultimately recovered water-based LCM (then a *used* product) was disposable at sea.

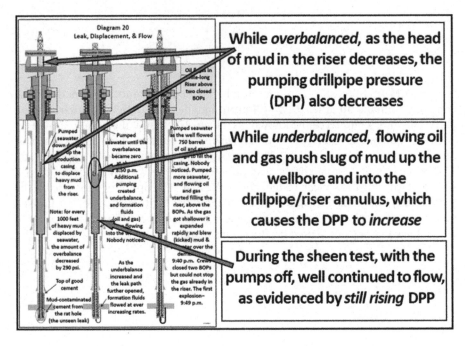

SLIDE 16

At that time, rig leaders were oblivious to the open path from the formation to the wellbore, and they start pumping seawater into the drillpipe (left-hand schematic, **slide 16**). Seawater that exited the bottom of the drillpipe displaced heavy mud from the riser (the drillpipe annulus). As mud was displaced from the top of the riser and replaced by seawater in the bottom of the riser, the circulating pressure (u-tube) on the drillpipe decreased with the reduced footage of annular mud; hence, the drillpipe pressure dropped by the minute.

(A full-sized **Diagram 20** is shown on page 278, herein.)

With the replacement of heavy mud in the riser by seawater, the total hydrostatic head at the bottom of the well was also decreasing. In fact, with each barrel of seawater pumped, and with each barrel of heavy mud discharged from the riser, and with each minute that passed, the total hydrostatic overbalance on the reservoir decreased from 1,000 psi—until it got to zero (0 psi)—neither overbalanced nor underbalanced. And the operator was still pumping seawater.

And with that next barrel pumped (creating underbalance), that's when the well commenced to flow (middle schematic, **slide 16**).

As the formation flowed, a column of hydrocarbons (oil with natural gas in solution) moved up the deep wellbore, lifting the mud above it. The uplifted mud rose up the casing and around the outside of the drillpipe, which *increased* the backpressure (u-tube) on the drillpipe. Data show the drillpipe pressure stopped falling and started rising—the demarcation between overbalanced and underbalanced—when flow from the reservoir commenced, at about 8:55 P.M.

The operator (apparently) did not see this change in drillpipe pressure and kept pumping seawater.

* * *

Seawater was pumped until the slug of LCM in the riser reached the top of the riser (just under the rig floor). When the rig crew saw the LCM, they turned off the seawater pumps. The LCM was water-based, and since all the oil-base mud in the riser was *above* the LCM, there should have been no more oil-based mud in the riser. A "sheen test" was then initiated to ensure there was no more oil-based mud. Test results showed a positive result (no sheen). Accordingly, rig leaders directed the crew to allow future returns from the riser (including the LCM) to go overboard (since there was no more oil-base mud).

But the rig data said there was a problem.

During the sheen test, while the seawater pumps were off, the well (according to NPT-2) should have been dead. Instead, additional mudlogging charts showed that the drillpipe pressure continued to rise while the pumps were off—*because the well was still flowing.*

Nevertheless, with a good NPT-2 and a good sheen test (and not noticing the well was flowing), the operator went back to pumping seawater, which further exacerbated the rate of flow.

The wellbore schematic on the far right of **slide 16** deserves its own slide.

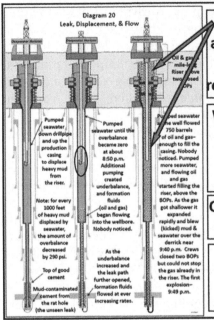

Flowed 750 barrels into casing and almost 1,600 barrels into the riser before shallowest O&G reached the bubble-point pressure

With gas below the bubble-point pressure, rapid expansion blew seawater, oil, and gas through the rig floor and over the derrick

Closed two BOPs, but O/G in riser continued to flow—and ignited.

Drillpipe was compromised by falling traveling blocks, which allowed flow to continue.

SLIDE 17

The well continued to flow (kick) hydrocarbons (right-hand schematic, **slide 17**), until the flow filled the entire 750-barrel wellbore, then filled the BOPs, and then started filling the 1,500-barrel riser. For reference, a 100-barrel kick would be considered *significant*.

Note: BOPs still open; operator still pumping seawater.

Somewhere along that rapidly accelerating path up the wellbore and riser, where there was less and less heavy mud to generate hydrostatic pressure, the shallowest gas (dissolved in the oil) reached the *bubble point*. That meant the dissolved gas finally experienced a hydrostatic pressure so low that it boiled from the oil and formed free gas, in the form of tiny bubbles[187].

And as any given bubble flowed (and was pumped) to an even shallower depth (lower pressure), the bubble further expanded. Perhaps slowly at first, then explosively fast. The collective mass expansion of bubbles displaced the shallowest fluids from the riser and thus further reduced the pressure on top of the already expanding gas. Once started, the process was autocatalytic (the evolved and expanding gas reduced the pressure, causing further expansion and allowing even more gas to evolve from the oil).

Though analog evidence was abundant, the first unquestionable evidence the well was flowing took the form of an eruption of fluids from the riser, through the rig floor, and over the derrick.

Records show "somebody" immediately closed an annular BOP and a variable bore ram (VBR)—see closed BOPs as shown in **slide 17**.

And herein lies the major difference between deep-water wells and all others, whether offshore or onshore.

When the deep-water Macondo annular BOP was again closed (recall it had been used throughout both NPTs), it worked exactly as designed and expected. It sealed the annulus outside the drillpipe. It stopped the reservoir from flowing (other than pressuring up the wellbore—creating another pressure-buildup curve—below the BOP).

But because the BOPs were on the seafloor a mile below the rig, all the crude oil and natural gas already in the 1,500-barrel riser continued to accelerate up the riser driven by the sub-bubble-point gas, which continued to expand violently.

The exiting oil and gas (from above the closed BOPs) engulfed the rig floor and derrick and blew laterally under the rig floor throughout the moon pool[188] area.

In seconds, engine-room suction fans inhaled a mix of gas and atomized oil, which fueled the diesel engines that drove the power generators. The generator room exploded (data showed 9:49 P.M.), which shut down power throughout the rig. Perhaps ten seconds later, the oil and gas under the rig floor exploded, taking out the moon-pool walls and entering the hallways, quarters, and galley. The entire rig, from substructure to the top of the derrick was engulfed in flames.

Good news: As bad as that was, the well was dead—no new flow could enter the deep wellbore or the riser past the closed BOPs (annular and VBR)[189], or enter the drillpipe, which was mechanically closed (still tied to the seawater pumps).

Bad news: But that changed when the conflagration on the rig floor (fed by burning oil and gas from the riser) eventually brought down the traveling blocks, which would have landed directly on top of, and thus opened, the top of the drillpipe. Such failed drillpipe opened a clear path from the reservoir, up the casing, through the ruptured drillpipe, into the riser, and up to the rig floor.

Data from a September 2010 DNV (Det Norske Veritas) inspection of the recovered BOP indicated several broken sections of drillpipe landed on top of the still-closed BOP. The broken (open-ended) drillpipe had thus provided a large conduit *through* the closed BOPs, and instantly allowed new formation fluids to enter the highly underbalanced wellbore and riser, further feeding the inferno on the rig.

Some experts argue that the blocks didn't fall, and that the powerless rig drifted off station, which parted the drillpipe (locked high in the blocks and low in the BOPs). The argument is academic without further data, but in any case, the then failed drillpipe became the open conduit right up through the closed BOPs.

---

[187] BUBBLES—BOYLE'S LAW (P x V = C)—No calculating necessary, but the simple equation shows there is a relationship between the volume (V) of a fixed mass of gas, and its pressure (P), in that their product is constant (C). In short, this means that as the pressure on a gas bubble decreases, its volume increases, and vice versa. A common example involves Mentos mints and Diet Coke, where the mixture erupts immediately after the first bubbles form.

[188] MOON POOL—Large opening through the lower deck—directly under the rig floor and surrounded by handrails—through which the BOP and other tools are lowered into and recovered from the sea.

[189] VARIABLE BORE RAM (VBR)—Rated to 15,000 psi. The two VBR rams have semi-circular faces that self-adjust around whatever size pipe they find. The enclosed drillpipe is not cut. Closing pressure does not need to be adjusted, as it does for annular BOPs.

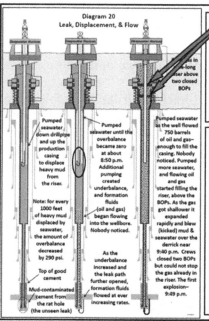

Diagram 20
Leak, Displacement, & Flow

Drillpipe was uplifted by the violent flow. Excess buckled DP was trapped between the closed BOPs, which prevented the blind shear rams from closing.

THE MACONDO BLOWOUT
- **Immediate explosions and fire**
- **11 deaths**
- **115 survivors evacuated the rig, many with extreme injuries**
- **Rig sank day and a half later**
- **Almost 5 million barrels of oil flowed into the Gulf of Mexico for 86 days before well killed**

SLIDE 18

The two closed BOPs (**slide 18**), in concert with the violent uplift of the flowing well, created a problem with unusual mechanical consequences, which significantly impacted the kill of the well. Specifically, though successfully used earlier for the positive pressure test, the blind shear ram (BSR)—located between the closed lower annular and the closed upper VBR (with open drillpipe between them)—would not close, and could not be forced to close, for weeks and months to follow. And the well continued to flow through the open drillpipe.

In short, when the BOP was recovered from the seafloor in September 2010 and subjected to a forensic examination by Det Norske Veritas (DNV), more drillpipe was found between the two closed BOPs than the distance between them. The drillpipe in the gap was deformed (buckled) and off-center, which had prevented the BSR from closing. Arguments persist among academics and experts as to whether the violent fluid uplift drove excess pipe into the gap or if pressure differentials (inside and outside the drillpipe) between the closed BOPs buckled the pipe.

Given the debate, the bottom line is this: The BSR would not close, and the entrapped drillpipe was subsequently found to be severely buckled.

The deformed, distorted pipe was so far off center the blades of the BSR could not function.

Also, the drillpipe inside the closed annular BOP was found to be externally fluid eroded, an indication of the violence of the flow even as the annular BOP was first closing.

A second serious impact of the BSR-buckled-drillpipe problem was that the emergency riser disconnect system depended on the BSR closing. That meant the auto-disconnect system at the LMRP (lower-marine-riser package) would not release. The LMRP included a remote-operated connector between the two annular BOPs, which, when released, would have disconnected the LMRP and the riser from the rest of the BOP stack. Therefore, the rig was stuck on location, unable to release from the riser or BOP stack.

* * *

The long-term environmental and financial consequences of the blowout are well documented, but key statistics define the tragedy.

Eleven men died. Survivors totaled 115, though more than half sustained serious and crippling physical injuries.

The *Deepwater Horizon* burned and sank a day and a half later. See the photo, **Diagram 21**, on page 291 herein.

The Macondo well flowed ("about" is a safe word) five million barrels of crude oil into the Gulf, before being killed on day 86. Events and procedures during those terrible 86 days are beyond the scope of this work, which remains focused on the *cause* of the tragedy.

* * *

So . . . where do we go from here?

Factors evidenced by data that
Contributed
to the *Cause* of the blowout

- **Rat Hole**
- **Float Collar**
- **Back-flowing the well**
- **Unseen forensic data**
- **LCM in the BOP**
- **Simultaneous operations**

SLIDE 19

With all the above helping us to understand the sequence of events that *caused* the Macondo blowout, let's look *back* at what we've seen as well as *forward* to how we can apply what we've learned to future activities.

Slide 19 lists factors that *contributed* to the blowout. Alone, any might seem innocuous. Yet, for Macondo, none was. And had any one of them been recognized and acted on with authority, the tragedy likely would have been averted.

(1) The footage of the *rat hole* was necessary as part of the procedure for evaluating the discovery and landing the long string of production casing. Yet industry experts realized decades ago that less-dense rat-hole mud could gravity segregate *upward* into heavier cement slurries in the casing and in the annulus (see **Q&A #3** for additional information). Such upward migration of mud would contaminate the cement. Mud-contaminated cement does not set, does not harden, and does not seal as designed. Preventing this known problem would have been easy—with zero rig time and zero risk—during the original design of the well.

Nevertheless, the design program for the well stated, "Do not need to set 16.5 ppg mud in rat hole as volume is only ~4 bbls."
That directive contributed to the cause of the blowout.

* * *

(2) The *float collar* played a role that could have been so different. By design, the float collar contains two high-pressure flappers that act as one-way check valves (preventing flow up the wellbore). The flappers are run in the open position so the casing fills with mud as it is being run. By design, the float collar must be "converted" for the flappers to activate as check valves. (See **Diagram 14**, page 202 herein.) The manufacturer tested the model programmed for the Macondo well, using Macondo mud, and confirmed the float collar would convert at a minimum throughput pump rate of 6 bpm. Data from the rig show the maximum pump rate at the float collar was 4.1 bpm. Therefore, the float collar did not convert, and the flappers never got the chance to act as check valves.

The open float collar contributed to the cause of the blowout.

* * *

(3) *Back-flowing the well* contributed on several fronts. Specifically, during the last 200-psi of the 2,400-psi NPT-1 drawdown, the contaminated cement in the lower annulus and the 180-foot shoe track finally yielded to the increasing *negative* differential pressure, and the well commenced flowing. That initial surge further cleared the mud-contaminated annulus-to-wellbore path, and the well rapidly flowed 15 barrels into the low-pressure wellbore during the next five minutes (5:55 to 6:00 P.M.).

The flow also lifted (back flowed) the top cement plug from the float collar, which subsequently allowed seawater injection into the kill line during NPT-2 (pushed the plug back toward the float collar).

Further, the continued flow of formation fluids into the underbalanced wellbore created the pressure-buildup curve (indicative of the reservoir's flow capacity). The buildup-curve pressures increased for thirty-five minutes and leveled-off at 1,400 psi, a measure of the underbalance between the flowing reservoir's formation pressure and the NPT-1 mixed fluid column (8,367 feet of seawater on top of 14.2-ppg mud).

The lack of recognition that the well was flowing during NPT-1 *contributed to the cause of the blowout.*

Records also reveal a significant in-court debate that centered on the possibility of a pre-NPT-1 breach in the casing below the float collar. This topic is to be covered in **Q&A #2**.

* * *

(4) *Unseen forensic data* is a big category. It includes all the flow-related data evidenced by the pressure chart for NPT-1 as well as other documented mud-logging flow-and-pressure charts. All these data were unseen, or more fairly, were neither seen nor understood to an extent sufficient to lead to a firm decision to stop testing and remediate whatever problem had allowed flow from the reservoir into the cemented production casing.

Further, NPT-2 provided two additional key pieces of unseen forensic data. First, there should have been a simple two-part drawing showing the following: (a) the NPT-1 wellbore, with its specific seawater-filled drillpipe and mud column, and its shut-in pressure (kick pressure) of 1,400 psi, side by side with (b) the NPT-2 wellbore, with its specific seawater-filled kill line and mud column, and its shut-in pressure (kick pressure) of—

Whoops, a simple calculation says—the NPT-2 drawing should have shown a 450-psi shut-in pressure (kick pressure) on the kill line.

It's the *missing* part that would have been so vivid in a simple drawing of the wellbore with balanced hydrostatic pressures up the drillpipe and up the kill line. Neither seeing nor missing the 450-psi kill-line kick pressure contributed to a falsely justified comfort with NPT-2.

By the same token, the data that showed the kill line to be full of seawater also showed that "something gave" when the kill line was pressured up. That data, apparently unseen, showed injection into the reservoir (in other words, communication with the annulus). Missing the assessment of "what gave" contributed to the comfort level in declaring NPT-2 a good, no-leak test.

The lack of assessment and response to the above noted forensic evidence *contributed to the cause of the blowout.*

* * *

(5) *LCM in the BOP* did not cause the blowout, but it plugged the kill line and allowed the NPT-2 test pressure to be dropped to zero (0 psi), which made NPT-2 look good—and that *contributed to the cause of the blowout.*

* * *

(6) *Simultaneous operations* (pumping mud overboard to a workboat, moving mud and cleaning mud pits, and preparing for the next day's operations) also *contributed* by making it difficult to accurately measure pit levels[190] *and mandatorily balance* barrels of seawater being pumped into the well with the barrels of mud, LCM, and seawater exiting the riser. To the extent possible with such incomplete data, post-blowout analysis and cross-correlation of charted data (beyond the simplicity of pit-volume totalizers and differential barrel counters with alarms) showed the well to be flowing (more barrels out of the well than being pumped in) before 8:55 P.M., but these data were either not seen or not acted on to such an extent as to recognize the flow, stop the work, and initiate well-control procedures—all of which *contributed to the cause of the blowout.*

* * *

Though each of the above *contributed to the cause of the blowout,* several factors were more directly linked to *causing the blowout.*

---

[190] PIT-LEVEL—Critical to drilling and all wellbore-related activities is an active record of the number of barrels of drilling mud in the system—both in the wellbore and in a number of interconnected mixing and storage pits on the rig. While drilling, the system is dynamic, with mud being pumped into the well and mud returning up the annulus (the riser). At other times the system is static— nothing moving. Whether static or dynamic, sensitive gauges and meters compare pit volumes and pump rates 24/7, with a simple reality—any loss of mud from the system likely means lost circulation, and any gain of mud by the system (called a pit gain) likely means a kick. The driller watches such readings (which are audibly and visibly alarmed), so he or she can take immediate action as necessary. The mudloggers monitor the same charts and alarms around the clock on behalf of the company man and toolpusher and immediately raise the alarm as warranted.

Factors evidenced by data that
Caused and Exacerbated
the blowout

- **Viable NPT results that confirmed a leak and the well's flow potential if underbalanced**
- **The lack of a primary cement-plug barrier before seawater displacement**
- **Viable pump-pressure data that confirmed the well flowed for an hour prior to the blowout**
- **Massive, unchecked flow (kick) that ultimately debilitated proper functioning of the BOPs**

SLIDE 20

The items listed in **slide 20** are strongly linked to the *cause* of the blowout:

(1) NPT-1 data showed that the Macondo wellbore was in communication with the annulus and that the well flowed when exposed to a pressure reduction of about 2,400 psi (underbalanced by 1,400 psi). Such confirmation should have (mandatorily) given the Macondo leaders no viable recourse other than to stop the temporary-abandonment procedure, investigate and identify the location of the leak, and remediate as necessary. Every action on the rig after the well was shut-in at 6:00 P.M. and after the leak/flow data were declared anomalous and further ignored—and every unseen intervention opportunity along the way, including data from NPT-2—drove the operation closer to the cataclysm that would follow.

(2) A critical barrier for any temporarily abandoned wellbore (for example, when preparing for a hurricane) is the 300-foot-long cement plug normally set a few hundred feet below the seafloor. It is designed with a single purpose in mind—plug the wellbore so *nothing* can get through the plug. Had the cement plug been set *after* NPT-2 (even if that

test was wrongly considered valid) and *before* the riser was displaced with seawater, the story would have ended quite differently.

Based on data that showed what had and had not been done prior to the blowout, a critical plan-changing decision had (apparently) been made to set the cement plug *after* the riser was displaced, ostensibly so the cement plug would be set in seawater rather than in oil-based mud. As part of the same decision, this would have delayed the setting of the lock-down seal ring (LDSR) but would have allowed it too to have been set in seawater rather than in mud.

On both counts, this meant the riser was displaced with seawater and the well blew out prior to setting the cement plug and the LDSR.

(3) Pump-pressure data gathered while displacing the riser with seawater provided *extensive* incontrovertible evidence the well became balanced (neither overbalanced nor underbalanced) and commenced flowing near 8:55 P.M. Had the evidence been recognized early on, immediate action would have allowed rig crews to shut in the well and kill a *low-volume, low-pressure kick*—prior to remediating the then-obvious casing-cement-leak problem.

(4) The Macondo blowout preventers (BOPs) have taken a lot of heat since the blowout due to this argument: The *blowout preventers* did not prevent the blowout. Point taken, but the topic is an important part of this presentation.

The normal pressure-testing and well-control functions of a BOP are well known (as taught in petroleum-engineering universities, on the job, and through mandatory, professional, well-control schools around the world). Specifically, as related to well-control kicks, the critical goal is to *minimize the volume* of uninvited formation fluids that enters the wellbore.

Accordingly, rig leaders and crew are skilled at recognizing *early* symptoms of kicks (flow from the well, gain of drilling fluid in the pits, etc.). After the *very first symptom* of a possible kick, prudent operator and contractor personnel take immediate necessary steps to allow closing the BOP as soon as possible. The immediate closing of the BOP minimizes the severity of the kick by preventing further influx of formation fluids and allows rig crews to gather necessary information to quantify the kick and kill the well.

Not so with Macondo. As shown herein, the steps that contributed to and caused the blowout took place *hours* before the BOPs were called into action. The well first flowed during the induced pressure drop associated with the NPT-1 simulation (yes, that was a 15-barrel *kick*). The well was shut in (with a valve at the cement unit, rather than with a BOP) near 6:00 P.M.

Later, after NPT-1 and NPT-2, with open BOPs and the help of seawater pumps, the well lost its overbalance (about 8:55 P.M.) and kicked and flowed for almost an hour. The flowing well filled the casing and most of the riser with nearly 2,000 barrels of crude oil and natural gas, and then blew out (about 9:45 P.M.) with violent, gas-driven energy through the rig floor and over the crown of the derrick—before the lower annular BOP was closed. A variable-bore ram BOP was also closed.

The first explosion occurred near 9:49 P.M.

As described above (following **slide 18**), the combination of the extremely long-duration kick and accelerating, voluminous flow contributed to lifting and buckling the drillpipe inside the BOP stack, specifically inside the BSR, located in the gap between the lower closed annular BOP and the upper closed VBR.

Note: Although the blind shear ram (BSR) had been used earlier in the day for the 2,700-psi positive-pressure test, the buckled drillpipe prevented the BSR from closing. Had the debilitated BSR been able to close and seal, the duration of the *in-progress* blowout would have been measured in hours and days rather than weeks and months.

Conclusions

- **Macondo blowout evidence is defined by basic petroleum-engineering, operating, and drilling concepts, training, and responsibilities**
- **Skilled application of such concepts would have made a difference at Macondo**
- **Also helpful would have been industry initiatives like drilling process safety, human factors, real-time data, safety & environmental management systems,**
- **But . . .**

SLIDE 21

Worldwide, degree programs most often referred to as *petroleum engineering* and *petroleum technology* are the only academic programs dedicated to teaching the engineering concepts necessary for drilling wells, controlling the complexity of subsurface fluids and pressures, completing wells for production, producing hydrocarbons, and managing the vast underground geologic complexes in which oil and gas are found. After graduation, petroleum engineers (and others, for sure) further mature through on-the-job training, specialized course work, graduate school, and years of experience to ensure a deep understanding of the concepts required for a career in the energy business.

For Macondo, as noted in **slide 21**, the engineering concepts that define the cause and manifestation of the Macondo blowout are not overtly complex. No differential equations. No spreadsheets. No sophisticated software packages.

Conversely, with a clear understanding of hydrostatics, mud weights, fluid pressures, cement, casing design, wellbore mechanics, well-control requirements, and formation flow criteria—basic Drilling 101[191]—no petroleum engineer[192] or experienced rig leader reading this document should be overwhelmed by any aspect of the Macondo blowout data.

However, knowing and understanding basic concepts is different from applying such knowledge to managing any and every well—as should have been done on Macondo, without question, without fail.

* * *

While *From the Podium* focuses only on the *Macondo data* that define the *cause* of the blowout, others—academic and engineering study groups, corporations, consultants, authors, technical experts, journalists, legal entities, and environmental groups—tirelessly investigate and pontificate on topics beyond the cause and effect aspects of the Macondo blowout, including human factors, group think, finger-pointing, company culture, training requirements, politics, metrics analysis, liability assessment, completion procedures, regulations, real-time data, process safety, root and latent cause analysis, operator and service-company relationships, toxic dispersants, product-service-resource initiatives, riser-gas detection, disaster management, non-Macondo historical catastrophes, equipment redesign, drilling reliability, drilling process safety, risk management, and other "could've, would've, should've" topics without limit. Entire books have been written on these topics and more.

Are such studies and related findings good? Oh, yes! They are *umbrella* issues because they pertain to Macondo and the entire industry. Each is a critical issue, being examined by the best of the best, and all for a good cause: to help minimize the chance of ever repeating a Macondo-type disaster.

But the above slide ends with the word *BUT. BUT* what?

---

[191] DRILLING 101—Slang term for the first-semester course in drilling engineering as part of petroleum engineering curriculum. Topics include drilling equipment, drill bits, rock mechanics, drilling fluids (mud), wellbore hydraulics, casing design, cement, and well control to name a few

[192] ENGINEER—uses math and science to make projects faster, safer, and more productive, cost efficient, and environmentally friendly. Petroleum engineers apply geo-science and engineering (mechanical, structural, civil, geological, electrical) to the petroleum industry, whether designing, drilling, and managing wells, or producing oil and gas (O&G), or managing O&G reservoirs.

Conclusions

- Macondo blowout evidence is defined by basic petroleum-engineering, operating, and drilling concepts, training, and responsibilities
- Skilled application of such concepts would have made a difference at Macondo
- Also helpful would have been industry initiatives like drilling process safety, human factors, real-time data, and safety & environmental management systems,
- But . . . HOW DO WE APPLY LESSONS LEARNED FROM MACONDO TO FUTURE WELLS?

SLIDE 22

Although involved experts strive for closure and applicability of the complex, seemingly open-issue "umbrella" topics, the question must be asked, as in **slide 22**: *How do we apply what we've learned from the Macondo disaster to future wells?*

The mix of senior leaders on Macondo, despite their collective education, experience, and responsibilities, failed to sufficiently action abundant real-time data in a way to prevent the disaster. Such failure is not acceptable. Would different leaders—whether clones of the Macondo team or hand-selected world-class deep-water drilling experts—have had different results? Likely, but we will never know.

Nevertheless, regardless of who was or could have been in charge, the following is offered as a tool that—had it been used on April 20, 2010 to assess and respond to real-time Macondo data—likely would have saved the day.

Process Interruption Goal

Process
- **Running casing**
- **Testing BOPs**
- **Installing a wellhead**
- **Drilling to next casing point**
- **Testing casing**

Interruption
- **Any unplanned/unexpected deviation**

Goal
- **Figure out what's wrong and fix it.**

SLIDE 24

(And, yes, the contents of **slide 23** have been included in **slide 24**.)

NASA and the commercial aviation industry use a proven concept that can be paraphrased as "process interruption" to solve unexpected, intractable, and otherwise lethal events. Certainly, a commercial flight to Dallas takes off, climbs, cruises, navigates, approaches, and lands. For a passenger to be able to say, "That was a good flight," every step in the process must have worked, in the right order, even if there were problems prior to disembarkation.

Sometimes, though, flights are interrupted by process mishaps. Think about Captain Chesley Sullenberger and the incident often called *Miracle on the Hudson*. Or *Apollo 13*, where an oxygen tank failed a mere 200,000 miles from earth. In both cases, the immediate and expert response to the respective process interruptions saved the day.

Now, the challenge—translate the concept from flying a space shuttle or an airplane to drilling a well.

An exploration well **(slide 24)**—from rig mobilization, through sequential drilled intervals, casing and cement jobs, evaluation, and rig demobilization—can be defined as a continuous sequence of processes that must be completed before we can say, "That was a good well." Here, though, a single process might be a *casing job*, followed by a process called *cementing*, followed by a process called *drilling to the next casing point.*

As with flying, every process in the scheduled well must be completed before the next step can be taken, and each step must be completed with success, even if there are problems along the way.

And that means, if any step in any process is interrupted by an unplanned or unexpected result (equivalent to engine failure, or loss of hydraulic power, or stuck landing gear, or an Apollo 13 audible alarm), then something is wrong—and the problem must be fixed.

A drilling example follows.

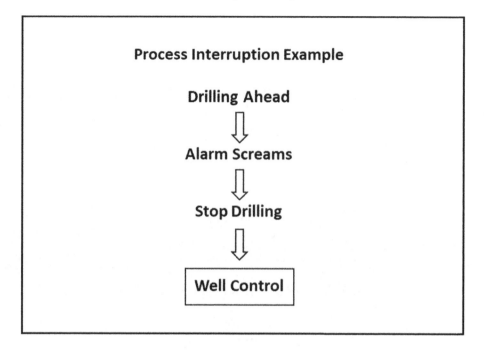

SLIDE 25

Few things are worse than doghouse coffee on an offshore rig.

But a screaming alarm (**slide 25**) on the rig floor can instantly overshadow vile coffee and cause the driller to stop drilling, shut down the mud pumps, pick-up off bottom, and close the annular BOP.

Well control for a full-fledged kick is perhaps akin to a pilot realizing he's just lost power. In both cases, it's time for immediate response and deliberate actions because the penalty for failure can be high. And on a rig, that's exactly what rig leaders and crews are trained to do.

But before going further, there's something wrong with the above rig-floor-alarm, stop-drilling, well-control example. Because it's too simple.

And that's because a rig-floor alarm while drilling at 15,000-feet below sea level requires the same in-depth scrutiny as does an alarm in the cockpit while flying at 15,000 feet above sea level.

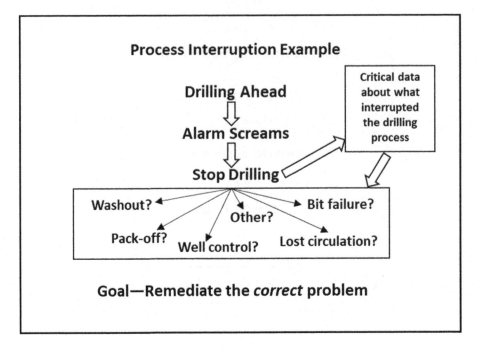

SLIDE 26

The problem is this: Several things, in addition to a kick (**slide 26**), can interrupt the process of drilling. Many are alarmed, but some are not, and even these need rapid response.

For example, an immediate drop in standpipe pressure may indicate a washed-out tool joint, where a few-seconds delay by the driller in picking-up off bottom and shutting-down the mud pumps may be the difference between a bit trip and a fishing job.

Further, a reduction of (or no) mud returns, or a drop in total pit volume, may be lost circulation—just the opposite of a kick.

All drilling parameters on the rig floor and at the bit—torque, drag, pump pressure, rate of penetration, mud weight, gas in returns, weight on bit, pit volume, mud properties, pick-up and slack-off, etc.—are measured and monitored for a reason. A change in a single parameter could mean nothing, or it could be something that demands immediate attention.

Hence, it's critical to all parties (and practiced throughout the industry) that whenever the drilling process is interrupted by an alarm—or by any audible or visual signal, event, or unexplained happening—the first step is to *stop drilling*. Only then can the time be taken to look at critical data and determine the source of the interruption so the problem can be identified and solved.

Hah! No big deal. That's what every rig around the world does while drilling.

But no drilling operation can afford to limit such prudence to just when the bit is on bottom.

For example, the Macondo disaster happened *not while drilling*, but during what should have been a straightforward, end-of-well, temporary-abandonment procedure.

Hence, we need to define a procedure, a protocol, that applies to every process throughout the entire well, during drilling of course, but *not just during drilling*.

And we will call the procedure our ***Process Interruption Protocol***.

It's the protocol (the procedure) we will use if any aspect of any process gets interrupted.

If any process related to the well
is interrupted

The
Process Interruption Protocol
must be . . .

- **Stop the process**

- **Resolve the Interruptive data**

- **Remediate the Problem**

SLIDE 27

Here, as shown in **slide 27**, the same procedure that works for the process of drilling (drilling gets interrupted . . . so we (1) stop drilling, (2) resolve the data, and (3) remediate the problem) needs to be applied to every process throughout the well.

If we are running casing (in other words, the process of running casing), and any step in the process gets interrupted (for instance, the casing gets stuck, or mud returns are lost, or the well kicks), then the *Process Interruption Protocol* says:

(1) Stop running the casing.

(2) Assess the data that accompanied the interruption.

(3) Remediate the problem.

Process Interruption Protocol—
Negative Pressure Test

PROCESS
- **Run the drillpipe**
- **Fill with seawater, close the BOPs**
- **Bleed trapped pressure to zero (0 psi), hold 30 minutes**

INTERRUPTION
- **Pressure would not bleed; the well made 15 barrels**

PROTOCOL
- **Stop the Process (NPT 1)**
- **Resolve the interruptive data (slide 29—1,2,3,4)**
- **Remediate the problem (fix the leak).**

SLIDE 28

What if, as per **slide 28**, the **Process Interruption Protocol** had been applied to the Macondo negative-pressure test?

The Macondo NPT process was straightforward.

(1) Run the drillpipe.

(2) Fill it with seawater.

(3) Observe the amount (2,400 psi) of trapped back-pressure.

(4) Close the annular BOP.

(5) Bleed small amounts of seawater from the drillpipe to incrementally reduce the trapped pressure.

(6) Continue bleeding seawater and reducing the trapped back-pressure until it gets to zero (0 psi).

(7) Hold the pressure at zero (0 psi) for 30 minutes, watching for any indication of a leak.

(8) If there is no pressure increase (no leak), declare the well secure.

(9) Conversely, if the NPT shows there is a leak, fix the problem.

* * *

But something happened (*the interruption*) during the Macondo negative-pressure test (NPT-1)—see items (5) and (6) above. The trapped pressure was manually reduced, as per the procedure, to about

200 psi. Nevertheless, while further manually bleeding seawater from the drillpipe to reduce the pressure from 200 psi to zero (0 psi):

(1) the drillpipe returned about 15 more barrels of seawater than expected, and

(2) the drillpipe pressure would not bleed to zero (0 psi).

Here, because the NPT was interrupted, the **Process Interruption Protocol** says:

(1) Stop the NPT.

(2) Resolve the interruptive data.

(3) Remediate the problem.

* * *

Wow, how clean and neat the problem would have been resolved, in gross contrast to one of the most lethal, costly, man-made, environmental disasters in history.

* * *

So, let's look back at the mud-logging chart that showed NPT-1 data.

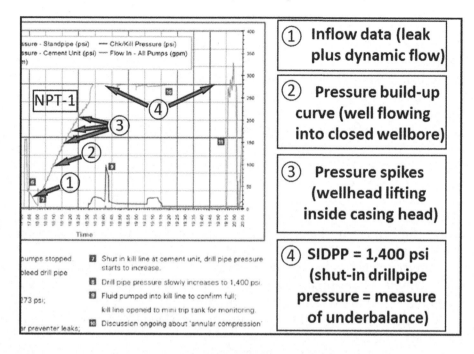

SLIDE 29

For the 8,367-foot-deep NPT-1, as shown in **slide 29**, the *interruptive data* presented itself from 5:55–6:00 P.M. (arrow from #1)

The **Process Interruption Protocol** says that when faced with *interruptive data* that was not part of the plan, *stop the process*, which means *stop NPT-1*. No more negative-pressure testing. No NPT-2. Now is the time to put the entire focus on the NPT-1 *interruptive data*.

A thorough examination of the *interruptive data* at the end of NPT-1 reveals, without ambiguity, that there are only two likely sources of the extra 15 barrels of seawater from the drillpipe:

(1) through a leaking annular BOP, or

(2) from the deep wellbore and annulus.

In looking at the *interruptive data*, we would have asked if the annular BOP closing pressure was again insufficient and if mud from the riser had again bled past the annular and forced 15 barrels of seawater up the drillpipe.

For Macondo, the data show otherwise. The riser was checked (after 15 barrels flowed back) and was found to be full of drilling mud. Therefore, no mud leaked from the riser past the annular BOP to lift seawater up the drillpipe.

See more in Q&A on this topic (**Q&A #1**).

As we continue to look at the NPT-1 6:00 P.M. *interruptive data*, the other alternative is that the reduced wellbore pressure—severely underbalanced by NPT-1—somehow invited flow from the annulus into the wellbore. *And since we ran the negative-pressure test to look for leaks— perhaps we've found one.*

This supposition would have matched the data, as such flow down the annulus and up the casing would lift seawater from the drillpipe, and the same flow would disallow us to bleed the drillpipe pressure to zero (0 psi). For these reasons (along with the immediately subsequent pressure-build-up curve and the 1,400-psi shut-in drillpipe pressure), the data show without ambiguity that the flow was from the deep reservoir, down the external-casing annulus, and up into the underbalanced (low-pressure) shoe track and wellbore.

Alternatively, as later argued in court, if the casing had been breached (for example, a hole in the casing just below the float collar), the flow path would have been directly from the pay zone into the side of the casing, rather than down the annulus and up the shoe track. This topic is discussed in **Q&A #2**.

Hence, had the **Process Interruption Protocol** been followed for Macondo, NPT-1 would have been officially stopped shortly after 6:00 P.M., and the *interruptive data* (bolstered by the build-up data that immediately followed) would have led to the right assessment. That correct assessment would have been that the *interruptive data* generated by the significantly reduced NPT-1 wellbore pressure did the following:

(1) proved communication with the annulus and

(2) invited the formation to flow into the underbalanced wellbore.

As a direct result of applying the **Process Interruption Protocol**, the simple conclusion at the time would have been this:

(1) There is a leak between the wellbore and the annulus.

(2) The problem must be investigated and remediated before continuation of the temporary-abandonment procedure.

In hindsight, and with weeks and months to look at data, it was discovered that float-collar and rat-hole problems caused the silent leak. But on the rig, in real time, those conclusions were not so obvious.

Nevertheless—

This clear-cut, unambiguous, data-based assessment of annular communication would have led the operator to mandatorily change the temporary-abandonment plan, and to commence identification and remediation of the leak, without the risk of a well-control incident.

Macondo—A Lesson Learned

Process Interruption Protocol—
STOP the process
RESOLVE the Interruptive data
REMEDIATE the Problem

Applicability:
Wells worldwide, any process, deep or shallow,
Onshore or offshore, design through abandonment

Goal:
To minimize the chance of ever
losing control of another well

SLIDE 30

As summarized in **slide 30**, the purpose of this presentation is to show that an engineering *assessment* of Macondo data justifies *application* of a simple, proven, problem-solving methodology (***Process Interruption Protocol***) that not only would have been lifesaving on Macondo but would also be relevant to other projects.

By identifying incremental processes throughout the drilling of any well, the lessons learned from Macondo—the ***Process Interruption Protocol***—would allow rig leaders to apply the technique and minimize the chance of ever losing control of another well.

Questions and Answers

Throughout this written presentation, I've endeavored to recall a litany of Q&A sessions with questions coming from every direction—young engineers, service company experts, corporate executives, litigators, petroleum engineering faculty, world-class offshore drilling engineers, and, of course, a diverse cross-section of other oilfield hands and petroleum engineers from around the world. And from those questions, I've bolstered and salted my generic presentation with answers and footnotes, as appropriate, throughout.

But there are other important Q&A topics, raised in technical sessions around the world, that I'll address here.

Q&A #1—The Bladder Effect

The possible phenomenon of "the bladder effect" was apparently first debated on the rig just after 6:00 P.M. as the causal reason for the well producing 15 barrels and the drillpipe pressure not bleeding to zero (0 psi). Neither side of the debate was supported by an engineering assessment of the data.

Regardless of the *basis* of the argument for the *bladder effect* (fluid leakage through the annular, or the annular rubber being pushed deeper into the BOP by heavy mud in the riser, or a bubble of gas under the annular BOP, or whatever, whether real or imaginary), the *result* of the argument being "won" by the operator was devastating—because it caused NPT-1 to be aborted and NPT-2 to be commenced.

Here, in the spirit of the **Process Interruption Protocol**, let us allow the hypothetical *bladder effect* to become a part of the *interruptive data*. That means the *bladder effect*, as related to the annular BOP, would need to be resolved from an engineering perspective.

Such resolution would have been simple. To do so, recall the that NPT was a simulation. The well was kicking *only* because the annular BOP was closed around the seawater-filled drillpipe. So, to get rid of the BOP-related bladder effect, let's take the annular BOP out of the process.

A three-step procedure to take the annular BOP out of the NPT equation would give us a quick resolution: (1) Open the annular BOP, and note the well is immediately dead (1,000 psi overbalanced), and then

(2) close the VBR (variable-bore ram), which contains no "bladder," and then (3) repeat the NPT pressure-reduction process through the drillpipe.

With the annular BOP (bladder) taken out of the equation, the repeated NPT through the drillpipe (using the closed VBR) would once again expose the deepest casing and annulus to the extreme drawdown (with zero chance of mud leaking down the riser).

With what result? (1) The path down the pay-zone annulus and into the wellbore would have already been flushed by the original 15 barrels of flow from the formation, (2) the well would have kicked sooner and harder (with higher flow rates), and (3) the build-up pressure would have more rapidly settled at the SIDPP (shut-in drillpipe pressure) of 1,400 psi. All of that would have led to the same conclusion, at the time—repeated here for clarity:

(1) There was a leak between the wellbore and the annulus.

(2) The problem mandatorily should have been investigated and fixed before continuation of the temporary-abandonment procedure.

Q&A #2—Casing breach below the float collar

This concept—a casing breach below the float collar—was argued in court by expert technical witnesses, each with a high incentive to win the breach/no-breach debate on behalf of his or her respective client.

Specifically—contrary to the data presented for the cement job described in **slide 8** (see page 363)—a claim was made that the casing had been breached (i.e., a wide-open hole in the casing) just below the float collar.

The claim was based on two technical reasons: (1) during the casing running procedure, the casing took load (increased drag) and had to be worked through the original under-reaming ledge at 18,130 feet (pictured in **slide 5**—see page 355); and (2) immediately after all the casing was run (and was rigged up for cementing), it took several attempts and more than 3,000 psi to finally break circulation—which was argued as evidence that the casing had just been breached (cracked, split, jumped a box) below the (later) pressure-tested top of the float collar and the plugged guide shoe.

The counterargument was that it took 3,000 psi to clear rock debris and previously drilled cuttings from the plugged guide shoe (a rounded device installed on the bottom end of the shoe-track casing).

The crux of the argument was that—given such a breach—all the original cement would have exited the casing at the breach (rather than at the bottom of the casing) and gone up the annulus (above the breach). Further, had this been the case, there would have been no annular or shoe-track cement below the breach, leaving the bottom half of the pay zone uncemented and exposed to the open wellbore (through the wide-open breach and up through the open float collar).

There was no hard data *presented* in court to counter either argument, and the judge was left to make his decision about how to proportionally allocate shares of fault (think dollars), accordingly.

Conversely, the following assessment allows the data to speak for itself. Specifically, the Macondo well flowed 15 barrels in five minutes, before the well was shut in. Afterward, the continued flow into the low-pressure wellbore was so slow that it took thirty-five minutes (recall the pressure-buildup curve) to pressure the wellbore to the equilibrium SIDPP of 1,400 psi.

That being said, had the flow been through (into) an open casing breach adjacent to the middle of the pay zone, as claimed, with more than 1,000 psi of underbalance—one inch from the proven-high-capacity *uncemented* Macondo reservoir—the resulting kick (into the casing and up through the open float collar) would have manifested explosively fast, measured in seconds.

That did not happen.

Conversely, the actual flow path was so arduous it took thirty-five minutes for the *flowing* produced fluid to pressure-up the inside of the casing to 1,400 psi. *Arduous* in this case means hydrocarbons were flowing down through 100 feet of mud-contaminated-cement in the ¾-inch-wide annulus and then up through the mud-contaminated cement in the 180-foot-long shoe track, before flowing through the open (not-converted) float collar and into the wide-open wellbore.

Bottom line: the slow-flow-rate kick data seen in the pressure-buildup curve (**slide 14**—see page 377) confirm *there was no breach in the casing below the float collar.*

Q&A #3—Spotting high-density mud in the rat hole

This would have been a zero-time, zero-dollar action. That is to say, while circulating bottoms-up after logging, just before pulling the bit for the casing job, a driller typically calls the mudroom to make a small (10-barrel) but heavy (17-ppg) mud pill. When ready, the driller flips a switch, picks up the 10-barrel pill, and counts strokes to pump it down to and out the bottom of the drillpipe before stopping the pumps, pulling the bit, and leaving the pill behind. With that simple procedure, the rat-hole gravity-segregation problem goes way.

This was not done on the Macondo well.

It's important to say here that any deep, hot, high-pressure cement job can experience problems. The Macondo production-casing cement job, as designed, had its own share of built-in *potential* problems—cement quality and set time, small cement volume, slow displacement rate, pressure-sensitive annulus, multiple workstring and casing sizes, and tight-fitting casing (7-inch casing inside an 8-1/2-inch hole). With so many *potential* problems that might affect final cement quality, one such almost-guaranteed (as opposed to *potential*) problem had an easy solution—prevention of cement contamination by spotting a heavy-mud pill in the rat hole. Without the dense mud pill, the resulting gravity-segregation began the moment the cement job was complete—long before the heavy cement (in the annulus and in the shoe track) had even the slightest chance of curing while located above the lighter-weight oil-base mud in the rat hole. (Recall earlier reference to API RP 65 Section 7.5)

Q&A #4—Use of dispersant after the blowout

Though my presentations targeted the *cause* of the blowout, months of public and media pressure about the environmental impact of the blowout increased public and industry concern across the nation and worldwide. I was asked about the Macondo dispersant during presentations at several locations, including the Evangeline (Lafayette, Louisiana) Sierra Club; Melbourne, Australia; and St. John's, Newfoundland (to name a few), which opened the door for the following answer.

BP's Macondo well flowed for 86 days and spilled 5 million barrels of crude oil. The spilled oil was "dispersed" with Corexit (produced by Nalco Holding Company) as preauthorized by the US EPA (Environmental Protection Agency) for oil spills in navigable waters. Environmental experts (including, for example, marine biologists and environmentalists from my alma mater, the University of Miami) have spoken **with vigor** against toxic, slow-acting, inefficient Corexit, and they have recommended non-toxic, faster-acting, more-efficient, enzyme-based bioremediation agents.

There may be many brands, but one is OSE-II. Such products, proven around the world, don't disperse the oil; they work by causing indigenous bacteria to get very hungry for oil.

The US EPA misclassifies such enzyme-products as bacteria-based (though there are no bacteria in the products) and has declared them unsuitable for oil spills in US navigable waters.

Post-Macondo hydrocarbon spills (for example, Santa Barbara County in 2015) have used and will continue to use the same ineffective and toxic Corexit dispersant unless the EPA can be convinced otherwise. For more technology-based information on this hotly debated topic, see www.ProtectMarineLifeNow.org.

Q&A #5—Macondo topics not covered herein

My point of view while studying Macondo often felt as if I were watching an autopsy by coat-and-tie attorneys who were pretending to be pathologists trying to determine not why the patient had died during a physical exam but how to allocate liability for the patient's death. And I didn't like what I saw. And I'd like to never see it again. So, I kept my assessment of Macondo simple.

Accordingly, my focus for this presentation has been on the following: (1) assessing data that define the definitive mechanical and operating steps and decisions that contributed to and *caused* the blowout, and (2) applying lessons learned from Macondo to future work.

Further, the following *additional* topics were not (could not be) included during my short verbal presentations, but they are likely of interest to a wider audience. They were not included in my presentation because none is considered to be a contributor to the *cause* of the blowout.

Each is addressed in *The Simple Truth.*

(1) drilling the Macondo pay interval below the lost-circulation zones (note: I address such geologic hazard drilling, specifically transition-zone drilling, in *The Simple Truth,* as per my 1976 [Not a typo!] SPE paper: "A Risk Analysis of Transition Zone Drilling");

(2) not circulating prudent/recommended volumes of mud to clear the wellbore before running casing;

(3) running a long string of production casing versus a liner with or without a tie-back;

(4) using nitrified cement;

(5) fluid losses recorded during the cement job;

(6) running fewer-than-recommended casing centralizers;

(7) not running a CBL (cement bond log); and

(8) other credible topics way beyond the scope of my one-hour presentation.

Additional Macondo-blowout topics not in my presentation continue to be covered in-depth by others, including (1) the highly credible "umbrella topics" listed in the text for **slide 21** (see page 398), and (2) all *post-blowout* decisions, activities, and operations that occurred aboard the *Deepwater Horizon* the night of April 20, 2010, and during the subsequent months-long killing of the Macondo well.

Because I was limited to about an hour of presentation time, my apologies (though without remorse) if I left out anybody's hot-button topic(s).

Q&A #6—Post-Macondo blowouts

Though my goal for Macondo was specific relative to *cause*, I was also convinced that the assessment of the data and application of results would ultimately help the industry's entire well-management community minimize the chance of ever losing control of another well.

Maybe that has happened. It's not improbable that somebody read my book or heard my presentation and made a related decision that kept a well safe. But there's no way to ever know . . . because a well *not losing control* is the norm.

Yet in the years since BP's Macondo blowout, there have been other significant blowouts around the world. I have not researched for even a minute any of those events, which I leave to others.

I can only hope that young engineers and skilled and experienced experts alike will do the following: (1) use every academic, intellectual, common-sense, and on-the-job-training tool they have so they can manage and take responsibility for every well they're ever involved with, (2) vigorously use the ***Process Interruption Protocol*** procedure as appropriate, and (3) look in-depth at each failure—by assessing data and applying lessons learned—always with the same goal: to minimize the chance of ever losing control of another well.

In Closing

*Only if we understand and care about
the cause of BP's Macondo blowout
will we know why it should not have happened
and why it should never happen again.*

Readers, especially those who have, or are studying to have, technical, operating, and/or management responsibilities in the oil and gas industry, please apply the lessons learned herein throughout the rest of your career, to every well, onshore, offshore, around the world;

Further, if you have found *The Simple Truth* and *From the Podium* useful, please pass them on, recommend them to others, and discuss them among your fellow students, faculty, and colleagues.

The Simple Truth (Book 1), standalone, is available from Amazon in paperback and Kindle.

From the Podium (Book 2) standalone, is available from Amazon and Ingram in paperback, and from Amazon in Kindle.

Deepwater Horizon 2020 (this book), which includes both the above books, is available in paperback and Kindle from Amazon.

ADDENDUM ONE—

Diagrams from

THE SIMPLE TRUTH

and referenced in

FROM THE PODIUM

The following list is of photos and full-scale schematics from *The Simple Truth: BP's Macondo Blowout*. Many of the *diagrams* were used (as noted) as the base drawings for slides in FROM THE PODIUM. The physical pages are not duplicated here, but the reference page numbers from THE SIMPLE TRUTH are note below.

One important exception: BP's Mudlogging Chart, from their Internal Investigation (Reported on their P. 88), is shown following this list.

The Simple Truth—Diagrams and Page Numbers
Also shows FROM THE PODIUM Slides and page numbers

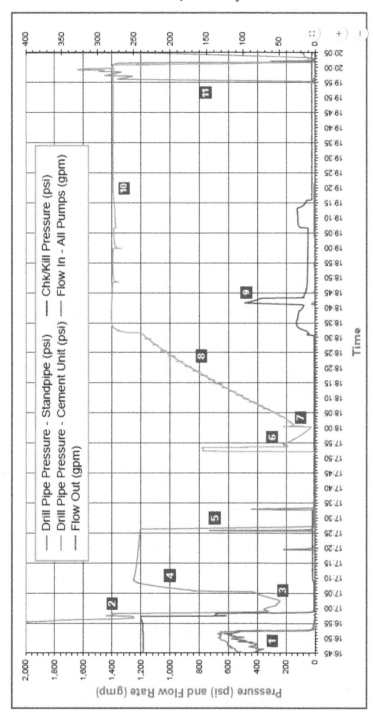

MUDLOGGING CHART
Book 2

ADENDUM TWO

DEEPWATER HORIZON

movie review

By John Turley

Media reviews of the movie *Deepwater Horizon* are plentiful. But for those in the industry who truly do care, a review of such a critical movie deserves more detail; hence, the following.

I'll use a personal experience as the basis for my review.

On October 13, 2016, I joined a hundred CSM (Colorado School of Mines) petroleum engineering students and faculty for a private screening of the movie. I had been invited to emcee the event, where I watched the movie for the third time and then led an energetic hour-long discussion and Q&A session.

Every attendee had a vested interest in the career-related movie, and each was intellectually capable of understanding every aspect of what appeared on the screen.

Afterward, audience comments ranged from "OMG" and "Unbelievable" to "How could anybody have survived?" Some comments were in the form of body language only, without words to describe tear-moistened emotions. I felt the same.

We took a break, then discussed three aspects of *Deepwater Horizon:* (1) the movie, (2) the people, and (3) the technology.

The Movie

First, the entire setting, including all acting family members, the rig hands, onshore and offshore facilities, and the massive *Deepwater Horizon* drilling rig, is done exactly right. Even as the story unfolds on the rig with 126 people aboard, we get to see good renditions of the control room, shops, the galley, offices, the rig floor, a workboat, and working personnel everywhere. Then, once the disaster unfolds, with mud, oil, and gas blowing violently over the derrick, followed by explosions and fire throughout the facility, the situation on the rig could not have been more horrific; nor could the visual effects have been more stunning, more realistic. For those who have ever been on, or who will ever be on, or who never want to be on, a drilling rig, whether onshore or offshore, the movie is a harsh view of a world we must strive to never see again.

With strong agreement among students and faculty, the bottom line for the film, as a good package of entertainment, is simply this: kudos; job well done.

The People

Also important to those who care are the relationships among the players, on several fronts. First, there's a well-portrayed rig worker (key to the story) and his wife and daughter as he prepares to go to the rig for his twenty-one-day hitch. Associated scenes do a good job of showing family dynamics and remind the audience that all persons out there, and those they leave at home, are real people with emotions and concerns and love for life.

On a different scale, the dynamics of relationships among leaders on the *Deepwater Horizon* are entirely different, albeit handled quite well in the movie. Though there are four key leadership positions on the rig (plus four visiting executives with minor roles), the conflict is simple: (1) the well belongs to BP, which pays all the bills, and BP's senior guys (*company men*) on the rig make all technical and operating decisions about the well; and (2) the rig owner, Transocean, has three senior leaders: (a) the *toolpusher*, who is in charge of the drilling rig and all its functions and personnel; (b) the *OIM*, offshore installation manager, who is responsible for the nondrilling facilities (in other words, the "hotel"); and (c) the *captain*, who is in charge of keeping the floating rig (considered by the USCG to be a vessel at sea) on station, hovering above the wellhead a mile below.

In a departure from reality, the movie OIM is given a major authoritarian leadership role throughout the movie, including critical rig-related matters (normally handled by the toolpusher), likely because the real-world OIM survives, while the toolpusher does not.

The rig status on the critical day is that the discovery well has been drilled, cased, and cemented. In preparation for temporary abandonment (*temporary* because it will take several years to evaluate and build the necessary deep-water facilities), the well must be pressure tested to ensure casing and cement integrity. The high-pressure test goes well. But the negative-pressure test (designed to manually reduce the wellbore pressure to ensure there are no leaks from outside the casing) fails to prove the well is secure and generates "anomalous" data. The predominant heated-argument on screen is the following: (1) the BP leaders (company men) agree that the test data were bad but argue they were bad only because of the "bladder effect." The movie does a good job with characters arguing about the technical aspects of the bladder effect (which, in the real world, does not exist, leaving an unnecessary open issue with the audience), and (2) every other non-BP leader, even the workboat captain, argues that the test data prove the well has a leak (information they would not know) and that the BP leaders don't want to admit the failure as it would lead to a major time-and-money cement repair job.

The audience does not know what's right or wrong, but by now they rank the BP rig leaders as bad guys, an apparent goal of the movie. The movie shows BP's fallback decision is to rerun the test a different way (using the kill line), which "successfully" shows the well has pressure integrity.

However, an argued one-liner in the movie proposes it's possible the second test was invalid because the kill line might have been plugged. In reality, it *was* plugged, and the second test was indeed invalid—with catastrophic results, though not mentioned again in the movie.

The falsely "successful" kill line test justifies for BP (and reluctantly for the other rig leaders, at least in the movie) the next step in the temporary-abandonment process—pumping seawater into the well to displace heavy drilling mud from the 5,000-foot-long drilling riser. Given that the well had a serious undetected casing/cement leak (a documented, albeit off-screen, failure of the company men to correctly interpret the negative-pressure test), such displacement of riser mud with seawater allowed the well to flow (also known as a kick, though unseen), even as more seawater was being pumped, exacerbating the accelerating flow.

The result was BP's Macondo blowout.

As soon as the well commences blowing out, rig personnel rightfully actuate a BOP unit (blowout preventer), but they panic verbally to one another, making the point that the BOP, in apparent total failure, does not stop the violent flow. The flow of oil and gas, like Old Faithful, finally explodes and burns, with blow-torch-like flames from the rig floor to the top of the derrick and throughout the living facilities—the cataclysm seemingly beyond belief, but very real.

Understandably, as the fire escalates, personnel conflicts go away, replaced by individual instincts for survival. The choices (well done in the movie) were few—fight your way through the fire and get to a lifeboat or jump overboard. However, as successfully portrayed in the movie and as supported by testimony during the USCG depositions after the disaster, serious injuries were abundant, as were individual life-saving acts of heroism worthy of military-type honors.

And though the viewing audience likely will not recognize on-screen, real-life names unless they live and work on the Gulf Coast, they will have watched eleven men, played by surrogate actors in the movie, just doing their jobs, on this, their last day.

Their bodies were never found.

From the student perspective, the film vividly portrayed the importance of understanding technical concepts and data, reacting to change, respecting authority, standing up to incompetence, and accepting and executing technical job responsibilities—without fail.

The Technology

The third aspect of the hundred-minute movie that needs clarification is the necessarily rapid coverage of abundant technical issues that took place during the rig's twelve-hour countdown to disaster. A number of issues were visual only, or introduced as one-liners, requiring attendees to ponder the significance.

For example, natural gas was seen erupting on several occasions from the seafloor around the BOP, increasing in frequency and violence proportional to the tension on the screen and the ticking of the clock. Not true. No gas erupted around the wellhead either before, during, or after the blowout. Sorry to say, but this was for show, and though it successfully looked ominous, it detracted from movie credibility.

There was also a conflict about a service company leaving the rig before running a CBL, or cement bond log. Every named player on the rig (and again, even the workboat captain) was astounded that BP had released (as per the movie) the service company without the CBL, while the BP leaders, when challenged, were confident with their decision.

The concern was that the cement outside the casing at 18,000 feet (not the structural-casing cement at 5000 feet, just below the seafloor, as wrongly shown in a diagram during the movie) could be bad, and the CBL would tell them so. Not true. The CBL does not test the cement. In reality, the tool is used in limited circumstances when there's been a significant problem during a cement job (and more so during standard completion operations). That was not the case on Macondo, where BP showed that the deep cement job, given enough time to set up, met the criteria for no CBL.

Conversely, the negative-pressure test directly tests the pressure integrity of the deep cement and the rest of the wellbore. Unfortunately, so much movie time was spent on the CBL debate that some attendees may have been *(wrongly)* convinced it was a leading cause of the blowout.

A key issue with the BOP involved the BSR (the blind shear ram). The BSR was located between two other BOP units that were closed immediately after the blowout started. Closure of the BSR was critical as part of a last-ditch emergency operation designed to release the rig from the BOP stack (to get the rig away from the well and the source of fuel to the rig fire). But a serious consequence of the massively flowing Macondo blowout was that the drillpipe between the two closed BOP units was so severely deformed that the BSR was unable to close. The movie tempts the audience with a "big red button" that would save the day. When the red button is finally pushed (after much debate), we see sharp (BSR) blades move toward one another—then stop. Consequently, the pipe is not cut, the well is not sealed, and the rig is stuck on location, burning on top of the fountain of oil and gas. There is no further mention in the movie about the BSR, other than that the BOP failed.

Summary

The CSM students in the theater were hungry for real data, wanted to understand the nuances of the one-liners, and did not want to be taken in by misinformation, all of which made for lively Q&A. And yes, because they *wanted* and *needed* to know, we thoroughly discussed "what caused the blowout," which, to be candid, was beyond the scope of the movie.

Nevertheless, though there are other technical subtopics worthy of debate, it's fair to say the *Deepwater Horizon* writers, producers, actors, and consultants did a respectable and credible job of creating dialog, building tension, revealing important issues and fears (*even before anybody on the rig knew there was any chance of a blowout)*, and then wrapping it up with spectacular visual effects. And that takes true creativity.

Bottom line. For the movie, the people, and the technology—job well done—albeit with a few caveats.

Deepwater Horizon is a must-see movie.

ADDENDUM THREE—

Author's Book Review

THE HOLE TRUTH

A Novel

Published 2019
By J.A. Turley

Professor Tony Zanatelli signs a summer-long contract to manage the drilling of a deep exploration well in the Gulf of Mexico. Guided by education and experience, and in spite of raging seas, raunchy geology, and tight-hole paranoia, he'll do the job well and make good money.

The project will cost millions, but hard data from the hole is the only way to know whether there's a treasure of oil and gas, or the well is a dry hole.

Regardless of results, Tony will deposit his summer salary, reunite with his wife, and share a wealth of new memories with his students.

Well—maybe.

Because the truth, the whole truth, about what resides deep in the bottom of the hole is targeted by those who will kill to prove neither life nor truth is bulletproof.

Excerpt follows

THE HOLE TRUTH

PROLOGUE
June 2005

FRANCINE ELIZABETH RACH, PhD, outdrove the arrogant bastard and his high-beam headlights as he chased her in and out of traffic southbound on Houston's Beltway 8. Looking ahead, she blasted her well-tuned Mazda RX-8 up the elevated US 59 overpass and backed off the gas only slightly as she approached the curve at the top, where he caught up, pulled alongside, and shot her in the face.

Had the brilliant geophysicist not been dead when her bright-red coffin spun out of control, flipped end to end, jumped the retaining wall, and fell to the roadway below in a ball of flame, she might have screamed his name.

Accident investigators took pictures, measured skid marks, interviewed shocked witnesses, and eventually added one more tragic, random, road-rage death to Houston vehicular statistics.

They were right only about it being tragic.

———————

Paperback and eBook editions of *The Hole Truth* are available through Amazon and in local bookstores through Ingram.

Comments Welcome:

John Turley at:
jatmessages@gmail.com

J.A. (John) Turley on LinkedIn

Website: JohnTurleyWriter.com

Made in the USA
Coppell, TX
25 May 2021

56319104R00236